Springer Water

The book series Springer Water comprises a broad portfolio of multi- and interdisciplinary scientific books, aiming at researchers, students, and everyone interested in water-related science. The series includes peer-reviewed monographs, edited volumes, textbooks, and conference proceedings. Its volumes combine all kinds of water-related research areas, such as: the movement, distribution and quality of freshwater; water resources; the quality and pollution of water and its influence on health; the water industry including drinking water, wastewater, and desalination services and technologies; water history; as well as water management and the governmental, political, developmental, and ethical aspects of water.

More information about this series at http://www.springer.com/series/13419

Martina Zelenakova · Petr Hlavínek ·
Abdelazim M. Negm
Editors

Management of Water Quality and Quantity

Springer

Editors
Martina Zelenakova
Technical University of Kosice
Kosice, Slovakia

Petr Hlavínek
Brno University of Technology
Brno, Czech Republic

Abdelazim M. Negm
Faculty of Engineering
Zagazig University
Zagazig, Egypt

ISSN 2364-6934 ISSN 2364-8198 (electronic)
Springer Water
ISBN 978-3-030-18358-5 ISBN 978-3-030-18359-2 (eBook)
https://doi.org/10.1007/978-3-030-18359-2

This Springer imprint is published by the registered company Springer Nature Switzerland AG
The registered company address is: Gewerbestrasse 11, 6330 Cham, Switzerland

Preface

The subject of water resources management is a very wide-ranging one, and only some of the most important aspects are covered in this volume. It soon became apparent that although a number of good books may be available on specific parts of the topic, no text covered the required breadth and depth of the subject, and thus the idea of Management of Water Quality and Quantity book came about. The book has been treated as the product of teamwork of 30 distinguished researchers and scientists from different institutions, academic, and research centres with major concerns regarding water resources, agriculture, land and soil, rainwater harvesting, and water quality and quantity.

The book can serve as a reference for practitioners and experts of different kinds of organisations with responsibilities for the management of water, land, and other natural resources. Equally, we hope that researchers, designers, and workers in the field of water management and agriculture covered in the book will find the text of interest and a useful reference source. The landscape which is sustainably managed conserves water, lowers the rate and volume of run-off water from rain, snowmelt, and irrigation, and helps reduce the amount of pollutants reaching surface water.

Water is an important medium regarding the transport, decomposition, and accumulation of pollutants, whether of natural or anthropogenic origin, which in excessive amounts represent considerable risks for all kinds of living organisms and thus also for human beings. The step towards effective protection of water resources is to know their quality. Systematic investigation and evaluation of the occurrence of surface water and groundwater within the Czech Republic and other countries as well is a basic responsibility of the state, as an indispensable requirement for ensuring the preconditions for permanently sustainable development as well as for maintaining standards of public administration and information. The fundamental requirement in this context is to optimise water quality monitoring and assessment and the implementation of necessary environmental measures.

This volume consists of 16 chapters. **"The introduction"** is written by the editors with inputs from all the authors. It is to introduce the book to the audiences where a brief summary of each chapter is presented. Chapter 2 titled **"Stormwater Management in Urban Areas"** was prepared by Jakub Raček and Petr Hlavínek from Brno University of Technology, Faculty of Civil Engineering, AdMaS Centre. It is devoted mainly to the management of stormwater in the cities. Chapter 3 titled **"The Green Roofs and Facades as a Tool of Climate Cooling in the Urban Environment"** written by Martin Šenfeldr, Petr Maděra, Pavla Kotásková, and Jitka Fialová from Mendel University in Brno, Faculty of Forestry and Wood Technology, Department of Forest Botany, Dendrology and Geobiocoenology and Department of Landscape Management and Miroslav Kundrata and Vlastimil Rieger from Czech Environmental Partnership Foundation. It introduces green facades and green roofs in the Czech Republic. The results from online monitoring of the vapour from green roofs and the cooling effect have been shown, and the roof and urban heat islands have been discussed. David Duchan, and Jaromir Říha from Brno University of Technology, Faculty of Civil Engineering, Institute of Water Structures, are the authors of Chap. 4 **"Infiltration of Rainwater in Urban Areas"**. In this chapter, the review of the most used facilities for infiltration is presented together with a brief description of the geotechnical investigation and hydraulic approach recommended in the Czech Republic. On the other hand, **"Stream Water Quality Modelling Techniques"**—Chap. 5—was also prepared by Jaromír Říha. In this chapter, practical examples of screening and detailed water quality studies applying modelling techniques are introduced. Chapter 6 titled **"Pharmaceuticals in the Urban Water Cycle"** was prepared by Adéla Žižlavská and Petr Hlavínek from Brno University of Technology, Faculty of Civil Engineering, AdMaS Centre. This chapter provides the existing knowledge about pathways of entry, occurrence, degradation, and behaviour of drugs in the water cycle, especially in sewage systems. The reduced quality and yield of water lead to an increase to reusing wastewater in the Czech Republic. Chapter 7 titled **"Biological Audits in the System of Water Treatment Control"** was prepared by Jana Říhová Ambrožová from Institute of Chemical Technology, Prague, Faculty of Environmental Technology, Department of Water Technology and Environmental Engineering. It discusses the significance of biological analysis in solving various problems during the treatment of raw water from reservoirs and streams. The concept of reused wastewater in buildings is based on the treatment of less polluted greywater. The greywater reuse mainly in urban areas is discussed in Chap. 8 titled **"Gray Water Reuse in Urban Areas"** by Jakub Raček. Chapter 9 titled **"The Necessary Documents for the Design Documentation for Water Supply and Sewerage Systems in the Czech Republic"**, which was prepared by Vojtěch Václavík and Tomáš Dvorský from Technical University of Ostrava, Faculty of Mining and Geology, Institute of Environmental Engineering, focuses on the explanation and extension of the knowledge from the area of design documentation for the field of sewerage systems, sewer connections, water supply systems and water-service pipes both for the professional and non-professional public. Chapter 10 **"Evaluation of Technical Condition of Sewerage Systems Operated by Municipalities in the Czech Republic"** was prepared by Petr Hluštík from Brno University of Technology, Faculty of Civil Engineering, AdMaS Centre, and Martina Zelenakova. This chapter deals with the comparison of the technical condition of sewerage systems in the Czech Republic operating by water companies and operated by municipalities themselves. Chapter 11 **"Numerical Modelling of the Fluid Flow at the Outlet from Narrowed Space for a Better Water Management"** was prepared by Vladimíra Michalcová and Kamila Kotrasová. It is dedicated to problems of numerical modelling of Newtonian fluid flows in changing flow space. Chapter 12 titled **"Numerical Modelling of Fluid Domain Flow in Open Top Channel"** was prepared by the same authors Kamila Kotrasová and Vladimíra Michalcova to provide a theoretical background for the influence of choosing mesh parameters on seismic response of fluid domains by numerical simulation of problems fluid–structure interaction during extremely loading. Chapter 13 titled **"Monitoring of Changes in Water Content in Soil Pores of Earth-Fill Dams"** was prepared in an international cooperation by Janka Pařílková and Zbyněk Zachoval from Brno University of Technology, Faculty of Civil Engineering, Institute of Water Structures; Milan Gomboš and Danka Pavelková from Institute of Hydrology, Slovak Academy of Science, Slovak Republic; Boriss Gjunsburgs and Jekabsons Gints from Department of Water Technology, Institute of Heat, Gas and Water Technology, Water Engineering and Technology, Riga Technical University, Latvia; Yanko Yanev and Daniela Toneva-Zheynova from Department Automation of Manufacturing, Faculty of Automation and Computing, Technical University of Varna, Bulgaria; Tymoteusz Zydroń and Andrzej Tadeusz Gruchot from Department of Hydraulic Engineering and Geotechnics, Faculty of Environmental Engineering and Land Surveying, University of Agriculture in Krakow, Poland. In this chapter, the team of the authors provides a view of international cooperation and its achievements when dealing with a project of applied research in the EUREKA programme, which was focused on the development and construction of monitoring technology enabling changes in water content to be monitored in a porous medium, exactly monitoring of water seepage in earth-fill dam. David Duchan, Aleš Dráb and Jaromir Říha from Brno University of Technology, Faculty of Civil Engineering, Institute of Water Structures, are authors of Chap. 14 **"Flood Protection in the Czech Republic"**. It is focused on flood risk assessment and management generally and specifically in the condition of the Czech Republic. Chapter 15 titled **"Small Hydropower Plants in the Czech Republic"** was also prepared by David Duchan and Aleš Dráb. The chapter provides an overview of the current use of hydropower potential in the Czech Republic in small hydropower plants. The last chapter, written by editors, contains **"Updates, Conclusions and Recommendations for Management of Water Quality and Quantity"** which closes the book volume by the main conclusions and recommendations of the volume, in addition to an update of some finding which may be missed by the contributors of the volume.

Special thanks to all who contributed in one way or another to make this high-quality volume a real source of knowledge and the latest findings in the field of Management of Water Quality and Quantity. We would love to thank all the authors for their invaluable contributions. Without their patience and effort in writing and revising the different versions to satisfy the high-quality standards of Springer, it would not have been possible to produce this book and make it a reality. Much appreciation and great thanks are also owed to the editors of the Earth and

Environmental Sciences series at Springer for the constructive comments, advice, and the critical reviews. Acknowledgements are extended to include all members of the Springer team who have worked long and hard to produce this volume and make it a reality for the researchers, graduate students, and scientists around the globe.

The volume editor would be happy to receive any comments to improve future editions. Comments, feedback, suggestions for improvement, or new chapters for next editions are welcomed and should be sent directly to the volume editors. The emails of the editors can be found inside the books at the footnote of their chapters.

Brno, Czech Republic — Abdelazim M. Negm
January 2019
Zagazig, Egypt — Martina Zelenakova
January 2019
Kosice, Slovakia — Petr Hlavínek
December 2019

Contents

Part I
Introducing the Book

Chapter 1
Introduction to the "Management of Water Quantity and Quality"

Petr Hlavínek, Abdelazim M. Negm and Martina Zelenakova

Abstract This chapter presents the main features of the book titled "Management of Water Quantity and Quality" and their related topics mainly in Czech Republic. Topics which are covered are divided into five themes including (a) storm water in urban areas, (b) water quality and reuse, (c) water supply and sewerage systems, (d) hydrodynamics modelling and (e) water structures. The main technical elements of each chapter are presented under its relevant theme.

Keywords Management quality · Quantity · Modelling · Water resources · Czech Republic · Pollution · Hydropower

1.1 Czech Republic: A Brief Background

In the Czech Republic is situated main European watershed separating the seas of the North, Baltic and Black Seas. The main rivers are Elbe (370 km), Vltava (433 km), Moravia (246 km), Dyje (306 km), Odra (135 km) and Opava (131 km). The long river is also Ohře (246 km), Sázava (225 km), Jihlava (180 km), Svratka (168 km), Jizera (167 km), Lužnice (157 km) and Berounka (139 km). The largest natural lake in Czech Republic is Černé jezero in Šumava.

P. Hlavínek
Faculty of Civil Engineering, AdMaS Centre, Brno University of Technology, 612 00 Brno, Czech Republic
e-mail: hlavinek.p@fce.vutbr.cz

A. M. Negm (✉)
Water and Water Structures Engineering Department, Faculty of Engineering, Zagazig University, Zagazig 44519, Egypt
e-mail: Amnegm85@yahoo.com; Amnegm@zu.edu.eg

M. Zelenakova
Department of Environmental Engineering, Faculty of Civil Engineering, Technical University, Košice, Slovakia
e-mail: martina.zelenakova@tuke.sk

M. Zelenakova et al. (eds.), *Management of Water Quality and Quantity*, Springer Water, https://doi.org/10.1007/978-3-030-18359-2_1

The climate is moderate in the Czech Republic, intermediate between the continental and oceanic types. Typical is the alternation of four seasons. It is characterized by predominant Western flux and intense cyclonic activity. The influence of the sea is manifested mainly in Bohemia, Moravia and Silesia, and continental climate influences are increasing. However, the greatest influence on the climate in the Czech Republic is the altitude and the relief.

Typical are abundant precipitation and transitions of frontal systems—annually they will pass through the Czech Republic on average 140 times. Most rainfall will fall in June or July, at least in January or February. The wettest place in the Czech Republic is the Jizera Mountains (especially the Bílý Potok area). The driest of Libědice in the district of Chomutov is situated in the collision shadow of the Ore Mountains. The average annual temperature is between 5.5 and 9 °C. The coldest month of the year is January, the warmest July. Tropical days are recorded on average 12 per year; tropical nights are very rare. Arctic days are usually 1–2 per year. The warmest places are the Dyje-Svratka and Dolnomoravsky valley, and then the big cities, especially Prague, where the temperature increases the dense housing development. The chilliest place is the peak of Sněžka. The windiest point of the Czech Republic is the peak of Milešovka. At the same time, it is the place with the greatest number of storms in the year.

1.2 Main Themes of the Book

The book covers the following themes: (a) storm water in urban areas in Chaps. 2, 3 and 4; (b) water quality and reuse in Chaps. 5, 6, 7 and 8; (c) water supply and sewerage systems in Chaps. 9 and 10; (d) hydrodynamics modelling in Chaps. 11 and 12; (e) water structures in Chaps 13 and 14. In the following sections, the main features of each chapter will be presented under its relevant theme.

1.3 Storm Water in Urban Areas

Chapter 2 focuses on storm water management in urban areas. The chapter provides the basic overview of storm water management (SWM) in urban areas in Central Europe connected with storm water (SW) run-off, and the potential solutions are defined. The current dry weather in the Czech Republic (CR) accompanied by extreme weather fluctuation provides the necessity to deal with reducing the surface run-off and then consider connecting the SW to the sewer system or the watercourse. SWM as a decentralized drainage system represents the point of natural processing of SW and its return in the natural water cycle. SWM in urban areas is connected with surface run-off quality, SW quality, SW management methods and SW reuse. And for conditions in the CR, the potential solutions are defined: evaporation, infiltration and regulated run-off to watercourses. These solutions can be implemented by the accumulation of SW and its use for irrigation and flushing toilets. The reduced quality and yield of surface and groundwater due to droughts and changing climatic conditions lead to an increase of reusing SW in the CR. Given the current continuing trend of reducing the cost of SW collecting, treating and water reuse technology, the SWM will be more often designed in practice.

Chapter 3 deals with the green roofs and facades as a tool of climate cooling in the urban environment. The chapter is focused on the introduction of green facades as well as green roofs in the Czech Republic. In the beginning, the history and development of green roofs and facades have been outlined, the type of green roofs and ways of facade greening, structures and systemic solutions, protection against slide and drainage have been described, and the functions and benefits of green roofs and facades have been estimated. Next part describes the measurement experiment of the water vapour from green roof and the quantification of green roof cooling effect. Finally, the Zero Carbon Facility has been described—an example from Brno (Czech Republic)—The Open Gardens. The results from online monitoring have been shown, and the roof and urban heat islands have been discussed.

There are many reasons for increasing green areas, especially in cities where the construction of buildings displaced all the greenery practically only into parks. Green roofs may become an almost integral part of roofs of new buildings in urban areas just because of air purification from airborne particles and temperature reduction, as well as the increase of humidity in the roof surroundings and partial water retention, especially during summer months. The roof with vegetation cover, besides the fundamental function, which is the protection of the construction and the indoor environment from meteoric water, fulfils many specific functions related to the improvement of the environment. The roof can be representative. It can be used for recreational purposes and at the same time fulfils the many ecological functions. The most important potential of green roofs though is their climate adaptation function. They help to moderate urban heat islands and measuring their retention capacity approved that even extensive green roof can significantly cut the peaks of run-off from extreme storm rains. Increased financial costs, whether on the load-bearing structure or the layers needed for the planting, will subsequently provide savings of the operating means. Well-designed layers of green roof significantly increase the durability of the membrane roofing as the vegetation, and vegetative layers protect against the weather effects the damp course of the roof or facade.

Greenery and substrates, especially on the roof, improve thermal insulating and reduce thermal extremes, and there is also smaller material expansion and thus the durability, compared to flat roofs without vegetation. A great advantage is a protection against UV radiation, which degrades the majority of materials, what contributes to extending the durability of the entire roof again.

During the experimental measurements, the authors quantified the green roof cooling effect using data acquired from meteorological station installed on the roof. The cooling effect was quantified as the ratio between the energy input by the solar radiation and the energy emitted on the evaporation of the lawn. The lawn evaporation was quantified using Bowen's ratio. Our results confirmed the importance of green roofs in the urban environment. The green roof of The Foundation Partnership

Brno was able to evaporate 1–6 mm.m^{-2} of water during the day. The process of evaporation depends on the actual conditions of potential evapotranspiration and on the amount of available soil water. Monthly sums of vapour of green roof ranged from 34 to 125 mm m^{-2}, and 46 to 69% of the solar energy input was consumed during evaporation. If this energy was not draining away by evaporation, it would be used to heat the city surface.

Chapter 4 discusses the different aspects of the infiltration of storm water in urban areas. In this chapter, the review of the most used facilities for infiltration is presented together with a brief description of the geotechnical investigation and hydraulic approach recommended in the Czech Republic. An analysis of factors influencing the process of infiltration is carried out, and related uncertainties influencing the design of the storage volume of the infiltration facility are discussed. The chapter focuses on the technically oriented problems affecting the infiltration of storm water in urban areas. A typical and new approach to infiltration of storm water is presented together with typical infiltration facilities. The chapter later focuses on the geotechnical survey, where necessary laboratory and field tests are described. From those tests can be obtained important soil properties for effective design of infiltration facilities. The hydraulic design of the infiltration facility from Czech legislative is introduced, where all uncertainties are combined in one coefficient. The new approach to hydraulic design with coefficients that include uncertainties in input parameters for economical and safe design is presented together with recommended values.

The reduction of surface run-off belongs to the topical tasks of urban water management. The efficient measures of storm water management are water retention and infiltration at the place of its origin, i.e. just within the built-up areas. It is necessary to install efficient arrangements at or close to individual estates such as private dwellings, blocks of flats, industrial and agricultural facilities. The effect on the technical drainage systems such as sewerage and open channels may be considered if the retention and infiltration capacities are efficiently planned, designed and performed.

It is recommended to perform careful and professional hydrological, geotechnical and hydrogeological survey and analysis before the design to determine soil profiles and infiltration rates. Hydrogeological investigations are required in cases when there is a likelihood of groundwater discharge or high seasonal water tables. Experience shows that saving money for the survey usually results in increased expenses for the design, construction and performance.

1.4 Water Quality and Reuse

Chapter 5 is devoted to discussing the groundwater flow problems and their modelling. In this chapter, the main groundwater-related issues and possibilities of their modelling are discussed. Special attention is paid to the water supply problems and groundwater protection, to the groundwater issues in urban areas during no flood periods and at the flood events. The assessment of groundwater impact on the hydraulic and civil structures is mentioned too. Numerous examples of groundwater flow modelling and its results are presented as well. The chapter focuses on the technically oriented problems affecting the groundwater flow characteristics. Typical problems in groundwater management are presented together with an overview and applications of groundwater flow models. One-, two- and three-dimensional models are briefly described. Related practical applications are listed and demonstrated in examples from technical practice. These are namely the influence of hydraulic and other structures on the groundwater regime, changes of the groundwater flow during floods and an impact of flood protection measures.

Groundwater flow models significantly contribute to the sustainable management of groundwater resources, on optimization of groundwater withdrawal and assessment of negative impacts of phreatic surface drawdown due to groundwater exploitation. They are necessary for contaminant transport modelling in aquifers. The models are also a crucial part of soil stability calculations. In this chapter, the summary of the models and assumptions for their use are shown.

As geological and hydrogeological conditions are complex, it is crucial to deal with the uncertainties in the entering data. One of the tasks of modellers and interpreters is to estimate the error and uncertainty in the results achieved. Moreover, the groundwater management and seepage modelling is a multidisciplinary issue at which significant role is played by geologists and hydrogeologists, water managers, hydraulic and civil engineers.

In the chapter, it is recommended to define objectives and formulate the problem carefully. Relevant interpretation and explanation of results must be presented for final decisions and technical proposals. All these items need experienced staff with good knowledge about the necessary data for the solution and a clear vision about the results and their application.

Chapter 6 discusses the different aspects of pharmaceuticals in the urban water cycle. Pharmaceuticals are an integral part of modern society and had helped humankind defeat many devastation population diseases. However, from the point of view of environmental protection and water control, they can represent a dangerous interference with the stability of the ecosystem. Medicines have been accompanying humans since the prehistory when they tried to reduce the pain or heal wounds with herbal extracts.

Pharmacology has undergone great development and has brought a lot of good to our society. On the other hand, however, the massive use of the drug (or hormonal contraceptives) and the relatively low degradability of many drugs has caused many of these substances to spread to the environment where the most frequent entry pathway is an outflow of sewage treatment plants.

Most drugs are present in the water cycle in the form of micro-contaminants and have a concentration range of $1\ \text{ng–}\mu\text{g}\ l^{-1}$, often only with very low detection levels. Moreover, due to the changing chemical, biological and physical conditions within the water cycle, there may be different syntheses that produce conjugates of these substances or different cometabolites of the original substances, and it is therefore very difficult to make correct analyses of the substances.

Conventional treatment of wastewater without specific conditions is very poorly effective and is capable of degrading only substances readily degradable such as some types of analgesics or substance with the high adsorption abilities. Pharmaceuticals represent a relatively wide spectrum of compounds that do not always have the same physicochemical properties, as a result of which we can say that the efficiency of different purification processes to the level of their degradation will be different and for this reason non-exist only one best treatment solution.

The persistent and bioaccumulate nature of many pharmaceuticals substances brings major risk to wild live plants and animals, except acute or long-term toxicity may influence their endocrine systems and even transcript of DNA or RNA. Such effects can cause mutagenic or teratogenic, cancer, changes in gender or adversely affect on the behaviour.

The immediate danger to humans is primarily in the bioaccumulation potential of these substances and their binding to a solid soil matrix where they can be leached into groundwater and contaminant drinking water sources.

The most frequently observed pharmacological groups in spill water are analgesics, antibiotics, hormonal drugs, pharmaceutical of mental illness or cytostatic agents.

Chapter 7 presents and explain the biological audits in the system of water treatment control. The whole water supply system from the catchment to the consumer should be studied and monitored as one continuum with many interrelations. This principle is one of the key points of water safety plans which are included in the new WHO Drinking Water Quality Guidelines. In the water supply system, it is possible to apply the risk analysis (HACCP; Hazard Analysis and Critical Control Points) that should consider the entire supplied area including the raw water source, collecting structures (reservoirs), water treatment (technology), distribution network together with the structures located in it (pumping stations, accumulation, water supply tanks) and the consumer endpoints in the network. It is needed to pay increased attention to the system of production, transport and accumulation of drinking water because the water quality may be affected at any point of the continuous water supply system, both internally and externally.

Microscopic analysis can help to understand the sources and consequences of eutrophication and the pollution of both treated and untreated water, thus affecting the work of the designers, construction engineers and operatives, water managers, water suppliers, health authorities and decision-makers involved. These methods can provide rapid results and reliable data, that are useful in selecting appropriate preventive measures or ecologically sustainable remedial actions. Water, in the sense of the possible prognosis of the condition and development of water quality, is very important and have a long-term history of practice in the Czech Republic. At the site, in connection with the issues of applying (hydro)biological audits in practice, it is certainly appropriate to state model examples of performed biological audits in the water supply. For illustration, audits of distribution networks were selected, where the supply was groundwater or surface water of various trophic levels, respecting the zoning of the tank and the possibility of the selection of a suitable profile, various arrangements of the distribution network and technological separation. The findings from the biological audits and the recommendations stated below have been accepted and gradually applied by the operators of water supply networks.

Microscopic analysis can provide information, which is inexpensively and rapidly obtained, on many drinking water quality failures, to supplement other commonly used analytical checks. This kind of information is difficult to obtain by chemical or microbiological analyses (cultivation techniques). For this reason, it is suitable for inclusion in the HACCP system. However, an essential precondition is to have people well trained in the determination of microorganisms. Biological and hydrobiological audits are discussed step by step in the context of risk analyses (HACCP), monitoring of water treatment technologies and the water safety plans used in the Czech Republic.

Chapter 8 explain in more details the reuse of the greywater in urban areas. The chapter provides the review of greywater reuse in urban areas, especially in buildings. In Central Europe, the reduced quality and yield of surface and groundwater due to droughts in the years 2015–2018 and changing climatic conditions lead to an increase to reusing wastewater (WW). The concept of reused WW in buildings is based on the treatment of low polluted greywater (GW) by greywater treatment system (GWTS) which produces white water. GW represents WW from showers, basins, washing machines, kitchen sinks and dishwashers. White water does not meet such strict parameters as drinking water and can be used for toilet flushing, irrigation or for other use. The WW reuse with GW in urban areas is implemented in the guidelines of countries: UK, Germany, Australia, Canada, USA and internationally by WHO. In the Czech Republic, reuse of GW has not yet been legislatively approved. The groups of GW treatment processes were reported: simple, extensive, chemical, physical, biological and MBRs treatment. The MBRs represent a modern treatment of WW and also are associated with biological treatment and separation of solid and liquid substances. GWTS has been provided data on MBRs technologies (microfiltration or ultrafiltration) that seem to be the acceptable solution in the building. Further, increase water and sewerage rates to allow a smooth recovery of predominantly obsolete utilities and the rehabilitation or construction of new water resources. This increase creates conditions for return on investment in the GWTS in the selected buildings.

1.5 Water Supply and Sewerage Systems

Chapter 9 presents and explains the necessary documents for the design documentation for water supply and sewerage systems in the Czech Republic. This chapter describes the preparation of documents for the individual stages of project documentation necessary for the designing of sewerage systems, water supply systems, sewer connections and water-service pipes. The initial part of this chapter describes the basic terms from the field of sewerage and water supply systems and the individual connections for better orientation and understanding of the topic. Attention is also paid to the basic requirements for the construction of sewerage systems and water supply systems resulting from the valid legislation, which serve as an important basis for both professional and non-professional public. It also provides the specifications of the technical requirements for the preparation of design documentation for sewerage (sewage system) and water system design, which are based on the normative documents valid in the Czech Republic and the European Union.

The following part of this chapter is focused on the design documentation and designing activities in construction related to water supply and sewerage systems.

The location of the building is a very important aspect of this chapter. The location of the construction of water supply and sewerage systems and their changes can be made only on the basis of a zoning and planning decision or a zoning and planning consent. A decision on the construction location determines the building site, places the designed construction, determines its type and purpose, the conditions for its location, the preparation of project documentation necessary for the issuance of a building permit, the announcement of construction and the connection to the public transport and technical infrastructure. Further attention is also focused on building permits, building permits versus public contract and building permits versus authorized inspector.

The following issues are described in this chapter as well:

- Use of water supply and sewerage system constructions—waterworks, the competent authority is represented by the Water Authority. It deals with the final inspection certificate, certificate of occupancy and notification of the use of construction. It also describes terms such as premature use, trial operation;
- Construction removal permit;
- Construction passport and documentation of actual execution;
- Obligations of the owners of water supply and sewerage systems (waterworks);
- Obligations of the owners of water-service and sewer connections;
- Unauthorized water withdrawal and unauthorized discharge of wastewater with respect to sanctions for such behaviour.

This chapter is intended for the professional and non-professional public interested in the preparation of design documentation for the construction of water supply and sewerage systems. After studying this chapter, students of universities focused on the issue of technology and water management can also extend their knowledge.

Chapter 10 focuses on the evaluation of the technical condition of sewerage systems operated by municipalities in The Czech Republic. The presented chapter deals with the comparison of the technical condition of sewerage systems in the Czech Republic concerning water companies operating water management infrastructure and operators—mostly municipalities themselves, i.e. the owners of sewerage systems.

The paper presents comparison and classification of various defects in the sewerage systems according to a norm EN 13508-2 standard used to assess the condition of the sewerage systems. The assessed localities are municipalities up to 2000 inhabitants in the Czech Republic, which themselves operate sewerage systems.

This chapter offers a solution for assessing the structural and technical condition of sewer systems using a multi-criteria analysis.

1.6 Hydrodynamics Modelling

Chapter 11 discusses how the numerical modelling of the fluid flow at the outlet from narrowed space is used for better water management. In the water management, events of significant changes in the running fluid flow space occur very frequently. Such change influences the characteristics of the flow field significantly and thus its effects on the environment. These might be effects on adjacent objects in close proximity such as walls of the designated space, or effects on objects bypassed with the running fluid. Such a situation frequently influences the surrounding area. The flow run may affect the terrain of the land or even the ambient climate in some cases. This chapter is dedicated to problems of numerical modelling of Newtonian fluid flows in changing flow space. Showing a particular task, which is solved numerically using CFD codes (Computational Fluid Dynamics) in ANSYS Fluent software, the reader becomes familiar with both, problems of the mathematical specification of fluid movements and principles of the correct choice of the numerical model. To resolve this task, four numerical models are selected.

The flow filed change is monitored in the changing flow space. Results related to flow velocity and turbulence intensity are presented. Some results could be verified by experimental measuring performed in an aerodynamic tunnel of the experimental research site of ITAM CAS, v. v. i., Centre of Excellence Telč.

As in any numerical solution, it is necessary to create a model for CFD (virtual prototype of the examined system). Subsequently, mathematical procedures are applied to it, so that selected data on processes running in the whole examined area are acquired from the specified boundary and initial conditions. At the same time, it is necessary to respect physical principles. Without the knowledge of turbulence physics, the researcher cannot be sure about the right model selection. For this reason, separate sections in this chapter are dedicated to methods of mathematical modelling and specification of selected numerical models.

Chapter 12 is devoted to provide a theoretical background for the specification of fluid effect on the liquid-filled rectangular channel during ground motion where ground-supported liquid-transporting structures are used to store and transport of fluid. Water flow in open channel deals with the equilibrium forces in the fluid. The fluid develops pressure on the channel. During an earthquake, the fluid exerts hydrodynamic impulsive and hydrodynamic convective pressures together with hydrostatic pressure on tank walls and bottom of the ground-supported fluid filling endlessly long channel. The numerical model of the seismic response of the fluid filling of endlessly long shipping channel was obtained by using finite element method in software ADINA. The results from the numerical parametrization: mesh density of 2D fluid region, mesh parameter “PATTERN” of 2D fluid region or 3D fluid region have influence for finally model of numerical simulation of water-filled channel grounded on soil.

1.7 Water Structures

Chapter 13 presented the various aspects connected with the flood protection in the Czech Republic. In this chapter, the main flood protection and flood risk-related issues are discussed. Firstly, the historical background in the Czech Republic is mentioned and discussed. Special attention is paid to the process of flood risk methodology and its applications at practical flood protection solutions. The present state of flood protection in the Czech Republic is briefly mentioned too. This chapter is focused on the flood risk techniques practised in the Czech Republic. The recent procedures used for the development of flood hazard and flood risk maps together with the development of danger and vulnerability maps are mentioned. The assessment of economic efficiency should be a necessary part of flood risk management plans. Therefore, methods for estimating flood damage and annual economic risk are mentioned together with multi-criteria analysis.

From previous floods is shown the importance of large dams with significant flood attenuation volume in flood routing. The most efficient is the management and real-time operation of the system of dams based on real-time rainfall measurement and data processing using adaptive models. The important issue is the technical surveillance of flood protection structures.

The chapter recommends focusing on further research into flood risk techniques to improve and refine results and outputs obtained. Those possible research topics should be: more detailed multi-criteria flood risk assessment, non-structural flood protection measures and the evaluation of their effectiveness, inclusion of risks from the exposure of inhabitants to flood hazards, assessment of environmental risks and risks due to the flooding of sensitive facilities and historical monuments, estimation of indirect losses and more comprehensive uncertainty analysis in risk management. The chapter shows that, in the present, it is important to continuously informing, educating and training of the population at flood-prone areas together with the flood rescue services.

Chapter 14 is concerned with the technical and architectural design of small hydropower plants in the Czech Republic. This chapter aims to present selected small hydropower stations realized in the Czech Republic for about 16 years, which are interesting both from a technical point of view and from the point of view of the architectural design. The chapter offers an overview of the current use of hydropower potential in the Czech Republic in small hydropower plants. Between 2002 and 2016, we have seen a relatively significant increase in the number of small hydropower plants in the Czech Republic by 162% and installed power by 93%.

The primary consideration in deciding on the realization of a SHPP is still the aspect of the economic efficiency of the investment invested. For this reason, most of the newly built hydroelectric power plants were implemented using existing weirs or dams. Besides the new buildings, a number of general reconstructions of the technological and construction part of the SHPPs were also carried out, aiming to increase the efficiency of these sources.

At present, most of the localities suitable for the economically efficient execution of the SHPPs are already exhausted in the Czech Republic. Most sites are now available offering under current conditions low economic efficiency for any investment in the SHPP. Further developments in the field of small hydropower plants can, therefore, be expected to be significantly slower than in the past 16 years. We expect the gradual redevelopment of new hydropower plants in previously unoccupied localities and the reconstruction of end-of-life historic hydroelectric power plants. Assessing the economic efficiency of a possible reconstruction of a power plant is based on a reliability assessment of its components.

Generally, in the Czech Republic, in the field of hydro-energy, the development of mainly pumping hydroelectric power plants in the future depends on the use of unstable electricity sources (wind, sun). The main obstacle to the implementation of pumped-water power plants is the fact that suitable locations are located in naturally valuable localities of the Czech Republic. It will be necessary to look for suitable technical solutions to minimize possible impacts on nature and landscape.

The book ends with the conclusions and recommendations, Chapter 15.

Acknowledgements The editors would like to acknowledge the authors of the chapters for their efforts during the different phases of the book including their inputs in this chapter.

Part II
Rainwater in Urban Areas

Chapter 2
Stormwater Management in Urban Areas

Jakub Raček and Petr Hlavínek

Abstract In an anthropogenically intact country, almost all stormwater (SW) is infiltrated, and it is absorbed by plants or evaporated. While SW in urbanized areas hardly finds an unaffected path to reach the natural water cycle, the natural water cycle has changed due to the buildings and roads associated with the growing population and also due to agricultural and forest management. This leads to a gradual change of underground water with a structural change and in extreme SW leads to local flooding. The current dry weather in the Czech Republic (CR) accompanied by extreme weather fluctuation provides the necessity to deal with stormwater management (SWM). Therefore, it is necessary to reduce the surface runoff and then consider connecting the SW to the sewer system or the watercourse. The implementation of the water policy of the European Community is subsequently the basis of legislation in the CR with technical measures for SWM: SW pretreatment, SW retention, and SW infiltration. The average demand for drinking water is over 100 L water per capita per day. However, about 50% of this need does not need to be drinking water. Thus, SW can be used as a replacement for toilet flushing, irrigation, washing, and other use.

Keywords Stormwater · Infiltration · Green roof · Water quality · Water reuse

---

J. Raček (✉) · P. Hlavínek (✉)
Faculty of Civil Engineering, AdMaS Research Centre, Brno University of Technology, Purkyňova 651/139, 612 00 Brno, Czech Republic
e-mail: racek.j@fce.vutbr.cz

P. Hlavínek
e-mail: hlavinek.p@fce.vutbr.cz

M. Zelenakova et al. (eds.), *Management of Water Quality and Quantity*,
Springer Water, https://doi.org/10.1007/978-3-030-18359-2_2

2.1 Introduction

In urbanized areas, stormwater (SW) does not return unaffected to the natural water cycle. This leads to a gradual change of underground water with a structural change in soil properties in urbanized areas, which is also related to the change of chemical and biological processes in the soil environment. Extreme SW leads to local flooding caused by the insufficient capacity of the sewage system and water streams. Thus, SW in urban areas has the impact on the pollution of watercourses due to the outfalls of the sewage system and the wastewater treatment plants (WWTPs). The current dry weather in the Czech Republic (CR) accompanied by extreme weather fluctuation in the years 2014–2017 [1] provides the necessity to deal with stormwater management (SWM). Therefore, it is necessary to reduce the surface runoff by taking into account the local conditions and then consider connecting the SW to the sewer system or the watercourse. When SW is not disposed of, but it is managed [2] with the principles of sustainable development of the urbanized area. It means SW is used in current buildings where the property owners are responsible for the quantity and quality of water from its own estate.

In selected countries of the European Union (EU), SWM has been described for many years in legislation with validation in practice with a comparison of environmental and economic indicators [3]. In CR, these principles are also successfully used, especially in newly urbanized areas. These newly urbanized areas further increase the dimension of sewers and restrain river beds, and it is necessary to purchase land for SW tanks and to force the owners to change the approach to designing measures for the SWM. Directive 2000/60/EC [4] defines the conditions for the implementation of the water policy of the European Community, and it is subsequently the basis of legislation in EU countries and also in the CR.

In Switzerland, the 1991 Federal Act of the Protection of Waters GSchG [5] requires the infiltration of SW. If this is not feasible, it is possible with the consent of the cantonal authority to regulate SW into the watercourses. Waters Protection Ordinance GSchV [6] of 2018 defines contaminated and uncontaminated wastewater (WW) suitable for the watercourse and its condition. Decree on soil pollution, VBBo [7] of 1998 describes the requirements and limits on the accumulation of pollutants in the soil after infiltration of SW.

In Germany, SWM is described in the WHG Water Act [8] of 2009 and refused SW rapid drainage in urbanized areas, and it is required to infiltrate on the property. Soil protection in SW infiltration is defined in the act to protect the soil BbodSchG [9], provincial laws and regulations. The Technical Directive DWA-Arbeitsblatt A 138 [10] of 2005 describes the procedures for the design, construction, and operation of SW infiltration devices. The Federal Decree NWFreiV [11] of 2000 and the Technical Specification TRENGW [12] of 2000 define the conditions for SW infiltration without requiring authorization. Recommendations for dealing with SW ATV-Merkblatt M153 [13] of 2007 assess the need for retention and SW pretreatment with respect to the junction to the watercourses.

In the CR, the main legislative regulation is Water Act No. 254/2001 [14] and this law sets out the obligation for new buildings and renovated buildings according to Act. No. 183/2006 [15] with SWM to implement the infiltration, retention, and regulated flow of SW. The Plan of Main River Basins [16] as a national planning document introduced a framework for implementation of sustainable SWM.

Act No. 274/2001 [17] defines the obligation for legal entities to pay for the drainage of SW into a sewage system with runoff at the WWTP. The SWM priorities are described in Decree No. 501/2006 [18] (amended by Decree No. 269/2009 [19]) and in Decree No. 268/2009 and are defined in the following order: infiltration, retention, and controlled flow rate into a watercourse, regulated bleed into a unified sewage system. According to the aforementioned priorities, infiltration should be addressed as a priority. If infiltration on the land is not possible because there are the clay soil and high groundwater levels, hydrogeology report according to Czech standard ČSN 75 9010 [20] is necessary for the design of SWM [21]. Czech technical standard TNV 75 9011 [22] of 2013 defines alternatives for decentralized drainage and also provides a central solution for larger urbanized units and individual areas. This standard is a guideline for the design of a technical SWM solution and the operation of collecting and retention objects including safety overflows.

The concept of nature-friendly SWM in urbanized areas represents designs corresponding to natural SW drainage before urbanization. Generally, SWM is a decentralized drainage system that is characterized by SW at the point of natural production and its use in the natural water cycle. Therefore, acceptable solutions for SWM are evaporation, infiltration, and regulated runoff to watercourses. These solutions can be implemented as the accumulation of SW and its use for irrigation and flushing toilets. Another implementation of SW is retention and slow, regulated drainage into a watercourse or in limited cases into the sewage system [23].

It is necessary to separate the low-polluted and high-polluted SW. Low-polluted SW is usually from roofs, car parks, and unfrequented roads. High-polluted SW requires pretreatment in the local treatment system or connection to the sewage system and subsequent treatment in the WWTP.

This chapter provides the basic overview of SWM in urban areas in Central Europe connected with SW runoff, and the potential solutions are defined.

2.2 Surface Runoff Quantity

At present, SW in urbanized areas hardly finds an unaffected path to reach the natural water cycle. In an anthropogenically intact country, up to 99% of SW gets infiltrated, and its absorbed by plants or evaporates [24]. The natural water cycle has changed due to the buildings and roads associated with the growing population. Moreover, the change also occurred due to agricultural and forest management. Paved areas such as roads, pavements, and urbanized areas prevent natural water infiltration into the native soils [25] and thus adversely affect groundwater replenishment. SW is polluted by substances from paved areas and thus represents relatively high pollution.

With the increasing amount of paved surfaces also the surface runoff increases while the value of groundwater recharge decreases. These situations cause extreme differences between excessive flood flows and drying watercourses which leads to flooding, erosion, water pollution, and reducing the groundwater level [26].

The increasing speed and volume of SW surface are characterized by the frequent occurrence of local floods [27] which are significant if the urbanized area is situated on a small watercourse [23]. In the CR were popular morphological changes associated with flow straightening and hardening watercourses which have a negative impact on the ability to transform the flood wave. Higher frequency of local floods due to landscape urbanization has a negative impact on the watercourse. Thus, these situations lead to hydraulic stress and pollution with effects on flora and fauna [28]. Hydraulic stress in the watercourse causes erosion and wash off aquatic organisms.

The hydraulic overflow of the sewer system is another consequence of excessive SW. In these situations, the pressure mode of the water flow in the sewer can cause the discharge to the terrain through the manholes and discharge into basements. The frequency of intense SW with the hydraulic overload of the sewer system is individual. Usually, these situations occur in the spring and summer months. Often it is local flooding, so there is no flooding of the entire urban area. It is usually the combined sewage system containing both municipal WW and SW which represents an increased flow and often causes the hydraulic overload of the WWTP. Then turbidity effluent from WWTPs is common for biological activation WW treatment tanks [26].

Disruption of the natural hydrological regime in cities has a negative impact on the environment during dry periods as well. During dry periods, vegetation is low and does not absorb solar energy which leads to further drying of the soil. Soil dryness and deficiency of capillary water have a negative impact on the development of the root system of vegetation. Also, the dry vegetation cannot transpire and thus does not much positive impact on the quality of life in urbanized areas [23].

2.3 Stormwater Quality

In the atmosphere, SW comes into contact with various chemicals. Its quality in this environment is influenced by air pollution. SW also combines with CO_2 in the air. After passing through the Earth's atmosphere, SW has a pH value of about 5.5; in some locations, it is below 4.5 [29]. SW pollution occurs in the following ways:

- combining with dissolved and undissolved substances at atmospheric level;
- accumulated pollution at the surface of the territory during dry period that is mixed with SW;
- pollution arising from the contact of SW with materials on the surface where the water falls.

Table 2.1 Average composition of selected SW parameters in the atmosphere in the CR from 2000, 2006, 2012 [29, 40]

Parameter	Unit	2000	2006[a]	2012[a]
pH	–	4.65	4.57	4.91
NH_4^+	$mg\ L^{-1}$	0.94	0.82	0.81
NO_3^-	$mg\ L^{-1}$	2.74	3.80	2.11
SO_4^{2-}	$mg\ L^{-1}$	2.17	1.69	1.37
F^-	$mg\ L^{-1}$	0.02	0.01	0.02
Pb	$\mu g\ L^{-1}$	4.30	4.30	1.31
Cd	$\mu g\ L^{-1}$	0.22	0.13	0.04

[a] Average composition for Prague 4-Libus

2.3.1 *Pollution at Atmospheric Level*

Pollutants in the atmosphere are one of the causes of SW pollution, especially in large cities and industrial areas. During the rain, the atmosphere is being treated, and the pollutants are trapped in SW. The chemical composition of precipitation depends on the composition and air pollution in the lower and middle level of the atmosphere. This composition may fluctuate considerably in terms of rainfall, duration, and intensity. Most polluted is the precipitation in industrial areas, the least in mountainous areas. According to Table 2.1, the long-term pollution monitoring of the atmospheric precipitation in the CR shows that there is a gradual reduction of pollution.

Regarding the anthropogenic sources of pollutions, acids and acidic substances are predominantly higher than alkali substances. Acids are predominantly present in sulfur compounds, especially in SO_2 and H_2S. Furthermore, acids are contained in nitrogen compounds such as N_2O, NO, and NO_2 from the combustion of fossil fuels, motor vehicles, and microbial denitrification in soil and water. Chlorine compounds are from the combustion of PVC plastics in incinerators. Ammonium ions in fertilizers and carbonates used in agriculture are the pollutant source of alkaline substances. Other atmospheric pollutants are heavy metals and organic substances [26].

2.3.2 *Accumulated Pollution During Dry Period*

The important parameters to determine SW pollution are the period of drought, the amount and intensity of atmospheric SW, and the volume of rain runoff. Higher concentrations of pollution are determined at the start of drainage by the first flushing effect. Thus, mobilized dry deposition of corrosion pollutants and impurities are leached at the start of the rain [26].

SW is the only natural way to clean roofs. SW from the roof is characterized by high pollution of acidic substances (CO_2, SO_2) and variable content of organic substances (pollen, branches, dust, bird droppings). Thus, SW from the roofs may contain pathogens from the bird droppings. Microbial contamination has the effect of microbial contamination in tanks and drainage pipes and SW collection. The odor of water is dependent on the value of contamination which requires treatment and disinfection in case of use of SW for irrigation or flushing the toilet. SW without treatment and disinfection from the roof drainage is not recommended for use in washing laundry [30]. However, SW from roofs is generally less polluted than SW from roads.

SW pollution from roads and paved areas is affected by automotive transport, erosion of paved areas, garbage, animal excrement, vegetation, industry, use of salt in winter, and others.

Chemical characterization of SW depends directly on the runoff surface process and various types of micro-pollutants which are released into the water due to anthropogenic activities in the catchment area [31]. These micro-pollutants are called xenobiotics and include hazardous substances such as pesticides, detergents, lays, pharmaceuticals, metals, and many others [32].

2.3.3 Pollution Arising from Stormwater Contact with Materials

The quality of SW is also dependent on the type of surface from which SW runs. SW pollution comes from direct contact with water, especially with roofing and piping. Parts of the buildings are exposed to sun, water, frost, and freeze in the form of particles that form a significant part of the pollution in SW. The extent of this pollution depends on the building and technical condition of the buildings and the material used. Roof and eaves can be labeled as conventional pollutants. The metal parts of the roof corrode and release toxic substances represented by heavy metals: Cu, Cr, and Zn.

Asbestos cement releases lindane after a long dry season into SW, which is a hazardous substance with permanent health effects and has a negative impact on the environment [26].

2.4 Stormwater Management Methods

SWM methods are represented by two main ways: measures to minimize surface runoff and measures to reduce runoff pollution [23].

Measures to minimize surface runoff are performed in place of SW. These measures represent minimizing paved areas in the design of urbanized areas. And for existing paved urbanized areas, measures are needed to disconnect these areas. Another measure is represented by the design of permeable areas and green roofs of buildings. Retention and use of SW for irrigation, toilet flushing, and cleaning in buildings are other measures to minimize drainage.

Measures to reduce runoff pollution are performed by regular and effective road cleaning to minimize pollution accumulation at the surface of the area during a dry period. It is necessary to reduce the salting of roads in the winter, the use of herbicides and pesticides. It is also necessary to avoid contact with SW pollution source as in improperly designed roof material or industrial areas for storage of chemicals [26].

SWM methods are partially dependent on specific local conditions. However, the fundamental solution is performed in the design of a decentralized SW system. The decentralized approach is addressed by the design of small areas, thus for individual plots or buildings. Alternatively, it is possible to design a centralized system for SWM which is represented by a common SW retention and infiltration for more areas, plots, and buildings. Pretreatment of SW is required before the infiltration or drainage into the watercourse. Pretreatment is designed according to the hydrogeological survey, local disposition parameters of the urbanized area, and economic conditions [26].

Based on the regulation to select SWM methods, infiltration of SW in a decentralized system should be considered a priority. Retention with controlled runoff is permitted under unsuitable hydrological conditions in the urbanized area. The last option is the controlled flow into the sewage system at WWTP.

The SWM design is dependent on the type of situation, especially whether it is a new or old approved area. Predetermined design criteria for SWM are necessary in the case of newly built urban areas. For existing urbanized areas, the designer suggests the most acceptable solutions in symbiosis with general requirements. General SWMs can be defined as follows [23]:

- conversion and reconstruction of existing drainage facilities;
- implementation of green roofs for buildings that have necessary constructional assumptions;
- perform decelerated retention in individual properties;
- new pretreatment of SW at the ends of the SW drainage sewage system;
- use existing road ditches that can be newly used for retention or ditching of SW;
- implement local measures to reduce drains on existing paved areas; local SW needs to be attached to grassland;
- propose measures for SWM for major routes;
- impermeable surfaces should be replaced with permeable infiltration.

For the design of SWM are necessary design data, in particular the type of catchment in urbanized areas (residential, mixed, industrial, etc.), map data (layout, urban plan, drainage plan, altimetry, etc.), geological and hydrological data (soil structure, soil infiltration capacity, distance of groundwater level, old environment pollution, etc.) [24].

According to the hydrological conditions related to hydraulic design rainfall calculations, it is necessary to define [26]:

Table 2.2 Overview of construction and technical measures for SWM

Stormwater pretreatment	Stormwater retention	Stormwater infiltration
Settlement shaft	Stormwater retention reservoir	Surface infiltration
Inlet shaft with sludge bottom	Filter well	The infiltration of water retention above ground
Geotextile filter bag	Retention on green roofs and water retailing roofs	Infiltration of groundwater water retention
Lightweight separator	Swimming pool with biota	Multicomponent infiltration elements
Settlement pond	Retention in the industrial areas	
Vegetation pond		

- types of impermeable surfaces to determine the runoff coefficient of the relevant surface;
- a survey into the quality of surface water and groundwater;
- a concentration of SW pollution;
- the determination of the property rights of the property and the approved areas.

According to Czech standard ČSN 75 6101 [33], the design flow of SW *QD* depends on the rainfall in the relevant locality, size of the area, and the runoff coefficient by the following calculation:

$$Q_D = F \cdot \psi_S \cdot i \quad (\mathrm{l\,s^{-1}}) \tag{1}$$

where Q_D represents the design flow of SW ($\mathrm{l\,s^{-1}}$), F is the total catchment area of the sewer, horizontal projection (ha) and ψ is runoff coefficient as an index number (–) [33] and i represents the intensity of design reserve rain required periodicity ($\mathrm{l\,s^{-1}\,ha^{-1}}$).

Typically used construction and technical measures are possibly divided into three parts (Table 2.2): SW pretreatment, SW retention, and SW infiltration.

2.4.1 Stormwater Pretreatment

SW pretreatment is difficult to design uniformly; thus, an assessment is made for individual construction and technical measures. It is important to know the contamination and the desired pretreatment efficiency. Generally, the pretreatment of SW is used following treatment principles: sedimentation, filtration, adsorption, chemical process, and biological process [26].

The sedimentation process has a high removal rate of pollution as many pollutants are solid particles and are therefore captured by the sedimentation process at the bottom of the device. Dissolved pollutants are trapped in the establishment of floating substances. Low water turbulence is important to increase sedimentation efficiency. Design and maintenance are necessary to prevent sediment expansion due to impact loads. It is important to minimize turbulence at the inlet and outlet of settling shafts and tanks [34].

The filtration effect occurs in a flat overflow over natural surfaces, alternatively via water-permeable paving, mineral concrete, and drainage asphalt. The treatment effect for removal of undissolved substances is high [34]. There is a strong accumulation of solids on the surface and in the upper 300 mm of soil. Suspended particles from the effluent are mechanically bound in the soil. In soils with fine pores, SW can be filtered and the fine particles (<0.2 mm) [35]. Filtration capability is determined by the diameter of the pores of the water path and their continuity. Due to the accumulation of particles, the filtering ability is reduced over time.

Adsorption is investigated by electrostatic or covalent forces on charged and uncharged surfaces. Heavy metal pollution is bound by adsorption on the exchangers. In adsorption, it is necessary to distinguish between non-specific adsorption that exists on the surface of the exchangers due to Coulomb forces and stronger specific adsorption. The force of the connection increases from Cd through Ni, Zn, Cu, and Pb. Pb shows the strongest specific adsorption in the soil. Substances bound by specific adsorption occur in the soil relatively tightly bound. For non-specific adsorption, substances are released into the soil. The long-term stabilization of metal ions occurs due to the incorporation into the crystal lattice of clay minerals. The most important cation exchangers in soils are clay minerals and humic substances [34].

Chemical processes are important for the deposition and degradation of pollutants in the soil. Free dissolved oxygen in the soil acts on metal oxides and metal hydroxides to cause their disintegration. To a limited extent, heavy metal sulfides are formed. From inorganic substances, complex compounds of heavy metals are important for treatment processes [34].

Biological treatment is carried out by microorganisms. Bacteria and fungi in the soil remove organic pollutants, thus converting them into inorganic substances. The final products from removing substances are carbon oxides and water. Oxygen content in soil is important for treatment efficiency. Also, plants can remove pollutants in the soil, especially Cd, Cu, Ni, and Zn are captured by their roots. In the long term, high concentrations of heavy metals can permanently damage the microbial life in the soil. Heavy metals in the soil immediately reduce the breathing ability of microorganisms. The toxicity of heavy metals to the microorganisms decreases on Cd-Ni-Zn-Pb [34].

SW biological pretreatment is achieved usually in conjunction with mechanical processes. Soil and substance treatment capability is based on physical, chemical, and biological processes.

Soil and substance treatment capability is based on physical, chemical, and biological processes. SW pretreatment is achieved usually in conjunction with processes such as settlement shafts, inlet shafts, geotextile filter, the lightweight separator, the settlement pond, and the vegetation pond.

The settlement shaft is designed to SW pretreatment with a high level of sedimentation substances at low water inlets. The shafts have a very low surface area requirement. SW is mechanically treated by sedimentation in shaft elements with a concreted bottom. Floating and light substances are collected on scumboards. The disadvantage is the turbulence that prevents the settling ability. Turbulence can be reduced by designing striking boards [26].

The inlet shaft with the sludge bottom is represented by a combination of the sedimentation and infiltration shaft. The bottom of this shaft is impermeable. This bottom is a settling space which is used for pretreatment of precipitation waters with a high proportion of settling substances. Shafts do not have high surface area requirements, and investment costs are low. The disadvantage can be limited maintenance. The treatment capacity is high due to sedimentation. A further increase in the reduction of the proportion of suspended particles in the filter gravel can be done through the casing of the shaft with geotextile filter.

The geotextile filter bag is for pretreatment of effluents with predominantly undissolved substances, thus in urban areas with small available areas. The special geotextile is sewn with respect to the size of the pumping shaft in the bag. SW is treated due to the mechanical filtering properties of the fabric with a throughput coefficient $k_f = 10^{-3}$m s^{-1} [26]. The advantage of such fabric is multiple usability and easy installation. The shafts have small surface requirements. The disadvantage is the low entrapment efficiency of the solutes.

The lightweight separator is used in addition to securing WW from technological processes for pretreatment of SW from paved areas with a high probability of contamination with petroleum substances. The treatment capability is high for light fabrics, but overloading is not permissible [34]. Lightweight separators for liquids such as petrol, diesel, and fuel oil generally contain a sludge compartment and a separation section Class I or Class II. Separators may contain sampling space. In some cases, lightweight sorbents may also have a sorption part [26].

The settlement pond is used for pretreatment of SW with a high proportion of settling substances in the outskirts of urban areas with a large number of undeveloped areas. Settlement pond contributes to improving the climate in urban areas. The settlement ponds are impermeable with respect to the ground. The settlement pond has a mechanical and biological function as a natural treatment pond. Settlement pond has high treatment ability for suspended and dissolved substances. It also has high treatment efficiency. With sufficient residence time, it has high sedimentation properties for $d > 0.1$ mm particles. Biological purification is carried out in the tank using aerobic and anaerobic processes [34]. The disadvantage can be a large area and necessary maintenance. It also poses a danger to children playing, so fencing is recommended here.

In the vegetation pond, SW flows mainly horizontally toward plants planted within a soil body which is sealed with respect to the subsoil. The vegetation pond is suitable for further purification of highly organically contaminated SW. The vegetation pond can be integrated into the urban area as a biotope and in residential areas [34]. It has high treatment ability of dissolved solids. The treatment capacity is high due to anaerobic and aerobic degradation processes; mechanical pretreatment is carried out at the same time due to the filtration of harmful substances in the soil body and the physical and chemical binding (adsorption) of pollutants to the soil particles. In winter, the biological treatment capacity drops by about 20%. The maximum degree of efficiency is achieved only with continuous inflow. Therefore, it is necessary to have retention space. It is advisable to install a treatment device for sedimentation of undissolved substances although it may have large area requirements [26].

2.4.2 *Stormwater Retention*

Rapid and immediate drainage of SW from paved areas causes extreme flows in sewers and watercourses, but also causes flooding of infiltration objects, so that the infiltration speed does not match the inflowing amount of SW. Appropriate retention measures and adequate constraints of drainage from the retention facility may be avoided by undesirable flooding effects of rainfall [28].

Large drainage areas can be solved by retention, thus the accumulation of SW and its controlled release into the watercourse. Retention tanks replace the natural retention properties of the landscape.

Retention tanks regulate large SW, rain drains. However, the protective function is predominant. The design of retention tanks must be preceded by detailed surveys. Protective retention tanks can be divided into retention tanks, retention reservoirs with a well-defined protection area, anti-erosion tanks, SW tanks, and others [26].

A simple and cheap solution could be a retention facility at a suitable surface, thus the roof or temporarily flooded parking area [26]. The SW retention tank can become an aesthetic element of the landscape architecture and possibly used as a biotope or even a pond for bathing. Decentralized SW retention can be characterized by the following structural and technical measures: SW retention reservoir, filter well, retention on green roofs and water-retaining roofs, swimming pool with biota, retention in industrial areas. Important elements of retention are device for limitation, drainage regulation: vortex regulator, float regulator, pipe shrinkage, filter bed, and others [26].

The SW retention reservoir is shown in Fig. 2.1, and it is for the retention of rainfall on the property. It is necessary to have a free area, and regular maintenance is required. If the tank is aesthetically well solved, it becomes a valuable architectural element of the garden. Investment costs are generally low [26].

The filter well is characterized by a drainage system that is impervious to the subsoil. The draining SW is fed into the control shaft through the attic passage in which the sliding device cuts the drain or completely shuts down. The measure can be used independently of soil permeability and eventual contamination. The filter well is suitable for pretreatment of heavily polluted water.

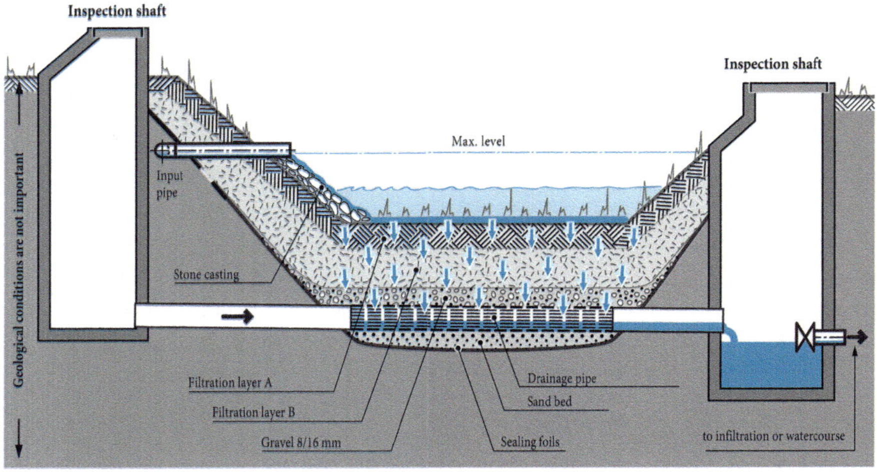

Fig. 2.1 SW retention reservoir [37, 38]

Retention on green roofs and water-retaining roofs are characterized by a green lawn with a soil substrate or a storage tank on flat roofs. The maximum amount of runoff is reduced by evaporation and slowing runoff. A static assessment is required in the design. Green roofs have only a retention effect. Thus, SW absorption is carried out on the soil substrate. Greenery can be extensive, where the vegetation layer is a mixed substrate of unpretentious plants after foundation, and it is not necessary to care for this greenery. The greenery on the roof can also be intensive when a green garden is established, lawns are grown, but also bushes and trees, and therefore, it is usually necessary to install the irrigation system. Depending on the type of green roof, SW may be retained in the soil body (intensive greening), but at least 30% (extensive greening). The green roofs represent biological and mechanical treatment, so green roof filters polluted matter through plants and soil. Greening the roofs can achieve improving the microclimate and correspondingly shaping the positive SWM impact of the urbanized area. Water-retaining roofs can be used where greening is undesirable. There is no water treatment effect in this type of retention [26].

Swimming pool with biota consists of a deep bathing area and a shallow part that is covered with plants. The biotope is a naturally self-treating water surface that functions as an ecosystem with a permanent quality of water suitable for bathing. In order to achieve water treatment efficiency, enough space is needed to regenerate water. Landscape architects recommend a minimum area of 50 m^2. Based on the climate conditions in the CR, the regeneration part mainly consists of aquatic, marshy, and moisture-bearing plants, such as orobi and reeds. The resulting level of treatment is simple water treatment like by filtration. If we manage to achieve automatic self-treatment process, it can ensure healthy swimming without chemicals [26].

Retention in industrial areas represents a limited runoff during short-term flooding. The depth of the water in these areas is within a few centimeters. Retention in industrial areas is easy to implement that partially restricts the operation of these areas. In the case of pipe connections to the docking facility or to the recipient, it is necessary to assess the cloud contamination of the rain drainage. Security overflow is essential [36].

2.4.3 Stormwater Infiltration

In the past, SW was naturally absorbed and supplemented the deficiency of groundwater. At present, most SW is discharged through combined sewage systems into watercourses. And this water does not return to groundwater on our local groundwater resources. In order to improve the water balance, it is necessary, according to local conditions, to reduce the surface runoff by local uptake, and only in the second step to connect the SW to the sewage system.

Infiltration of SW is not possible to implement in the protective zones of water sources, high groundwater levels and where impermeable or very low permeable subsoil (rock, clays, cracked environment) occurs. The level of pollution of SW is an important limitation for infiltration. Infiltration of SW must be ensured so that it does not affect the stability of building structures and buildings. The assessment of the suitability of the absorption should be done by the hydrogeologist according to local conditions [26].

In the natural water cycle, the infiltration zone occupies the function of permanent and reliable protection of lower water horizons. The protection of the infiltration zone is carried out by multiple physical, chemical, and biological processes and is heavily influenced by transport events and hydrogeological conditions. The intensity of individual natural processes in the underground is different. In the groundwater zone, filtration, adsorption, ion exchange, precipitation, and biological degradation are predominant. These processes are more intense in the soil layer for vegetation than in the lower layers of the infiltration zone. The dissolution and dilution processes are important in the water zone. These processes are mutually associated and may be related to natural changes in the environment. During the infiltration of SW on the soil, surface substances are separated by tiny particles and their sorbed substances. Another part of the substance is transported to the soil to a depth of 10–100 mm, and the dissolved substances are transported partially down to the level of groundwater. The deficiency may be a reduction in the efficiency of natural treatment capabilities associated with underground water quality, or the release of pollution accumulated in the top zone and their deeper movement. These negative effects can be reduced to a permissible level by technical measures [26].

For subsurface infiltration objects, the treatment capacity of biologically active soil is excluded, and the vulnerability of groundwater is considerably higher. Subsurface infiltration is therefore suitable only for SW runoff with low pollution. Therefore, the more polluted SW must be treated properly before being in the subsurface [28].

For the design, construction, and operation of the acquisition device, it is necessary to respect the regulations of the laws on the protection of water and soil protection.

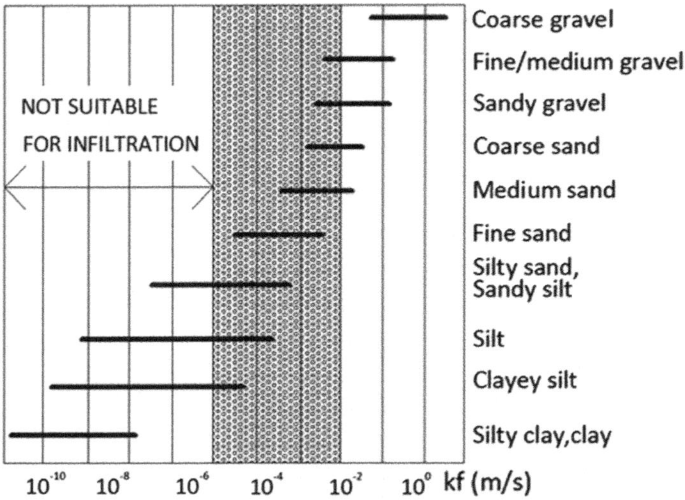

Fig. 2.2 Interval of the recommended coefficient of permeability for infiltration (indicated by a gray color) [39]

Protective measures reduce the release of substances from SW at runoff, reduce the input of substances from paved areas, modificate SW before the infiltration, and limit infiltration of effluents from contaminated areas.

In the areas of soil sedimentation, landslides due to erosion, in undermined areas, and possible washout effects, it is necessary to examine the effect of infiltration on geological stability. It is necessary to carry out an assessment when there are soils in certain localities that become unstable after being soaked with water.

An important condition for SW infiltration is the permeability of the soil and underground rock layers. Soil penetration is predominantly based on the grain size, the grain size, and the slope curve; soil structure and water temperature are also decisive for soils. It is expressed by the coefficient of permeability k_f in m s^{-1} [26].

For incoherent rocks, the coefficient of permeability kf is generally between 1×10^{-2} and 1×10^{10} m s^{-1}. Values kf lower than 1×10^{-6} m s^{-1} can only be used for infiltration with accumulation (controlled retention), so it is necessary to design a security overflow. At a coefficient of permeability more than 1×10^{-3} m s^{-1}, the SW overflows in the small capacities so quickly down to the groundwater level that it is not possible to provide a residence time sufficient for purification by chemical and biological processes. In Fig. 2.2, the gray-marked area represents the recommended interval for infiltration coefficient of permeability.

If the kf values are less than 1×10^{-6} m s^{-1}, the infiltration time is extended. It can initiate anaerobic processes in the unsaturated zone that could negatively affect its ability to capture and transform pollution. The infiltration design must assess the possibility of infiltration of space and distance from buildings and property boundaries [26].

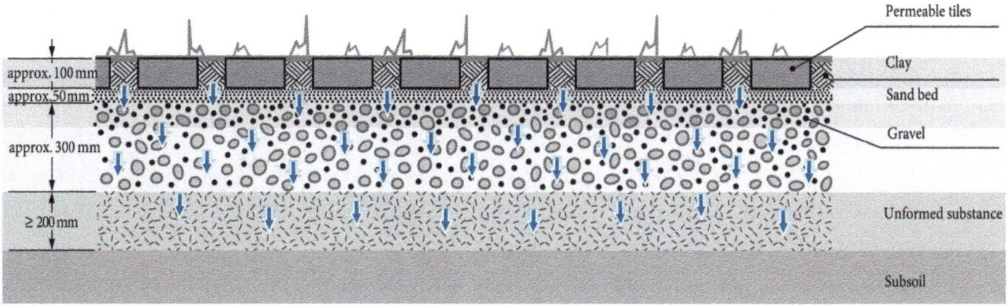

Fig. 2.3 Surface infiltration by permeable tiles [37, 38]

To design SW infiltration facility, it is usually necessary to use design rainfall with a lower occurrence period of 0.2 or 0.1 year^{-1}. During design rain, no flooding of the device should occur. For the predictive duration of the draft rain, the ATV-Arbeitsblatt A 138 [10] suggests that the design of the flat overflow should be based on the duration time $T = 10$ min. It is recommended that this drainage system, which drains through flats and areas where water can be retained, is increased to 15 min [26].

Surface infiltration is shown in Fig. 2.3, and it is infiltration through a permeable, paved, or vegetation surface area that does not capture SW. Infiltration capacity of the soil should be greater than the expected rain runoff. Frequent low hydraulic permeability is directed toward relatively high requirements on the contact surface. When the infiltration capacity of the soil layers to receive SW is not sufficient, it is necessary to drain the plain by drainage pipes. An advantage is the treatment ability of SW in the cover layer of soil with vegetation. Required infiltration is performed by the following surfaces: grassy areas, grassed gravel areas, grassing blocks, permeable tiles, permeable asphalt, and concrete.

Grass areas have high retention and evaporation capacity with high biological treatment efficiency. Low investment costs, easy maintenance, and control are benefits. The surface is formed by grass or higher vegetation, and the surface is suitable to bend to increase infiltration.

Grassed gravel surfaces have a high retention and evaporation capacity and are capable to biologically treat the pollution. These areas are inexpensive and suitable for passenger car use. The upper layer consists of a mixture of gravel and soil, which is composed of grass and gravel. Due to its load capacity, it can be used for parking areas.

Grassing blocks are represented by pavements with grassed joints, whereby the vegetation blocks have a high treatment effect. The blocks can be plastic or concrete. Grassing blocks can form reinforced communications to garages and residential buildings or as parking areas.

The perforated tiles are characterized by concrete paving with drainage joints passing through the SW through joints into the subsoil. Concrete paving from porous concrete is created by using a narrow grain fraction to form continuous cavities that convert to underground water. Passable tiles are suitable for use on walkways, bicycle paths, parking, pedestrian zones, and access roads.

Permeable asphalt and concrete are formed by adding much less of fine particles to the mix than in conventional concrete or asphalt. For permeable concrete, the pore volume is 15–22% of the total volume. For classical concrete, the pore volume is only 3–5%. Permeable asphalt or concrete is placed on a sandy subgrade in which SW is retained until it is absorbed into the soil.

The infiltration of water retention above ground is appropriate where there is not enough space for surface infiltration. In this method of infiltration, it is necessary to assess the aesthetic aspect of these objects. The infiltration facility can be supplemented with appropriate plant growth and permanent water surface, which improves the aesthetic appearance of the urbanized area. Devices can be divided into infiltration ditches and infiltration tanks.

The infiltration ditches can be grassed, covered with vegetation, or covered with gravel. Infiltration ditches should be dimensioned to allow only short-term infiltration. Longer drift is unacceptable, as this increases the unwanted surface clogging. Therefore, a maximum height of 300 mm is recommended. Bottom trench levels of ditches should be designed and operated horizontally to achieve the most uniform distribution of water for infiltration.

The infiltration tank is shown in Fig. 2.4, and it is characterized by infiltration through the vegetation layer of soil in the ground tank. For infiltration of tanks, the ratio between the bound impervious surface and the infiltration surface is generally greater than 1:15. This high hydraulic load in relation to the requirement of relatively rapid emptying of the tank requires sufficient and permanently secured permeability of the subsoil. It is important to correctly estimate the timing of the tank collation. Fragmentation is primarily caused by substances that carry SW drainage which is deposited in the infiltration zone and forms sealing deposits. Deploying the bottom of the tank toward the inflow site prevents the bottom from clogging. Thus, there is an increased sedimentation near the inflow. If infiltration of the tank does not have a pre-sedimentation, it is necessary to consider the design of the bottom permeability reduced to one-fifth. This takes into account the design collimation during tank operation. The use of infiltration tanks is only recommended in larger drained areas, usually from the area of 1 ha, or in new urbanized plots with free area. The advantage is high infiltration efficiency. The disadvantage is the bottom trapping in inappropriate operation as well as the necessity of fencing, as the filled tank poses a threat to the children playing.

Infiltration of groundwater water retention is used when there is an insufficient area for over-ground SW infiltration so that SW can infiltrate underground via infiltration in the grooves, piping infiltration, shaft infiltration, or infiltration in plastic pumping blocks.

Infiltration in the grooves is shown in Fig. 2.5. It represents the infiltration of SW in a trench filled with gravel or other porous material. In the groove area, the water is accumulated and passed to the lateral level with the intensity corresponding to the permeability of the surrounding soil. The size of the groove is generally limited by

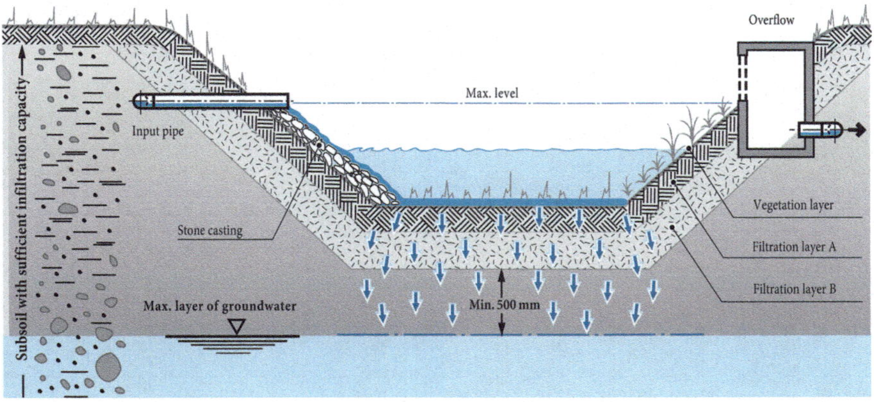

Fig. 2.4 Infiltration of water retention above ground—infiltration tank [37, 38]

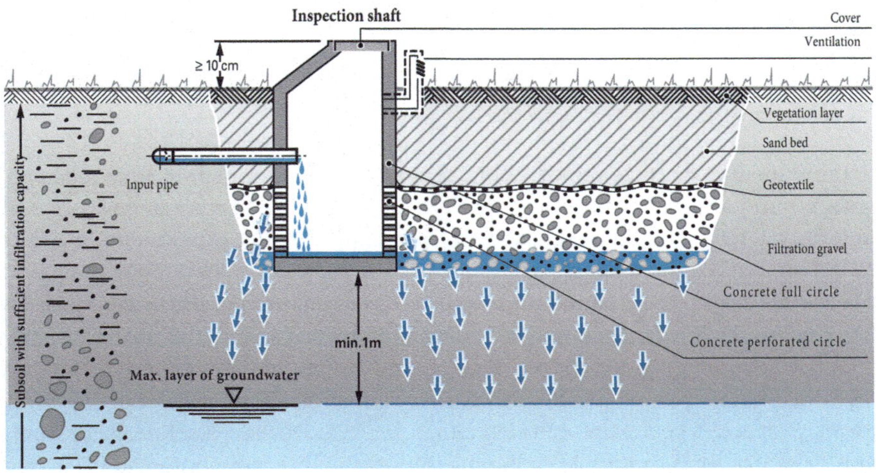

Fig. 2.5 Infiltration of groundwater water retention—infiltration in the grooves [37, 38]

the height of the water table. The construction of the groove must ensure that the water is evenly distributed along the length of the groove.

If SW is infiltrated through the vegetation layer of soil, high cleaning efficiency is achieved. If SW is connected directly to the lower permeable layers, it is necessary to install pretreatment device.

Pipeline infiltration is carried out in a perforated pipeline placed in a bed of gravel or other porous material. The pipeline is covered with terrain similar to or the same material. The diameter and perforation of the infiltration pipeline should have the adequate hydraulic capacity. The advantage of piping infiltration is the rapid distribution of point inlets.

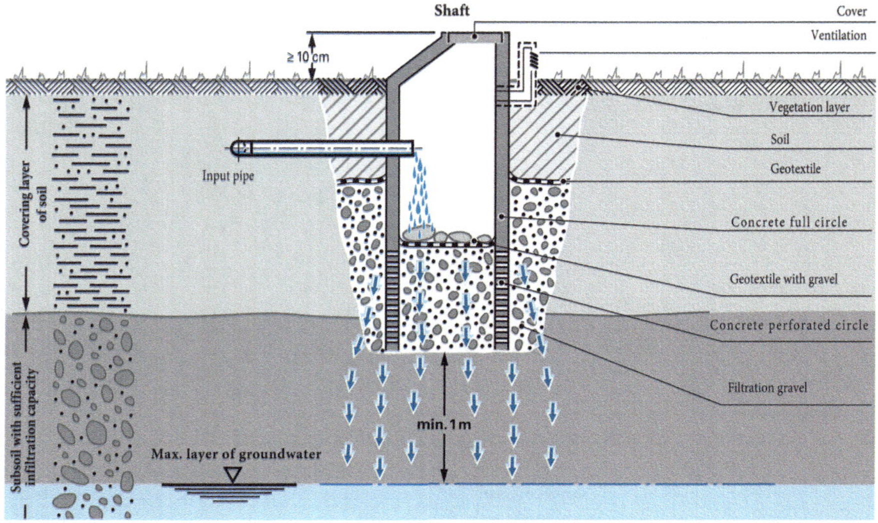

Fig. 2.6 Infiltration of groundwater water retention—infiltration in shafts [37, 38]

Infiltration in shafts is shown in Fig. 2.6, and it is performed for point and subsurface infiltration. SW first flows through an artificially formed filter layer. The infiltration shaft is usually composed of concrete planks. The minimum shaft width is 1000 mm in internal diameter.

Infiltration in plastic plumbing blocks is an alternative to classical infiltration drainage from aggregate and gravel or plastic drainage pipes. Infiltration cages and tunnels can be interconnected to create a high-capacity input field. The infiltration cages are located in an excavation which is laid out with a 150 mm layer of 16/32 mm gravel, wrapped in geotextile, and after being connected to the water line, it is covered with gravel of 16/32 mm and then with soil up to 150 mm above the cage. The soil layer is suitable for geotextile insulation from the gravel layer. From classical gravel drains with an absorption capacity of 30–35%, infiltration cages have an absorption capacity of about 3 times greater, up to 95%. Infiltration block guarantees high load capacity thanks to its construction. The excavation distance from adjacent buildings should be at least min. 2 m for buildings that are waterproof and 5 m for buildings that are not waterproof. Multicomponent infiltration elements are dealt with: a combination of plunging in a plume with a pipeline or a rinse, a combination of pipelining or piping [26].

2.5 Stormwater Reuse

In the past, SW was collected from paved surfaces, accumulated in collecting tanks for watering. SW is still perceived today by the general public as a slightly polluted source of water for further use.

The average demand for drinking water is over 100 L water per capita per day. However, about 50% of this need does not need to be drinking water. Thus, SW can be used as a replacement [26] for toilet flushing, irrigation, washing, and other use.

In the case of using SW to water the garden, lawn, or car wash, simple treatment processes are sufficient: filtration and sedimentation. It is necessary to ensure that larger particles such as leaves and branches do not get into the collecting tank. These particles would cause operational problems such as clogging pipes and header tank and could also damage the pumping SW for use.

SW usually requires only rough mechanical pretreatment to remove solids such as leaves and the like. It is necessary to consider fecal contamination by birds and other animals, so it is advisable to supplement the system with hygienic disinfection.

SW for washing requires effective filtration. Insufficient SW treatment results in stains on the laundry. This fact can be eliminated by two separate water connections to the washing machine, which most of the modern washing machines are equipped. Thus, for prewash, main wash and rinse use SW. Drinking water from the water system is needed only for the last rinse.

SW is advantageous for flushing toilet, as it is soft and no stone is deposited. Also, flushing toilet consumes along with bathing the most water in the household. Flushing toilets does not require high-quality water, thus using drinking water is unnecessarily wasteful.

As for watering the garden, SW contains a low salt level, so there is no soil salinisation. In addition, it does not contain chlorine although there are even plants that do not tolerate SW, such as Canadian blueberries. In addition, drinking water is too valuable to water the garden.

SW can be used for car washing and cleaning where there is no need for drinking water. In all these cases, a large amount of water is required, and it is economically and environmentally advantageous to use SW rather than drinking water.

Filtering SW leads to the removal of impurities and the residual bacteria. The rain drain is fed into the tank. SW should be stored in a cold place and not exposed to direct sunlight. Tanks on the surface of the terrain are mostly cheaper but are exposed to the influence of temperature, light, and eventual pollution. For these reasons, tanks located in the ground are recommended. The location of tanks in cellars is not recommended because the cellar temperature does not exceed 18 °C, thus as not to endanger the development of microorganisms. It is also recommended for hygienic reasons not to store the water in the storage tank for too long.

2.6 Conclusion

SWM as a decentralized drainage system represents the point of natural processing of SW and its return to in the natural water cycle. This chapter provides the basic overview of SWM in urban areas in Central Europe connected with surface runoff quality, SW quality, SW management methods, and SW reuse. And for conditions in the CR, the potential solutions are defined: evaporation, infiltration, and regulated runoff to watercourses. These solutions can be implemented by accumulation of SW and its use for irrigation and flushing toilets.

2.7 Recommendations

The concept of SWM in urbanized areas presents designs corresponding to natural SW drainage present outside the urbanized areas. The reduced quality and yield of surface and groundwater due to droughts and changing climatic conditions lead to an increase of reusing SW in the CR. Given the current continuing trend of reducing the cost of SW collecting, treating, and water reuse technology, the SWM will be more often designed in practice.

Acknowledgements This chapter has been worked out under the project No. LO1408 "AdMaS UP—Advanced Materials, Structures and Technologies," supported by Ministry of Education, Youth and Sports under the "National Sustainability Programme I."

References

1. Czech Hydrometeorological Institute (2018) Dry period 2014–2017, evaluation, impacts and measures. 92 p. ISBN 978-80-87577-81-3
2. Stránský D, Kabelkova I, Vítek J, Suchánek M (2014) The concept of stormwater management in the Czech Republic: current situation
3. Sieker F et al (2002) Regen(ab)wasserbehandlung und -bewirtschaftung unter Berücksichtigung der Anforderungen nach §7a WHG und einer möglichst ortsnahen Versickerung. Umweltbundesamt, Berlin
4. Directive 2000/60/EC of the European parliament and of the council of 23 October 2000 establishing a framework for Community action in the field of water policy. 2000. Official Journal L 327, 22/12/2000 pp 0001–0073
5. Federal Act of the Protection of Waters (2017) Bundesgesetz über den Schutz der Gewässer, (Gewässerschutzgesetz, GSchG). https://www.admin.ch/opc/en/classified-compilation/19910022/201701010000/814.20.pdf
6. Waters Protection Ordinance Gewässerschutzverordnung (GSchV) (2018). https://www.admin.ch/opc/en/classified-compilation/19983281/201806010000/814.201.pdf
7. Verordnung über Belastungen des Bodens (VBBo) (2016). https://www.admin.ch/opc/de/classified-compilation/19981783/201604120000/814.12.pdf
8. Gesetz zur Ordnung des Wasserhaushalts (Wasserhaushaltsgesetz—WHG) (2009). http://www.gesetze-im-internet.de/whg_2009/WHG.pdf

9. Gesetz zum Schutz vor schädlichen Bodenveränderungen und zur Sanierung von Altlasten (Bundes-Bodenschutzgesetz—BBodSchG) (1998). https://www.gesetze-im-internet.de/bbodschg/BJNR050210998.html
10. Arbeitsblatt DWA-A 138, Planung, Bau und Betrieb von Anlagen zur Versickerung von Niederschlagswasser (2005)
11. Verordnung über die erlaubnisfreie schadlose Versickerung von gesammeltem Niederschlagswasser (Niederschlagswasserfreistellungsverordnung—NWFreiV) (2000). http://www.gesetze-bayern.de/Content/Document/BayNWFreiV/true
12. TRENGW—Technische Regeln zum schadlosen Einleiten von gesammeltem Niederschlagswasser in das Grundwasser (2000). https://www.umwelt-online.de/recht/wasser/laender/bay/trengwz2000.htm
13. Merkblatt DWA-M 153 (2007) Handlungsempfehlungen zum Umgang mit Regenwasser
14. Act No. 254/2001, on water and amending certain acts (Water Act). Czech Republic (2011)
15. Act No. 183/2006, law on Territorial Planning and Building Code (Building Act) (2006)
16. The Plan of Main River Basins of the Czech Republic (2007) Ministry of Agriculture of the Czech Republic. Governmental Resolution No. 562
17. Act No. 274/2001, on public water supply and sewerage systems and on the amendment of some other acts (Water and Sewerage Act) (2001)
18. Decree No. 501/2006 on general land use requirements (2006)
19. Decree No. 269/2009 amending Decree No. 501/2006 on general land use requirements (2009)
20. ČSN 75 9010 Infiltration of rainwater device (2012) p. 44. Czech Republic
21. Samek O (2013) Motivace k hospodaření s dešťovou vodou. TZB-info. https://voda.tzb-info.cz/destova-voda/9961-motivace-k-hospodareni-s-destovou-vodou
22. Standard TNV 75 9011 Management of stormwater (2013) p 65
23. Stránský D, Kabelková I, Vítek J, Suchánek M (2007) Association of wastewater treatment experts of the Czech Republic. Background to the concept of stormwater management in urbanized areas
24. Hlavínek P (2015) Stormwater management in Urbanized areas. In: Stormwater management examples from Czech Republic, Slovakia and Poland. Switzeland: Springer Hydrogeology, 212 p. ISBN 978-3-319-25833-1
25. Paul MJ, Meyer JL (2001) Streams in the Urban landscape. Annu Rev Ecol Syst 32(1):333–365
26. Hlavínek P, Prax P, Sklenárová T, Dvořáková D, Polášková K, Kubík J, Hluštík P, Beránek J (2007) Stormwater management in Urbanized areas 1. Brno: Ardec, 164 p. ISBN: 80-86020-55-X
27. Tetzlaff D, Grottker M, Leibundgut Ch (2005) Hydrological criteria to assess changes of flow dynamic in urban impacted catchments. Phys Chem Earth 30:426–431. https://doi.org/10.1016/j.pce.2005.06.008
28. Krejčí V, Hlavínek P, Zeman E et al (2000) Drainage of urbanized areas: a conceptual approach. NOEL, Brno. ISBN 80-86020-39-8
29. Pitter P (1999) Hydrochemie, ICT, Publishing: UNITISK, s.r.o., 568 p. ISBN 80-7080-340-1
30. Klein. Aspekte der leitungsgebundenen Regenwassernutzung, Wasserversor- gungsverband Euskirchen-Swisttal, 10 p. http://www.wasser-eu-sw.de/pdf/regenwasser.pdf
31. Hlavínek P, Žižlavská A (2015) Classification, resources and elimination of xenobiotics in stormwater. In: Stormwater management examples from Czech Republic, Slovakia and Poland. Springer Hydrogeology, Switzeland 212 p. ISBN 978-3-319-25833-1
32. Fatta-Kassinos D, Bester K, Kümmerer K (2010) Xenobiotics in the urban water cycle. Mass flows, environmental processes, migration and treatment strategies. Environ Sci 16(1):507
33. Czech standard ČSN 75 6101 (2012) Sewer systems and house connections. 41 p
34. Geiger W, Dreiseitl H (2001) Neue Wege für das Regenwasser—Handbuch zum Rückhalt und zur Versickerung von Regenwasser in Baugebieten. Oldenbourg Wissenschaftsverlag, Hardcover/gebunden. ISBN: 3486264591
35. Scheffer F, Schachtschabel P (1992) Lehrbuch der Bodenkunde. Enke, 1992, Hardcover/gebunden. ISBN 3432847734

36. AGW (1996) Amt für Gewässerschutz und Wasserbau des Kantons Zürich. Die Versickerung von Regenwasser auf der Liegenschaft
37. VSA—Verband Schweizer Abwasser—und Gewässerschutzfachleute (2002) Regenwasserentsorgung: Richtlinie zur Versickerung, Retention und Ableitung von Niederschlagswasser in Siedlungsgebieten, Ent-wurf
38. Krejčí V, Kabelková I (2002) Use of retention and infiltration of rainwater. Chapter 11
39. Markovič G, Zeleňáková M, Vranayová M, Kaposztásová D (2015) Research of infiltration facility efficiency and quality of rainwater harvested from surface runoff in real conditions. In: Stormwater management examples from Czech Republic, Slovakia and Poland. Springer Hydrogeology, Switzeland 212 p. ISBN 978-3-319-25833-1
40. Cenia (2013) Czech environmental information agency. Statistical yearbook of the environment of the Czech Republic 2013, Ministry of the Environment. 414 p

Chapter 3
The Green Roofs and Facades as a Tool of Climate Cooling in the Urban Environment

Martin Šenfeldr, Petr Maděra, Pavla Kotásková, Jitka Fialová, Miroslav Kundrata and Vlastimil Rieger

Abstract The aim of this monograph chapter is to introduce green facades and green roofs in the Czech Republic. In the beginning, the history and development of green roofs and facades have been outlined; the type of green roofs and ways of facade greening, structures and systemic solutions, protection against slide and drainage have been described; and the functions and benefits of green roofs and facades have been estimated. Next part discusses measurements of the vapor from green roofs and the cooling effect. Finally, the zero carbon facility has been described—example from Brno (Czech Republic)—The Open Gardens. The results from online monitoring have been shown, and the roof and urban heat islands have been discussed.

Keywords Roof sheathing · Climbing plants · Succulent · Vertical gardens · Zero carbon · Cooling effect · Evaporation · Water cycle · Urban heat island · Green infrastructure

---

M. Šenfeldr (✉) · P. Maděra
Department of Forest Botany, Dendrology and Geobiocoenology,
Faculty of Forestry and Wood Technology, Mendel University in Brno, Brno,
Czech Republic
e-mail: martin.senfeldr@mendelu.cz

P. Maděra
e-mail: petr.madera@mendelu.cz

P. Kotásková · J. Fialová
Department of Landscape Management, Faculty of Forestry and Wood Technology,
Mendel University in Brno, Brno, Czech Republic
e-mail: pavlakot@mendelu.cz

J. Fialová
e-mail: jitka.fialova@mendelu.cz

M. Kundrata · V. Rieger
Czech Environmental Partnership Foundation, Brno, Czech Republic
e-mail: miroslav.kundrata@nap.cz

V. Rieger
e-mail: vlastimil.rieger@nap.cz

M. Zelenakova et al. (eds.), *Management of Water Quality and Quantity*,
Springer Water, https://doi.org/10.1007/978-3-030-18359-2_3

3.1 Introduction

The roof protects the under-roof spaces against weather. The basic function is to protect the construction and the indoor environment from precipitation water, to protect and secure the condition of the indoor environment. The roof can be operationally used if it is designed for that purpose [1].

Roofs that are not designed for that purpose fulfill only the basic function. They are expected only for the movement of instructed persons on the roof, who is ensuring the control and maintenance of the roof itself and its supplementary constructions.

If the roofs are designed for operational use, they can be used as roof terraces, roof gardens, car parks, heliports, and more. It requires the adequate design of the structure and construction of the roof and special drainage solutions. Moreover, it requires the solutions of the security measures, because the so-called uninstructed persons move on the roofs.

As ČSN 73 1901 Roof Design—Basic provisions stated, the roofs must meet basic requirements such as mechanical resistance and stability, fire resistance, protection of the indoor environment from noise. Hygiene, health, and the environment protection (exclusion of occurrence of molds, water leakage and exclusion of damp of building structures, and subsequent deterioration of the indoor environment by moisture) should be followed. Besides the energy saving and thermal protection, but also the aesthetic requirements for the appearance of the roof must be met. The bearing structure of the roof should be designed in such a way that its durability is the same as that of the building construction. Reliability is ensured by adequate technical solutions and maintenance of roofs. The system of inspections, controls, maintenance, and renewal of the roof, including drainage elements, must be determined by the designer.

One of the possibilities of building roofing is a green roof, a roof with a vegetation cover, where the upper layer of the roof consists of soil and vegetation in the form of mosses, herbs, grasses, shrubs or trees. These types of roofs are called green, greening roofs, vegetation roofs, grassed roofs, roof gardens, etc. It can also fulfill the ecological function except for the representative, mainly nowadays.

There are many reasons for increasing green areas, especially in cities where the construction of buildings displaced practically all the greenery only into parks. Green roofs may become an almost integral part of roofs of new buildings in urban areas just because of air purification from airborne particles and temperature reduction, as well as the increase of humidity in the roof surroundings and partial water retention, especially during summer months [2].

The green roof may not be just a matter of flat roofs, i.e., with a slope of up to 5°, as it may seem, but it is possible to green the roofs with a slope of up to 60°. In slopes of 20°, the substrate must always be secured against landslide, in the case of smaller slopes with respect to substrate coherence. However, the roof with a slope of 2°–5° is considered optimal [3].

The concept of a green facade (vertical gardens, vertical walls, green walls, or living walls) is used to grow plants located in a vertical plane so as to cover walls, pillars, or columns of buildings.

3.2 History and Development of Green Roofs and Facades

The tradition of green roofs can be found both in the Nordic countries, e.g., Iceland, USA, Scandinavian countries and in regions with very different climatic conditions, e.g., Tanzania. They are mostly without the possibility to be walked on, and their meaning was purely practical, i.e., thermal insulation and heat or cold accumulation. In the harsh climate of Iceland, houses with the walls and the non-supporting part of the roof from two to three layers of peat turf were built. This provided adequate protection against heat leakage, and these houses did not need to be heated, because the necessary heat was produced by the residents of the house [4].

The first preserved mentions about green roofs have been found in the Middle East during excavations of Nineveh (Assyria and Babylon areas, north of today's Iraq) from the period of the reign of King Solomon (929–917 BC. N. L.) [3].

However, the world's second wonder—the Semiramis Hanging Gardens in Babylon—which were probably founded in the sixth century BC, is better known. It was a terraced roof garden with a square ground plan. The weight of the soil, not to mention the trees that grew out of it, was an extraordinary load. Roof gardens were supported by colonnades—arches supported by stone columns. Admirable is mainly the support and irrigation system. A sophisticated irrigation system with spiral pumps powered by human power was used here. The water was further distributed by a system of channels to vegetation. The water was pumped from the nearby river Euphrates [5].

The roof gardens have spread from the Middle East to Southern Europe during the period of the Greek and Roman Empires. During this period, the rich Roman burghers of the higher classes, the so-called patricians, built these gardens on their houses and palaces to put their wealth on display. Because poorer people cannot afford them, they placed pots on the terraces [3].

The construction of roof gardens almost stopped for 400 years after the demise of the West Roman Empire [4]. The green roofs were built in the Byzantine Empire again in the palaces in the Middle Ages. Their popularity grew in France and especially in Italy since the mid-eleventh century, but the biggest boom occurred during the Renaissance, thanks to patrons for example in Rome, Venice, and Verona. In the seventeenth and eighteenth centuries, the roof gardens extend into German and English cities. There is also theoretical work on this issue, and in some Italian urban states, the regulations on the construction of green roofs are adopted [3].

An important milestone not only for the structure of green roofs, but also for the entire construction industry, was the year 1867, when Joseph Monier patented his invention—reinforced concrete [3]. The building physical properties of reinforced concrete allowed the development of green roofs. The new material helped to solve a number of existing problems. Roof gardens still had high demands on the static solution, but the reinforced concrete ensured the permanent strength and stability of the building and extended the service life of all building elements [6].

Since the beginning of the last century, the theme of green roofs has been dealt with, for example, by the Swiss architect Le Corbusier, who published five points of modern architecture already in 1926. The second point is about roof gardens on flat roofs. Le Corbusier emphasized that green roofs increase the green area of the city, and their soil protects the top floor of the house from moisture [7].

Green roofs are more often used due to the development of building chemistry and plastics after the Second World War. More and more significant buildings are being designed with roof green after 1950 [6].

Green roof gardens are created not only as a protection against external influences but also as a means of ecology in the 70s and 80s in the 20th century in the German-speaking countries. German experts deserved their research on the development of new materials for greening roofs, such as the protective layers against root penetration, or drainage prefabricated from polystyrene foam. They also developed the principles for establishing green roofs, even for greening of roofs with the slope. The companies specialized in green roofs are established in parallel with the research and they become a regular part of designing new objects or renovations. Since 1993, a program for the implementation of roof gardens for agriculture, which was ended after the economic crisis in the year 2008, worked in Russia (http://ccisf.org/cci/) [3].

The green roofs are popular in Germany or Australia. The Canadian city of Toronto has adopted a regulation for mandatory green roofs in large commercial, administrative, and residential buildings already in 2009, wrote Guardian. France decided that each new building in the commercial zone would have to have a "green" roof in the spring of 2015. The developer will be able to choose whether to place photovoltaic panels on the roof or to cover the roof with plants [8].

The green roofs, or rather the roof gardens, begin to appear in the Czech Republic in the second half of the nineteenth century. The first sign is the roof of the castle stable in Lipník nad Bečvou, which is still functional. It has undergone two large reconstructions—the first in 1911 and the second not long ago—in 2005–2006 [3].

Functional and no problem realization of the green roof is allowed by the development of hydro-isolation technologies. Increased interest in green roofs can be seen here after 1989, when it is more often possible to get acquainted with the practical experience of building green roofs from abroad. Ecological educational centers, which also focus on the issue of green roofs, are established in our country.

Climbing plants are one of the possibilities of greening the facades; their history dates back to ancient times. Olšan [9] states that the oldest iconographic sources about climbing plants on the walls are wall paintings in tombs. Grapevine and ivy were used to climb various pergolas along the paths or were planted on sunny walls in ancient Greece and Rome. In the gardens of the late Middle Ages (France, England, Netherlands, and Central Europe), there were plants tied to support constructions on the walls in the gardens. These were fruit trees most often. This type of plant cultivation was referred to as the Mauerspaliere in Germany.

Olšan [9] states that interest in the use of climbing plants in the composition declined at the time of Baroque. Climbing trees were planted to buildings and elements of only a small garden architecture. For example, fences, which could be "embroidered" by climbing plants, were formed along the lower part of the high green walls and hedges. The assortment was not too wide; woodbine, roses, honeysuckles, ivy, and peas were mainly used. Since the first half of the nineteenth century (periods of neoclassicism and romanticism), climbing trees have found greater use in natural landscape gardens and parks. The variety Rosa x francofurtana became the dominant growing climbing variety, which spread rapidly throughout Central Europe. This plant was mainly used to climb the arcades and was later pushed by new varieties of climbing roses. To this period, ivy, woodbine, and Aristolochia can be included. The second half of the nineteenth century was greatly influenced by the theoretical work of German garden architects and practitioners of Gustav Meyer and Hermann Jäger.

Vertical gardens are most often associated with the French botanist, scientist, and art designer Patrick Blanc (1953). He became famous for the vertical green walls called le mur végétal or vertical garden (plant wall). The green wall of the Hotel Pershing Hall, Paris, from 2001, is the first big realization of the Blanc system. Until then, it was applied only to relatively small areas. A total number of 320 plant species were used to create this over thirty meters high natural tapestry with a variable look. It is a living green composition that connects all six floors of the building [10].

It is possible to consider as a predecessor of vertical gardens in the Czech Republic a rock created more than 40 years ago in the Arboretum of Mendel University in Brno (see Fig. 3.1). Vertical outdoor gardens are rather rare in our country.

3.3 Types of Green Roofs

Standard ČSN 73 1901—Roofs design—the basic provisions distinguish only two groups of green roofs—classical growing layering with intensive greenery and economical growing formation with extensive planting. However, in practice and also in professional literature Čermáková and Mužíková [3] a finer division is used. According to the degree of necessary gardening interventions and the origin of the plant community, we distinguish biotope, intensive, semi-intensive, and extensive green roofs. In the design of the roof, the climatic conditions of the site should be taken into account.

3.3.1 Roof with Intensive Planting

We speak about intensive planting when the depth of the substrate is at least 300 mm, but the exact distribution is different in the literature. The formation can be up to

Fig. 3.1 A predecessor of vertical gardens—a vertical rock in the Arboretum of Mendel University (*Source* Kotásková)

1000 mm in depth, and it allows planting not only of lawns, perennials, grasses, bulbs, annuals, but also of shrubs or trees with flat root systems [6].

This type of greening is, therefore, more demanding for maintenance and also for the purchase price. Intensive green roofs are designed with such vegetation, which requires gardening maintenance to keep its character. If this maintenance is neglected, the planting vegetation will die and, as a consequence of this, there will be a significant reduction of the value of the entire roof covering—deterioration of aesthetic perception, reduction in retention capacity, erosion of the substrate, and associated loss of proper function of the roof. It is recommended to provide these roofs with a subsurface irrigation system.

This type of roof can only be used on flat roofs. The advantage of this type of greening is that it is a real roof garden with the possibility of using it for relaxation (see Fig. 3.2). Deeper layers of substrate and appropriate vegetation enhance the function of the green roof. Greater loads are to be expected; the weight of these roofs when they are fully saturated with water can reach 300 kg/m^2 or more [3].

3.3.2 Roof with Extensive Planting

Extensive greening differs from the previous type by a thinner layer of the substrate (approximately 60–200 mm). Thanks to it a smaller surface weight should be taken

Fig. 3.2 Roof with intensive planting (*Source* Kotásková)

Fig. 3.3 Roof with extensive planting (*Source* Kotásková)

into account. This does not place such high demands on the support construction of the roof. The surface weight in the fully saturated state is in the range of 60–300 kg/m^2 [3]. This type of roof is usually not intended for a walk, and there is no need for artificial irrigation. The disadvantage is the limitation of the choice of vegetation, because the plants growing on these roofs must be able to withstand difficult conditions such as drought or wetting. That is why we can often see succulents on these roofs, such as stonecrops, houseleeks, or xerophilous grasses and perennials (see Fig. 3.3). The care lies only in the occasional removal of undesirable air raids. This type is also suitable for reconstruction.

3.3.3 Roof with Semi-intensive Planting

This is the transition between intensive and extensive planting. The thickness of the substrate is between 150 and 300 mm. Planted species usually do not require much care, except for ensuring sufficient moisture, especially after the realization of planting. This type of roof is usually designed for grassy surfaces suitable for walking. Therefore, in our conditions, semi-intensive greenery requires minimal maintenance and especially sufficient moisture and sufficient thickness of the substrate.

3.3.4 Biotope Roof

Biotope green roofs represent an approach in which the roof layering is not planted, but it is gradually populated by the air raid plant species from the immediate vicinity. It is a roof with extensive greenery of natural biotopes. These roofs are left largely without care. It is the most ecological and economical type of planting [11].

3.4 Ways of Facade Greening

Pejchal [12] lists systems connected with loose soil and not connected with loose soil (vertical gardens).

3.4.1 Systems Connected with Loose Soil

Facades with systems connected with loose soil are those where the plants are planted directly into the soil close to the facade. Climbing plants can be found among herbs—both annuals and perennials, as well as among trees. Self-climbing and also plants not able to self-climb (liana) are used, which belong to a group of climbing plants.

Plants, which are not able to climb themselves, require support for their growth. Some of them are spinning around their support (e.g., *Lonicera caprifolium, Wisteria sinensis, Thunbergia alata, Humulus lupulus*), another are tendril plants—these use tendrils to grip to support (*Vitis vinifera, Parthenocissus vitacea, Clematis vitalba, Lathyrus odoratus, Cobaea scandens*) or buckling, which are attached by side shoots (*Rosa canina*), spines (*Rubus fruticosus*), thorns (*Lycium barbarum*) or leaves (*Tropaeolum majus*) [12].

Burian [13] reports that self-climbing plants have a relatively narrow assortment. We distinguish the rooting ones (*Hedera, Campsis, Hydrangea, Schizophragma*) and tendrils (*Parthenocissus*) (see Fig. 3.4).

Fig. 3.4 Facade system connected with loose soil—self-climbing (*Source* Kotásková)

3.4.2 Systems Not Connected with Loose Soil

Systems that are not connected with loose soil are the new way of greening the facades. They are further divided into the shelf, modular, and planar systems. Vertical garden growing systems can be simply divided into two groups, systems using the substrate and without using the substrate.

Shelf systems have pre-hanging containers or troughs on the wall, in which the plants are grown.

Modular systems allow, with the help of prefabricated elements hanged on the support construction of the wall, to be covered all over its surface. The supporting construction is mostly composed of hanging profiles that are directly screwed to the facade and cassettes made of metal or plastic mesh with growing substrate or substrates slabs made of modified foam substances or mineral fibers.

Planar structures have vegetation systems made of materials supplied with "standard meters". Plants can be planted after they are installed on the wall. The support construction consists of supporting pillars with horizontal and vertical bars. These supporting pillars are either anchored directly to the wall of the building or in a certain distance to the ground. Textile systems based on the principle of hydroponics are then attached to the construction. They usually have two layers of synthetic high-absorbent fabric (felt) that are attached to a non-absorbent plastic support plate. Plants are planted into them.

Textile and substrate systems represent a compromise between cassette systems (more reliable but heavy) and textile systems (less reliable but light). In textile and substrate systems, a classical soil substrate, moss, mineral substrate (lava), or stone (mineral) wool may appear. The substrate is covered with non-woven fabric with slots or circular gaps for planting on its surface. In terms of overall construction and fastening, it is largely identified with textile systems, the difference is mainly in the substrate, which is placed in the slots, circular holes, or pockets [14].

For these types of facades, irrigation must be provided. The most reliable method is a drip irrigation system where water, including the necessary nutrients, is distributed through hoses into the whole system. It is started up either by humidity sensors or timers.

3.5 Structure of Membrane Roofing of Green Roofs

The design of the structure is based on the supposed use of the roof, the slope of the roof plane, and the limit load of the substructure. Typical is the structure of the green roof with the classical order of the layers where the damp course is placed above the thermal insulation. It consists of layers that in their combination ensure correct function of the chosen solution. They are:

- supporting structure;
- gradient layer;
- thermal insulation;
- waterproof membrane;
- protection barrier;
- drainage layer;
- filter membrane;
- vegetative layer;
- vegetation.

Some layers may be omitted in some cases, or one layer may fulfill multiple functions at the same time (e.g., hydrophilic mineral wools combine substrate, dry area, and hydro-accumulation layers into one product).

3.5.1 Supporting Structure

The bearing layers may be formed by an area-covering bearing structure or a structure of rod and area-covering elements. For a flat roof, the bearing structure is made up of a ceiling structure above the last floor.

3.5.2 The Gradient Layer

The gradient layer is formed from embankments, light concrete, concrete, plastic, or mineral fiber components. It can also be formed with a bearing structure. The gradient layer of the thermal insulating materials can fill the thermal insulating layer at the same time.

3.5.3 Thermal Insulation

For the design of the thermal insulation layer of the green roof, it is important that the thermal parameters of the entire roof structure including the vegetation layer meet the normative requirements (see ČSN 730540-2: Thermal protection of buildings. Requirements) [15].

Thermally insulating layers are formed from foam glass thermal insulation boards, extruded polystyrene, as well as porous or gap materials with limited water absorption capacity.

3.5.4 Waterproof Membrane

The damp proofing is the most important layer of the roof structure, as it ensures that the water does not penetrate into the roof and under roof structures. In the case of vegetation roofs, the quality of the used material and the design is even more important, because the hydro-isolation will be difficulty accessible for inspection and repair. In the event of a leak, it is practically impossible to locate exactly the site (the water can flow a few meters from the perforation site) where the failure occurred and the whole vegetation formation has to be removed, what is very costly. The design of damp proofing is based on the ČSN P 73 0606—Damp proofing of the buildings, where the minimum slope for dewatering elements is 1.75% and for safe drainage 3–5%. Checking the waterproofness of the damp proofing is carried out both visually and by flood testing, before the other layers are applied, in order to remove possible deficiencies [3].

Around the permeable structures and exposed areas, it is advisable to design gravel bands or tiled pavement. For correct function, it is important to make perfect bonds, including overhangs, the use of foils with resistance to root growth. It is also needed to pay attention to the compatibility of damp proofing with other layers of the roof structure; otherwise, we need to put the separating layer between them [16].

Stabilization of damp proofing against wind suction can be done by mechanical anchorage, bonding, stabilizing layer, or a combination of these methods. Stabilization by imposed load in the case of green roofs is ensured by the weight of the vegetation layer and at the edges of the roofs where the largest wind suction occurs, the imposed load is done by, for example, the belt of aggregate with fraction 8/16 with the minimum thickness of 40 mm or the fraction 16/32 with a minimum thickness of 50 mm, or concrete products.

The materials used for the damp proofing of vegetation roofs are the most frequently modified bitumen felt and hydro-insulating foils. Oxidized bitumen belts can only be used in case of specially sealed joints, and a separate protective layer with root growth resistance must be installed. Long-term tests have shown that different plants have been rooted in bitumen isolates, because microorganisms living on roots can decompose these substances. Root growth resistance is of modified bitumen belts ensured by adding of additives or copper sheet in this case [3].

3.5.5 Protection Barrier

The coating serves primarily to protect the damp proofing against its mechanical damage. However, it can also fulfill root growth resistance function and assists in the accumulation of less water. Its function can also be taken over by a dry area of profiled foil with overlaps (at least 250 mm) and glued joints. However, this solution should be verified by the producer.

Non-woven fabrics (minimum plan weight 300 g/m^2), cement or concrete screed, hydrophilic mineral wool, as well as sheets and foils used in the dry area are used as the coating. When using a cement-based screed, it is advisable to limit its absorption capacity and to ensure leaching of carbonate by using a damp course spattle or sealing coat. The cement screed should have a thickness of at least 30 mm, and it must be separated from the damp proofing by a separation and dilatation layer (e.g., PE foil) [3].

3.5.6 Drainage Layer

This layer is used to drain excess water to the roof drains to prevent the substrate from overflowing. Plant roots need to absorb oxygen for plant nutrition, what is impossible, when it is immersed in water and plants die. However, the dry area may also serve to accumulate water in the case of long-lasting droughts, and in this case, it is called the dry—hydro-accumulation layer. It provides a minimum amount of water in the vegetative layer for growing plants and reduces the flow of rainwater in short intense rainfall.

The materials used for this layer may be loose (Liapor, brick recyclate) or profiled foils or boards. For loose materials, the ability to retain water (15–20%) is desirable, so these materials must have predominantly open voids and varied texture. The use of natural aggregates is therefore inappropriate. Profiled sheets and profiled polystyrene boards entrap the water for a season without rain and can hold 5–10 l of water per m^2 of their surface. Perforated foils are also available on the market, where the perforation on the top of the protrusion also serves to aerate the root system of the plants [3, 6].

3.5.7 Filter Membrane

The filter bed (also separating) serves to prevent the dry area from clogging through the washed substrate elements. The filter fabric is in some cases directly embedded directly on the profiled foil, which can reduce the time needed for placing. For flat roofs with a substrate thickness of up to 250 mm, fabrics with a basis weight of 100–200 g/m^2 are sufficient. The filter bed is laid at all points of contact with the substrate with other materials.

3.5.8 Vegetative Layers

The roof substrate can be understood as a substitute for common soil for the living space of the plant community. For the greening of the roofs, it is not an ideal solution to use only soil, especially clay, because of its worse composition and higher weight. If we want to use the topsoil that was detached before the building of the house, it is advantageous to lighten it with mineral fillers or sand. Mineral fillers include broken Liapor, pumice stone, broken porous bricks. It is recommended to lighten the topsoil with 30–60 volume percent of mineral fillers with a fraction of 0–16 mm. The choice of the substrate and its thickness is then chosen according to the composition of the planned vegetation and the bearing capacity of the substructure [3].

3.6 Systemic Solutions for Green Roofs

3.6.1 Urbanscape

Urbanscape is a system suitable for extensive green roofs. It is a composition of the layers, where the substantial layer is made of a product called Green Roll, which substitutes the natural substrate. It is an optimal basis for plant growth. It is a mineral wool material that produces a stable felt that is highly absorbent and can retain more water than traditional or even modified substrates [11]. The retention of the same amount of water requires a much thinner layer than the substrate. The material has a low bulk density, so it is statically (it does not burden the roof too much) also a good solution for roof reconstruction.

So the urbanscape system includes a pre-cultivated green vegetation layer with stonecrops (Sedum), a Green Roll mineral wool, a plastic drainage film that drains excess water during the rainy season, while ventilates the plant root system and a protective foil that protects the damp proofing against rooting.

3.6.2 *Vedaflor Systems*

The system offers three variants of green roofs—for extensive but also for semi-intensive and intensive planting. Variants differ in the thickness of the professionally tested ecological substrate [11]. The thinnest layer of the substrate allows growth of stonecrops; the medium thickness is able to ensure the growth of grasses and herbs; and at the greatest thickness, it is possible to design roof gardens.

The damp course is made of bituminous melting tapes or plastic foils. Furthermore, the system includes fleece with retention and protective function, a filtering function, and a drainage function [11].

3.6.3 *Isover System*

The hydrophilic mineral wool is the main layer of the green roof like in the urbanscape system, which partially substitutes the layer of the substrate depending on the thickness of the vegetative layers. These are hydro-accumulating boards that significantly reduce the weight of the vegetative layers while it ensures the retention capacity of the roof. Plants can be rooted in these plates without any problems.

For roofs with low extensive vegetation, the thickness of boards 50 mm and substrate 30–100 mm is recommended. It is also necessary to use the drainage foil as protection against overflow, to increase the drainage capacity of the drainage profiled foil. Roofs with semi-intensive layering have a vegetative layer made up of a 100–200 mm substrate and 100 mm hydro-accumulation plates. Roofs with intensive planting have a total vegetative layer thickness of 300–1000 mm. Mineral wool boards should form up to half of it. The sufficient thickness of the mineral wool layer is 100–300 mm [11].

3.7 Protection Against Slide and Drainage

It is not usually necessary to provide the substrate against the landslide at a slope of up to 20°. It depends not only on the slope but also on substrate thickness, length of the sloping roof area and substrate coherence. If the vegetation is not completely rooted, it could slide at 15°. One of the measures to prevent the landslide is to place the beams under the membrane roofing. The beams should have round edges to prevent damages of damp proofing on the sharp edges. Another way is a solution with different fabrics, plastic beams, or plastic or steel mats. Using the entire skidproof system is possible [17].

3.8 Draining of the Green Roofs

There are two ways to get the water on the green roof—firstly is it meteoric water (rainfall and snowfall, eventually hail) and secondly the water through irrigation. Excessive water, it means this, which is not entrap by substrate or hydro-accumulation layer, must be removed from the membrane roofing.

The draining of the green roof may be outside and inside the building. The external draining is usually designed by the sloping to the edge of the floor plan equipped with a hanging gutter (see Fig. 3.5).

The internal draining is solved by troughs or perforated pipes that are plugged into the inlets or directly with inlets. A perfect connection of the damp proofing to the roof inlet must be ensured. It is necessary to separate the inlet from the vegetation, for example, with a roof adapter, so that it does not glaze. Suitable treatment is to pour aggregate around and separation from vegetation with filter geotextiles.

In order to ensure draining of the meteoric water at the sloping roofs in a sufficient degree, it is usually necessary to lay along the eaves edge around 300 mm wide strip of aggregate with the drain pipe. The eaves edge must be provided with a sufficiently rigid edge profile to provide shear forces from the layers of aggregate and soil (see Fig. 3.6).

Fig. 3.5 Edge of the roof with semi-intensive planting (*Source* Kotásková)

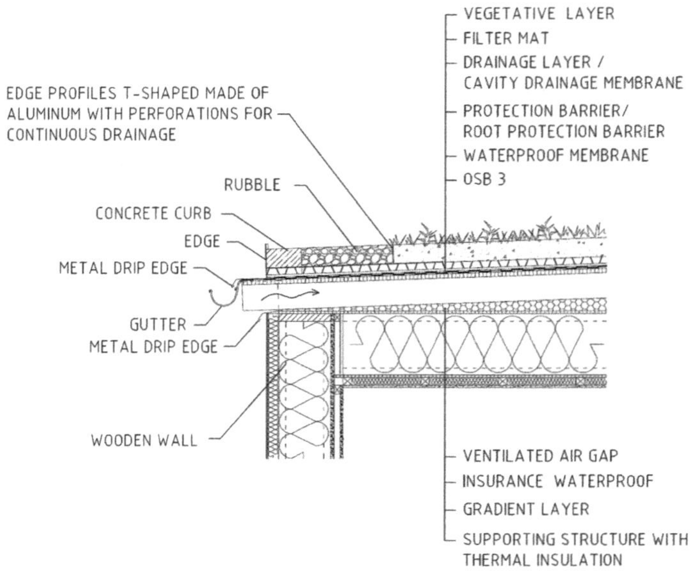

Fig. 3.6 Structure of the green roof and solution of the edge of roof ([95] modified by Kotásková)

3.9 Functions and Benefits of Green Roofs and Facades

Together with the basic functions that each roof has to meet, the green roof has many specific functions related to environmental and microclimate improvement, and its aesthetic character is inimitable:

- create, on the same site on which the building stands, new areas of green and outdoor living space;
- create new green areas for rest and recreation;
- reduce the share of concrete and paved areas;
- improve the look of cities and landscapes;
- enhance the living and working environment.

Within a landscape in villages or in adjacent areas of the cities, it is possible by the use of green roofs to partially hide the building, which may not be recognizable from the surrounding terrain from certain points of view. For even better results, we can inspire with houses that have vegetation not only on the roof, but they can also have the walls land up with the soil [4].

Green roofs function [16]:

- reduces the spreading rate of open areas and the share of paved areas;
- produces oxygen and binds carbon dioxide;
- filters particles of dust and dirt out of the air and absorbs pollutants;
- prevents overheating of roofs and thus swirling of the dust (see Fig. 3.7).

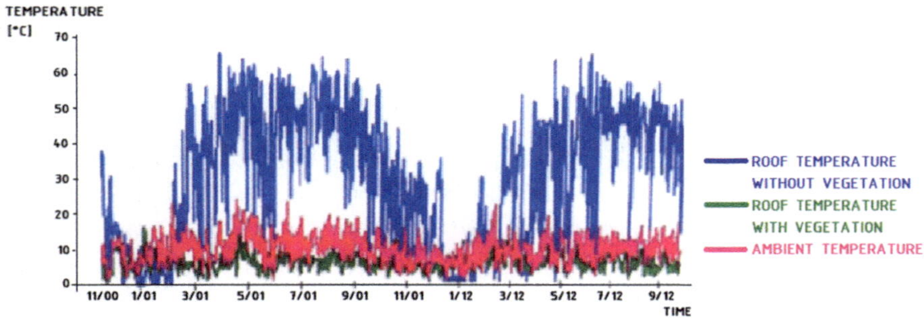

Fig. 3.7 Course of the temperatures of the roofs with and without vegetation depending on the surrounding temperature ([6] modified by Kotásková)

The measurement showed that at a temperature of 30–35 °C there is the maximum temperature of 20 °C at a substrate depth of 100 mm. In winter, at night temperatures around −10 °C, the temperature between 0 and −1 °C was measured in the substrate at a depth of 50 mm [6].

- reduce temperature fluctuations caused by rotating of day and night;
- reduce air humidity fluctuations.

Advantages of green roofs:

- have almost an unlimited lifetime in the case of a professional design;
- have a thermal insulation effect;
- protect the attic from intensive solar heat in the summer;
- acts as insulation from a sound;
- are considered to be non-flammable;
- slow down effluent of meteoric water, thereby reducing the use of the public sewer.

As Minke state out below [16]:

- spread the aromatic scent;
- create an environment for insects;
- are aesthetic, induce a positive state of mind and a sense of relaxation in man.

3.9.1 Influence on the Microclimate of the Building

Large concrete and asphalt areas have a high accumulation of heat and mostly low reflectance, thus contributing to the increase of temperature (the so-called thermal island), which has complex impacts on the environment in which we live. The difference between the temperature in the center and the outskirts of the city may be up to 11 °C in the summer months [2]. Increased temperature causes raising of the dust and pollutants from the ground. They then fly through the air and cause more difficult breathing and increased dustiness in habitable rooms, which is also undesirable even for cleaning. Dust particles and higher air temperatures are the reason of up to 15% decrease of sunshine and over 30–100% increased the occurrence of fog [2]. These aspects also cause a greater occurrence of storms and contribute significantly to smog situations [3].

Vegetation growing on green roofs limits thermal extremes by heat exchange between the plant and the surrounding environment until the equilibrium state is reached. The energy required to convert one liter of water into steam is approximately 2.2 kJ [16].

The layering of green roofs influences the microclimate in buildings, both by regulating the temperature, as described above, and by increasing the humidity around the roof. The increasing of the moisture is realized by transpiration of plants, evaporation from the soil, and evaporation of condensed dew on the surface of the plants. These aspects depend on the composition of the vegetation, where the ideal is a vertical layering of vegetation, dense green with large areas of leaves [3].

3.9.2 Protective Function

The green on the roof causes less temperature fluctuations, which lead to the smaller extensibility of the materials and thus increase the service life compared to flat roofs without vegetation. A great advantage is the protection against UV radiation, which degrades the majority of materials, and this again contributes to extending the service life of the entire roof. Therefore, we cannot forget the fact, that it is possible to use damp proofing with lower resistance to UV radiation (the damp proofing is exposed to its impacts only during the construction before it is covered with other layers). On the other hand, it is necessary to apply damp proofing with a certificate against the growth of roots. The vegetation layering also partly works as an element, which reduces the risk of flooding and sewerage network loads, because the drainage from these roofs is reduced up to 50% and it happens gradually, what is appropriate during the torrential rains.

3.9.3 The Function of the Green Facade

Its function is not only aesthetic, but has many other benefits that improve the environment of the residents, such as the greening of the places where is not ordinary greenery or where it was destroyed by the building. Every piece of greenery, especially the irrigated one, brings people a fresh feeling, extra in the summer months when the concrete surfaces in the cities are hot.

Green facades create a thermal insulation of the cladding of the building. It protects against cold in winter, and on the contrary, it prevents heat radiation and thus overheating of rooms. They reduce the dustiness in the surroundings, like the green roofs. The vegetation on the facade also absorbs the noise from the surrounding area.

The substantial positive is the social aspect of the placing of vertical gardens. In the places, where the facades were installed, the vandalism was greatly reduced and the safety of the place overall improved. It is obvious that the spaces created by these facades affect the behavior of people. People naturally respect this unusual part of nature. I think, it is good to place vertical greenery even in places where there is not only a problem with the built-up area but also a problem with disrespect and ignorance of the residents in the environment. The facades could serve to the natural upbringing of children and to the respect for nature in general. Thanks to the ability to green the city, the vertical garden can be one of the many sub-tools to achieve sustainable urban development [18].

Negatives, on the other hand, include the concentration of insects in the leaf stand, the effect of sticky roots on the facade. For example, such an ivy has more aggressive roots and can disrupt the facade. In disputed cases, for example, part of the tenants is for, and part is against, it is advisable to choose climbing plants that wrap around the prepared support tendrils that need support for their growth. Today, there are a lot of systems on the market, which can help to fix these plants [19].

3.9.4 Cooling Functions of Trees and Green Roofs in the Urban Climate

Almost every city on earth is warmer than the surrounding area by 1–4 °C and acts as a "thermal island" [20]. The effect of the thermal island of cities lies primarily on the high portion of the built-up area on a small proportion of vegetation. Buildings, asphalt, concrete, and other dark and hard surfaces absorb solar radiation, what causes that these surfaces are about 10–20 °C warmer than the surroundings and subsequently released the heat to the atmosphere [21–27]. In addition, these surfaces quickly drain water into the sewerage system and do not allow its gradual evaporation. Thereby, it would naturally cool down the microclimate of the city.

The vegetation responds to solar radiation by evaporating of the water and thus by cooling the surface of the leaves. The energy required to convert liquid to steam is referred as specific evaporative heat. The amount of 2450 kJ of energy is needed to evaporate one kilogram of water at 20 °C. This energy is consumpted from the incidental solar radiation and thus cannot participate in heating the surface and then radiating the heat to the surroundings. Another way of reducing the negative effects of the effect of the city's thermal island lies in the absorption of water in plants—this water very well accumulates the incidental solar radiation. Photosynthesis also plays an important role because it consumes 2.83 kJ of energy for the production of each glucose molecule. The very specific physical and chemical properties of water are keys in terms of cooling function, in particular, high specific evaporation heat of water and very high specific thermal capacity (three times greater than most other materials such as iron and aluminum rocks). These extraordinary characteristics of water are caused particularly as a result of the presence of so-called hydrogen bonds. Atoms bound in the water molecule are not arranged linearly (in one line), but the chemical bonds between the atoms make an angle of approximately 105°. The mentioned nonlinearity of the molecule owes the water molecule for its polarity, for the existence of hydrogen bond, also called hydrogen bridges.

Vegetation elements in cities gain increasing importance as climate changes, mainly because of their ability to eliminate the negative effects of global warming. The building of green roofs, facades, and planting trees in cities are a key adaptation strategy against the negative effects of heat islands, reinforced by the influences of global warming [28, 29]. Many cities have developed heat island strategies, and green roofs and facades are their integral part as it is, for example by Vienna [30], Melbourne [23, 31, 32], Kuala Lumpur, et al. Singapore, and Hong Kong [26], Athens [25]. The vulnerability of European cities due to climate change is convincingly evaluated by Tapia [28]. According to data from 2006, green roofs had the largest representation in Germany, with a green roof of 10% of all buildings [33].

Green (vegetative) surfaces in cities can play an essential role in the water (and thus energy) balance of the urban landscape only if they are sufficiently represented. While today we can quite exactly quantify the increased costs of their construction related to buildings (residential, administrative, and industrial), the quantification of their environmental functions (hydrological is one of the most important) has not yet been examined in the Czech Republic in detail. The most common studies related to the cooling effect of green roofs and facades logically come from warm areas such as the Mediterranean [34, 35] or tropics [36], but studies from Central European are no exception [37].

3.10 Measurement of the Vapor from Green Roof—The Case Study from Brno

A case study focused on measuring the quantity and dynamics of vapor from green roof was carried out in a vegetative period of 2017 in the Open Garden of the Foundation Partnerství Brno. The purpose of the case study was to determine what part of the energy of the incoming solar radiation was used for green roof evaporation. The evaporation was determined based on Bowen's ratio.

3.10.1 Materials and Methods

The measurements were realized on a green roof located in the center of Brno. It is an intensive green roof with the area 463 m^2. Typical species for extensive planting were used for greening (e.g. *Dianthus carthusianorum, Festuca rubra commutata, Festuca ovina, Festuca glauca, Stipa pennata, Silene vulgaris, Lychnis viscaria, Knautia arvensis, Sedum acre, Agrimonia eupatoria, and Bromus erectus*). The substrate was composed of 70% of topsoil with the addition of sand (fraction 0–2 mm) in a ratio of 3: 1 and 30% of drainage particles (milled mineral fiber, slag, milled brick recycled −0 to 16 mm).

A meteorological station was installed on the green roof to measure climate variables. The following variables were observed within the meteorological station: wind speed and direction, global radiation, precipitation, temperature and soil humidity. The temperature and air humidity sensors were installed at the height of 0.5 and 2 m above the ground for the purposes of Bowen ratio calculation. A bearing construction and an instrumental box with accessories are the part of the station. The data from all sensors were continuously stored in the data logger memory at ten-minute intervals.

The following variables entered into the calculation of the green roof vapor using Bowen ratio: global radiation (W/m2), air temperature at 2 m above ground (C°), air temperature at 0.5 ms above ground (C°), air humidity in 2 m above ground (%), and air humidity at 0.5 m above ground (%). All calculations were realized for data at hourly intervals. The pressure of saturated vapor at a certain temperature was calculated in the first phase of the calculation, and it was calculated separately for temperatures at heights of 0.5 and 2 m above ground level—formula 1 according to Allen et al. [38]:

$$e_s(T) = 0.61121.\ \mathrm{e}^{\frac{17.502.T}{240,97+T}} \tag{3.1}$$

where

e_s saturation vapor pressure at temperature T (kPa)
T temperature (°C)

In the next step (see formula 2, [38]), the actual vapor pressure at the heights of 0.5 and 2 m above the ground was calculated using the previous calculation (3.1) and the relative humidity of the air:

$$e = \frac{e_s.\mathrm{Rh}}{100} \tag{3.2}$$

where:

e the real pressure of the vapor (kPa)
e_s the pressure of the saturation vapor at a certain temperature (kPa)
Rh relative humidity of the air (%).

Subsequently, Bowen's ratio was calculated according to the following formula (formula. 3, [39]):

$$\beta = \gamma \cdot \left(\frac{T_{2\mathrm{m}} - T_{0.5\mathrm{m}}}{e_{2\mathrm{m}} - e_{0.5}} \right) \tag{3.3}$$

where

β Bowen ratio
γ psychrometric constant = 0.066 kPa °C^{-1}
T_{2m} temperature at height of 2 m (°C)
$T_{0.5m}$ temperature at height of 0.5 m (°C)
e_{2m} vapor pressure at height of 2 m (kPa).
$e_{0.5m}$ vapor pressure at height of 0.5 m (kPa)

In the last step, the evaporation was calculated according to the following formula (formula 4, [39]):

$$E = \frac{1}{L}\left(\frac{\mathrm{Rn}}{1+\beta}\right) \tag{3.4}$$

where

E the evaporation of the lawn in (mm m^2 h^{-1})
L latent evaporation heat of water = 2450 kJ kg
Rn net radiation calculated as 5% of the global radiation
β Bowen ratio.

3.10.2 Results and Discussion

There is evident very intensive vapor from Fig. 3.8. It shows a gradually increasing trend of daily vapor values, reaching the peak in June (median 4.3 mm/m^2). This trend has gradually decreased since July to the minimum values reached in October (median 1.2 mm/m^2). Maximum values of daily vapors occurred in June and August with a value of 6 mm/m^2. On the contrary, the minimum daily values were reached in September and October with a value of 0.2 mm/m^2. The above-described trend of the quantity of vapor reflects the potential evapotranspiration, respectively, evaporative requirements of the atmosphere (the values of potential evapotranspiration are not presented in the chapter, but they were calculated) and at the same time the water supply in the soil.

Table 3.1 shows that most of the solar radiation was supplied in June, on the contrary, the least amount in the month of October. The green roof has evaporated most of the water in June (125 l * m^{-2}), on the contrary, the least in October (34.4 l * m^{-2}). The lawn has evaporated 570.2 l * m^{-2} during the monitored period, what is significantly higher than the sum of the vertical precipitation over the period (422.4 l * m^{-2}). The horizontal precipitation and probably also the supplies of soil water accumulated from the winter period also play a significant role from the point of view of the quantity of vapor from the green roof. The amount of 46–69% of the energy supplied by the solar radiation has been consumed on the vapor of the lawn. The values calculated by us correspond to the values obtained by other authors (e.g. [40]). According to

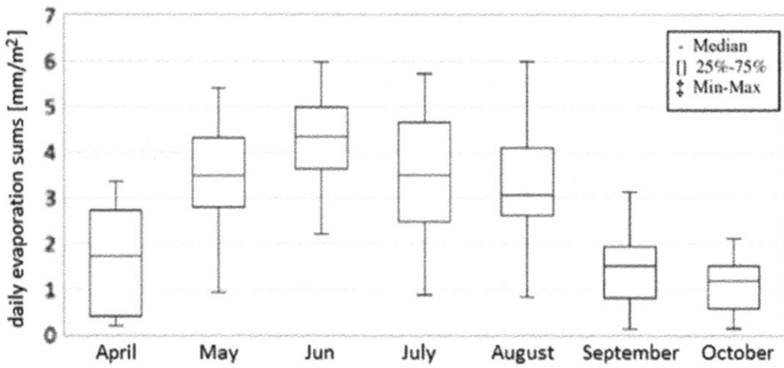

Fig. 3.8 Daily vapor sums of 1 m^2 of a green roof in individual months of measurement (*Source* own investigation)

Table 3.1 Ratio between the energy delivered by the solar radiation (Net radiation) and the energy emitted on the evaporation of lawn in the individual months of measurement (*Source* own investigation)

Month	Net radiation (W/m^2)	Net radiation (Ws/m^2)	Net radiation (kJ/m^2)	Evaporation (mm/m^2)	Evaporation (kJ/m^2)	Energy (%)
April	2998.0	259,022,984.0	*259,023.0*	48.7	119,197.5	**46**
May	4872.2	420,958,771.0	*420,958.8*	109.2	267,456.0	**64**
June	5412.8	467,662,810.0	*467,662.8*	125.0	306,303.2	**65**
July	4759.9	411,253,425.0	*411,253.4*	108.6	266,072.2	**65**
August	4172.7	360,525,358.0	*360,525.4*	102.0	249,818.2	**69**
September	2104.4	181,821,750.0	*181,821.7*	42.3	103,703.4	**57**
October	1590.0	137,375,585.0	*137,375.6*	34.4	84,263.3	**61**

Minke [41], in some cases, there is the possibility of 90% usage of solar radiation energy for the vapor.

The evaporation of vegetation of green roofs or facades is a significant measurable indicator, which is a key component of both water and energy (especially thermal) balances of green surfaces in the urban ecosystem. Most scientific studies are focused on the thermal balance of green roofs and facades or its components. Favorable influence on the cooling of the external surroundings is well documented [42], whether using thermal imaging [43–45] or direct measurements of temperature, both on the surface and at different distances from it [35, 37, 46–48].

Different simulation models and programs are often used [49–51], such as EnergyPlus, ENVI-met or DesignBuilder [52–54]. Gros et al. [55] have developed new indicators of the cooling output of green elements in the cities—energy performance index (EPI) and ambient temperature mitigation index (ATMI).

Favorable influence of green strata on the thermal regime inside the building is proven [36, 56–58], and it consists of the reduction of temperature extremes (fluctuation) due to thermal insulation properties, reduction of summer temperatures, and increase of winter temperatures [59–62], which in result leads to energy savings for both heating and air-conditioning [63]. Measurement and comparison of energy consumption dedicates to maintain appropriate indoor temperature for buildings with different types of roofs, including green roofs [64]. The cost–benefit analyzes were also carried out for this purpose [65, 66]. Also durability analyzes show that green roofs and facades are profitable in the long-term point of [67–69].

While the thermal regime is described in many studies in great details, the water regime of green roofs and facades is neglected in the current literature. Only a few studies also deal with the measurement of the air humidity above green surfaces [34, 70], with drainage of water from the torrential precipitation [71] or the transpiration of lianas of green facades [72]. Evapotranspiration is one of the most important causes of the cooling effect of green roofs and facades [29, 57, 71, 73], but only very few studies are focused on the direct measurement [74–76], which mainly use lysimeters [77].

However, green roofs and facades are considered a suitable contribution not only to the improvement of microclimatic conditions [34, 78–81], but also the aesthetic quality of buildings and their surroundings [82], decrease of noise [46], dust reduction and also air quality [83] and as such positively affect the health of the population [84–86] and are positively perceived by them [82].

The practice of using green roofs and facades in the Czech Republic is largely unsatisfactory. Vegetative (green) roofs as the most common application are perceived as a complicated solution associated with increased investment costs, difficult maintenance, and unclear overall effect. A multidisciplinary approach (team of plant breeders and climatic context, construction professionals, experts for evaluation of ecological contexts, etc.) combining theoretical reasoning and models with several levels of experimental methods and practical verifications is necessary for the clear identification of barriers defending to larger extension of green roofs and facades and searching of practically usable solutions. Due to the lack of technically relevant information for designers and investors, it could otherwise happen that fans of green roofs and facades will overestimate the thermal insulation effects and cooling effects and their opponents will underestimate it.

3.10.3 Concluding Remarks

Achieved results confirmed the importance of green roofs in the urban environment. The green roof of The Foundation Partnerství Brno was able to evaporate 1–6 mm/m^2 of water during the day. The process of evaporation depends on the actual conditions of potential evapotranspiration and the amount of available soil water. Monthly sums of vapor of green roof ranged from 34 to 125 mm/m^2, and 46–69% of the solar energy input was consumed during evaporation. If this energy was not draining away by vapor, it would be used to heat the city surface.

3.11 Zero Carbon Facility—Case Study of Environmental Partnership, Brno

3.11.1 Urban Context

Combating climate change in urban areas will always remain a complex combination of various tools: technical and natural infrastructure, soft political, planning, educational, and social, behavioral priorities or instruments. Often its basis could be just the recovering, interlinking, and re-packaging of existing city activities scattered in traditional sectors, such as greening public spaces or buildings, rainwater management, river restorations, public awareness, and emergency policies into one comprehensive climate adaptation strategy and implementation plan [87–89].

In recent decades, there has been no single blueprint or solution and sharing best practice seems to be the most efficient process for increasing the resilience of urban areas toward increasing weather extremes. Measuring the impact of various climate adaptations/mitigation tools, including their economic feasibility should be an integral part of their implementation (see, for instance, [90], or conclusions from the last global conferences in Rotterdam or Capetown [91, 92].

3.11.2 The Property of Open Gardens—A Case Study

The 15,000 m^2 large complex of the educational Open Gardens (see Figs. 3.9 and 3.10), the property of the Environmental Partnership Foundation in Brno, became a model zero carbon facility with energy and water balance proven by 2–5 years long continuous measuring. Data online are available on http://www.otevrenazahrada.cz/energie [93].

The Open Gardens infrastructure consists of:

- Historic six-floor building A with minimal environmental improvements—comparative baseline.
- 200 years old four floors (980 m^2) administrative building B reconstructed in 2012 into a passive energy standard one with 20 KWp PVE power plant on a flat roof.
- New three floors (1030 m^2) educational and dissemination building C with a green roof and meteorological station on top.
- Educational part of the garden (4000 m^2—including buildings) with nature science interpretation play stations for schools, underground tanks for collection of rainwater from roofs and sidewalks (total storage capacity is about 50 m^3), 8 geothermal drills, each 130 m deep, small reed bed water treatment biotope, and with experimental measuring of water sap on 4 maple trees (Mendel University since 2016).

Fig. 3.9 Aerial photograph of The Open Gardens (*Source* Nadace Partnerství)

– Small City Farm with community gardens, bees, animals, and outdoor kitchen with extensive 36 m large green roof (rainwater discharge monitored since January 2018) on remaining 11,000 m^2 of garden.

3.11.3 Monitoring Online

Monitoring of energy and water balance, including indoor and outdoor temperatures, is a baseline for the evaluation of the efficiency of climate adaptation/mitigation tools and also for comparing technological solutions with its natural green infrastructure impact.

For monitoring of the energy and water balance, 70 online meters were installed in Open Garden facility (the system was completed in 2016 with small improvements in 2017). The metering enables to measure not only the consumption/production of electric energy or the use of drinking water but also the energy for heating and cooling the buildings, rainwater discharge from three types of roofs, use of gray water for flushing and watering gardens, etc. Data collected after almost six years of operation of the facility enables us to analyze in detail the functioning of the buildings (used for optimizing facility management), their impact on the microclimate and return on investment of several smart technologies implemented on the foundation property.

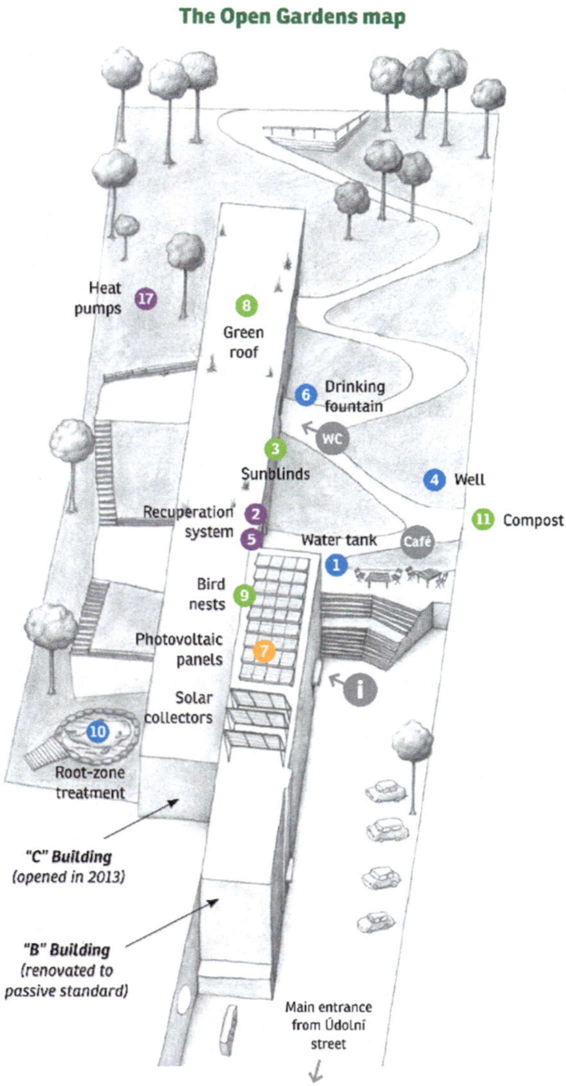

Fig. 3.10 Open Gardens map—distribution of smart technologies (*Source* Nadace Partnerství)

3.11.4 Zero Carbon Energy Performance

Well-insulated buildings (passive energy standard) powered by recuperation and other technologies minimize energy consumption by itself to the level below 20 kW/m^2/year. The first year after construction there was typically higher heat energy use (see Fig. 3.11).

Zero carbon impact is multiplied by connecting passive buildings to a geothermal source of energy for heating and cooling. Energy from eight drills of a total length

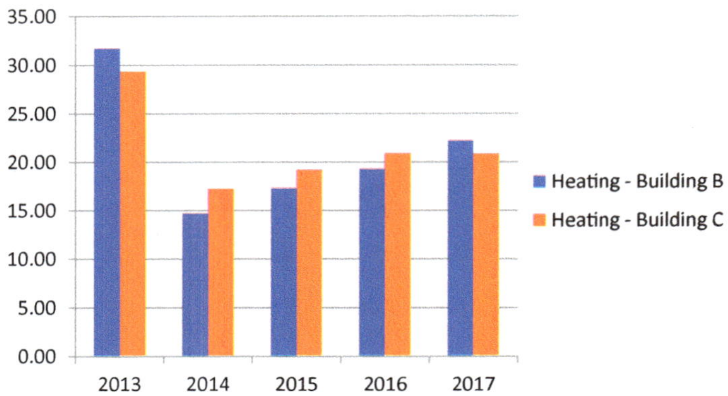

Fig. 3.11 Zero carbon energy performance (*Source* Own investigation)

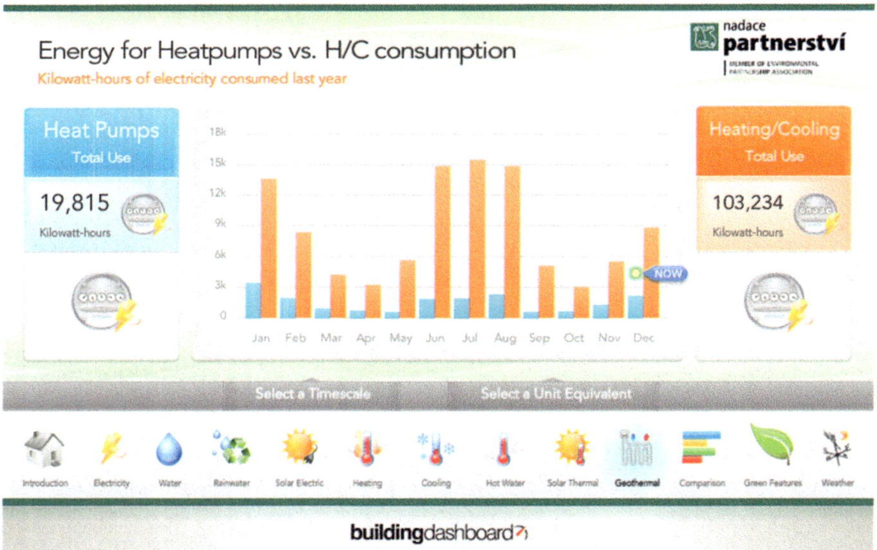

Fig. 3.12 Monitoring system showing SCP for the year 2017 (*Source* Nadace Partnerství)

of 905 m is transformed into cooling or heating energy via four heat pumps by using a minimum of electric energy (Seasonal Coefficient of Performance = 5–6) (see Fig. 3.12).

This very low amount of additional energy for heating and cooling combined with high efficiency of technology used for tempering two buildings (2000 m^2 floor area) enables a very high return on investment. The system of geothermal drills, heat pumps, and related networks distributing energy pays back in only 7 years (annual costs for heating and cooling are 1 EUR/m^2/year). Furthermore, given its positive influence on CO_2 reduction and replacement of increasing use air-conditioning units, which produce noise and contribute to the urban heat island, the heat pumps using energy from the ground have been proven as the most sustainable and environmentally friendly solution for urban areas.

Additional electrical energy for the offices is partly generated from the 20 kWp PVE located on the roof (produces 20% of the total need), and the rest is purchased from renewable energy provided by a distributor from the grid. The Open Garden facility covers 100% of its electric energy need from renewable sources.

3.11.5 Roofs and Urban Heat Island

The Open Gardens provide conditions to monitor the impact of various types of roofs and surfaces on the microclimate and to evaluate their potential to be used as climate adaptation elements in the urban areas. See infrared air snap from summer day, 29.8.2016—14:00, air temperature 28 °C, humidity 36% in Fig. 3.13a, b. The rainwater discharge from the roofs runs through flow meters before it is collected into the underground tank for further reuse. The discharge has been monitored online since April 2016 on the intensive green roof (15–25 cm thickness of soil) on the new building C and on the flat roof with a PVE plant on the reconstructed building B, both with the same acreage of 300 m^2. Since January 2018, we have measured discharge also from the extensive green roof (36 m^2 of Sedum matting on 7 cm thick mineral wool) on the outdoor wooden construction. The composition of species is as follows *Sedum sexangulare, Sedum spurium "Fuldaglut", Sedum reflexum, Stonecrop Angelina, Sedum lydium, Sedum lydium "Glauca," Sedum kampschaticum, Sedum hybr. "Immergrunchen," Sedum album, Sedum hispanicum "Minus", Sedum album coral carpet, Sedum acre.*

A significant water retention capacity of green roofs is apparent from the cumulative data. Even the extensive Sedum mat roof is able to reduce the discharge of rainwater by 10 times. It is evident that the extension of green roofs can significantly moderate peaks in the city sewage system during storms. The positive effect of the green roof on microclimates based on data from Open Gardens monitoring in vegetation season 2017 has been summarized by Svobodová [94]: The green roof is capable of evaporating 1–6 mm/m^2 during the day. Monthly sums of evaporation varied between 34 and 125 mm/m^2, which translate into 46–69% of solar energy depletion by evaporation. This energy otherwise would contribute to urban overheating. The comparison of discharge from three types of roof is given in Table 3.2.

Another important positive effect of the green roofs is their contribution to the "temperature stability" of the building. In summer, the layer of soil on the roof prevents overheating, and in winter, it improves insulation. See differences in temperatures measured on the surface and 10 cm under the surface in summer (Fig. 3.14a) and winter (Fig. 3.14b).

A higher need of cooling energy used in building B (without a green roof) compared to the neighboring building C with intensive green roof also indicates a positive influence of green roofs on temperatures in the upper floors (see Fig. 3.15).

Fig. 3.13 **a, b** Infrared air snap from summer day, 29.8.2016—14:00, air temperature 28 °C, humidity 36% (*Source* Mendel University, Brno)

Table 3.2 Comparison of discharge from three types of roofs monitored in the Open Gardens (*Source* Own investigation)

	Precipitation	Discharge of rainwater in liters		
	Total in mm	Flat roof B	Intensive green roof C	Extensive green roof
04-12/2016	368	26,000	35	N/A
2017	530	35,727	0	N/A
01-07/2018	231	25,163	0	763

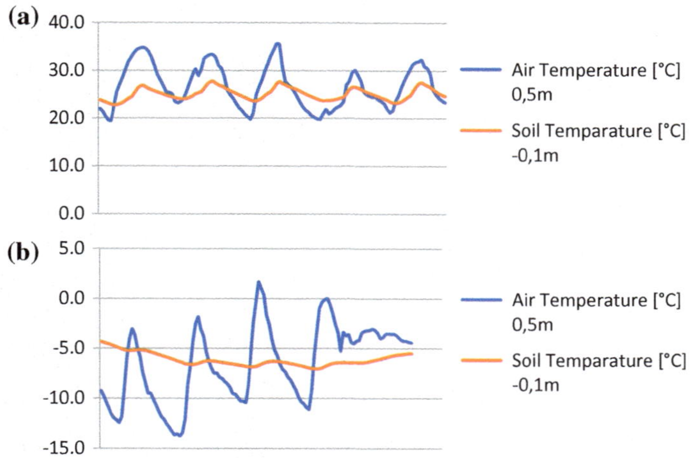

Fig. 3.14 **a**, **b** Temperatures on the surface and 10 cm under the surface in summer (**a**) and winter (**b**) (*Source* Own investigation)

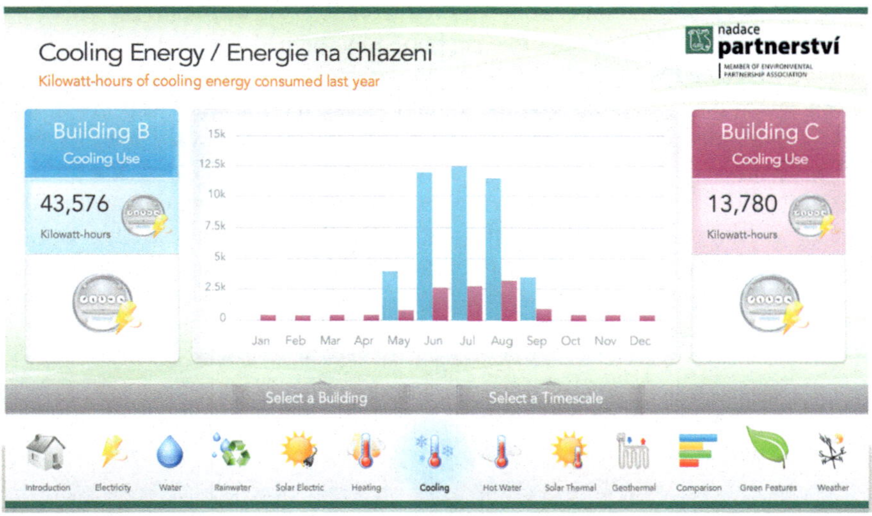

Fig. 3.15 Screenshot from the monitoring dashboard—comparison of energy used for cooling of two passive buildings in 2017: Build B with a normal roof, Build C with an intensive green roof (*Source* Nadace Partnerství)

3.11.6 Lessons from Water Management in the Open Gardens

The facility was conceptually designed in 2010 with the intention to test technologies for minimizing its ecological footprint. There were no hard data about the retention capacity of the green roofs available, and most of the water management components have been adopted during the operation of the facility based on experience. Water saving taps and toilet flushing systems enable to save 30% of drinking water. Use of rainwater for flushing toilets in building C as well as for watering the gardens enables to save almost 60% of drinking water. The original expectation is that rainwater discharge from the green roof will contribute to the storage significantly, but this was true only in the first two years of operation (2013, 2014) with "normal" distribution of temperatures and precipitations. Dry and hot years 2015–2018 did not allow almost any discharge from the intensive green roof, and all the precipitation capacity was accumulated in the soil and later evaporated through the plants. Thus, the underground storage tank is fed just by rainwater from classic roofs and from paved walkways in the property. It does not provide enough rainwater for the needs of the highly visited and intensively used environmental center and offices (30,000 visitors per year), and it needs to be supported by the underground water pumped from the well (see Fig. 3.16).

The return on investment into the system re-useable rainwater (50 m^3 storage capacity, pipe network, pumps, and ultraviolet cleaning technology) is about 10 years.

The obtained data show that in optimizing rainwater management the decision makers should take into account that green roofs have a significant effect on moderating urban heat islands, but they will not contribute by rainwater discharge to water management balance. It is more useful to collect rainwater from solid surfaces and classic roofs wherever possible.

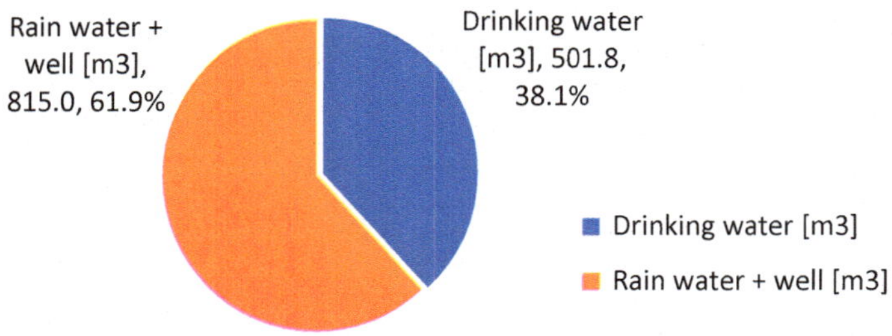

Fig. 3.16 Amount of the water from the pipeline, water from the well and rainwater (*Source* Own investigation)

3.12 Conclusions and Recommendations

The roof with vegetation cover, besides the basic function, which is the protection of the construction and the indoor environment from meteoric water, fulfills many specific functions related to the improvement of the environment. The roof can be representative. It can be used for recreational purposes and at the same time fulfill the ecological function.

The most important potential of green roofs though is their climate adaptation function. They help to moderate urban heat islands and measuring their retention capacity approved that even extensive green roof can significantly cut the peaks of runoff from extreme storm rains.

Increased investment, whether on the load-bearing structure or the layers needed for the planting, will subsequently provide savings of the operating costs. Well-designed layers of green roof significantly increase the durability of the membrane roofing as the vegetation, and vegetative layers protect against the weather effects the damp course of the roof or facade.

Greenery and substrates, especially on the roof, improve thermal insulating and reduce thermal extremes; there is also smaller material expansion and thus the durability, compared to flat roofs without vegetation, increases. A great advantage is a protection against UV radiation, which degrades the majority of materials, what contributes to extending the durability of the entire roof again.

By measuring on a particular green roof and calculations, it was verified that the vegetation reacts to solar radiation by evaporation of water, and thus by cooling of the surrounding environment, which is great benefit especially in the urban environment in mentioned summer months.

Acknowledgements This chapter was written within the project "Green Roofs and Facades as a Tool for Improvement of Thermal and Water Balance in Industrial Space" No. TH03030230 supported by the Technology Agency of the Czech Republic.

References

1. ČSN 73 1901 (2011) Roof design—basic provisions. ÚNMZ, Prague
2. Kotásková P, Štěpán J (2016) Green roofs of wooden constructions. In: Building partner 2/2016, pp 27–39. ISSN 1805-5958
3. Čermáková B, Mužíková R (2009) Green roofs. 1st ed. Praha: Grada, 246 pp. ISBN 978-80-247-1802-6
4. Frkal L (2007) Houses protected by soil. 1st ed. ERA group, Brno 94 pp. ISBN 978- 80-7366-095-6
5. Zamarovský V (1990) For the seven wonders of the world. Albatros, Praha, p 253
6. Šimečková J, Večeřová I (2010) Green roofs—Hope for the future. 1st ed. Union of Planting and Maintenance of Green, Brno, 40 p. Available from www: http://www.zelenestrechy.info/UserFiles/File/szuz_zelene-strechy_indd.pdf
7. Haas F (1983) Architecture of the 20th century: nationwide textbooks for college students at university, 3rd edn. SPN. Textbooks for universities, Praha

8. Guardian (2015) France decrees new rooftops must be covered in plants or solar panels [online]. [cit. 2018-03-02]. Available from: https://www.theguardian.com/world/2015/mar/20/france-decrees-new-rooftops-must-be-covered-in-plants-or-solar-panels
9. Olšan J (2011) History of trelages, pergolas and green walls (Historie treláží, pergol a zelených stěn). In: Green facade (Zelené fasády: odborný jednodenní seminář). Společnost pro zahradní a krajinářskou tvorbu, Praha 152 pp
10. Růžička V (2011) Beautiful vertical gardens saving homes (Krásné vertikální zahrady zachraňují domy). [online]. [cit. 2016-03-08]. Available from: https://bydleni.idnes.cz/vertikalni-zahrada-0aw-/architektura.aspx?c=A110624_111117_architektura_web
11. Vrabcová A (2017) Detached house with a green roof project (Návrh rodinného domu se zelenou střechou). Bachelors thesis. Brno, Mendel University
12. Pejchal M (2011) Plants for "vertical gardens" outdoors. (Rostliny pro "vertikální zahrady" ve venkovním prostoru). In: Green facade (Zelené fasády: odborný jednodenní seminář). Praha: Společnost pro zahradní a krajinářskou tvorbu. 152 p. pp 1–6
13. Burian S (2011) Use of climbing plants (Využití pnoucích rostlin). In: Green facade (Zelené fasády: odborný jednodenní seminář). Společnost pro zahradní a krajinářskou tvorbu, Praha, 152 p
14. Přerovská Z (2013) Vertical gargens in exterior and interior (Vertikální zahrady v exteriéru a interiéru). Diploma thesis. Mendel University, Brno
15. ČSN 73 05 40-2 (20012) Thermal protection of buildings—Part 2: requirements. ÚNMZ, Prague
16. Minke G (2001) Green Roofs: planning, realization, practice examples. 1st ed. Ostrava: HEL. 92 pp. ISBN 80-86167-17-8
17. Balík L (2009) About the green roofs (O zelených střechách) [online]. [cit. 2018-05- 06]. Available from: https://www.zelenastrechacz.cz/zelene-strechy
18. Ulrychová M (2009) Mur vegetal—Patric Blanc [online]. [cit. 2018-03-02]. Available from: http://www.greenlab.cz/cs/clanky/mur-vegetal-patrick-blanc/
19. Růžička V (2015) Green facades in several ways (Zelené fasády na několik způsobů). [online]. [cit. 2016-04-08]. Available from: https://mujdum.dumabyt.cz/rubriky/zahrada/zelene-fasady-na-nekolik-zpusobu_1954.html
20. Oke T (1973) City size and the urban heat island. Atmos Environ 7(8):769–779
21. Taha H, Sailor D, Akbari H (1992) High-albedo materials for the reduction of building cooling energy use [online]. 1992, 17 [cit. 2018-04-10]. doi: 10.2172/7000986. Available from: https://www.osti.gov/biblio/7000986
22. Forman RTT (2014) Urban ecology. Cambridge University Press, Science of Cities, p 462
23. Norton BA, Coutts AM, Livesley SJ, Harris RJ, Hunter AM, Williams NSG (2015) Planning for cooler cities: a framework to prioritize green infrastructure to mitigate high temperatures in urban landscapes. Landsc Urban Plan 134:127–138
24. Krüger EL (2015) Urban heat island and indoor comfort effects in social housing dwellings. Landsc Urban Plan 134:147–156
25. Varras G, Chiotelli K, Fragaki V, Karras G, Tsantopoulos G (2016) Potentials and prospects for the expansion of green areas on buildings in the metropolitan area of Athens. Acta Hortic 1108:331–337
26. Aflaki A, Mirnezhad M, Ghaffarianhoseini A, Ghaffarianhoseini A, Omrany H, Wang ZH, Akbari H (2017) Urban heat island mitigation strategies: a state-of-the-art review on Kuala Lumpur, Singapore and Hong Kong. Cities 62:131–145
27. Dhalluin A, Bozonnet E (2015) Urban heat islands and sensitive building design—a study in some french cities' context. Sustain Cities Soc 19:292–299
28. Tapia C, Abajo B, Feliu E, Mendizabal M, Martínez JA, Fernández JG, Laburu T, Lejarazu A (2017) An indicator-basedvulnerability evaluation for European cities. Ecol Ind 78:142–155
29. Sheweka SM, Mohamed NM (2012) Green facades as a new sustainable approach towards climate change. Energy Proced 18:507–520
30. Brandenburg Ch, Damyanovic D, Reinwald F, Allex B, Gantner B, Czachs Ch, Morawetz U, Kömle D, Kniepert M (2015) Urban Heat Islands. Strategieplan Wien, Wiener Umweltschutz-abteilung—Magistratsabteilung, p 22

31. Rayner JP, Raynor KJ, Williams NSG (2010) Façade greening: a case study from melbourne. Australia Acta Hort 881:709–713
32. Murphy SM, Rayner JP, Hall G, Francis J (2016) Growing green: developing industry guidelines for green infrastructure. Acta Hortic 1108:291–296
33. Köhler M (2006) Long-term vegetation research on two extensive green roofs in Berlin. Urban Habitats 4:3–26
34. Pérez G, Rincón L, Vila A, González JM, Cabeza LF (2011) Behavior of green façades in mediterranean continental climate. Energy Convers Manag 52:1861–1867
35. AboElata AAA (2017) Study the vegetation as urban strategy to mitigate the urban heat island in the mega city of Cairo. Process Environ Sci 37:386–395
36. Sunakorn P, Yimprayoon Ch (2011) Thermal performance of biofacade with natural ventilation in the tropical climate. Process Eng 21:34–41
37. Martin M, Afshari A, Armstrong PR, Norford LK (2016) A new validation protocol for an urban microclimate model based on temperature measurements in a Central European city. Energy Build 114:38–53
38. Allen RG, Pereira LS, Raes D, Smith M (1998) Crop evapotranspiration- Guidelines for computing crop water requirements-FAO Irrigation and drainage paper 56. Fao, Rome 300(9):D05109
39. Duffková R, Kučera, J (2005) Methodology of analysis of water stress of grassland. In: Rožnovský J, Litschmann T (edn) Seminar "Evaporation and evapotranspiration", Brno, March 23, 2005, ISBN 80-86690-24-5, pp 59–66
40. Kučera et al (2011) Evapotranspiration of selected types of agricultural and forest stands. In Hydrology of Small Basin 2011
41. Minke G (2001) Green Roofs: planning, realization, examples from practice. 1st ed. HEL, Ostrava. ISBN 80-86167-17-8
42. Afshari A (2017) A new model of urban cooling demand and heat island-application to vertical green systems (VGS). Energy and Buildings, in press
43. Spala A, Bagiorgas HS, Assimakopoulos MN, Kalavrouziotis J, Matthopoulos D, Mihalakakou G (2008) On the green roof system. Selection, state of the art and energy potential investigation of a system installed in an office building in Athens Greece. Renew Energy 33:173–177
44. Zölch T, Maderspacher J, Wamsler Ch, Pauleit S (2016) Using green infrastructure for urban climate-proofing: an evaluation of the thermal mitigation measures at the micro-scale. Urban Forestry and Urban Greening 20:305–316
45. Niachou A, Papakonstantinou K, Santamouris M, Tsangrassoulis A, Mihalakakou G (2001) Analysis of the green roof thermal properties and the investigation of its energy performance. Energy Build 33:719–729
46. Wong NH, Tan AYK, Chen Y, Sekar K, Tan PY, Chan D, Chiang K, Wong NCH (2010) Thermal evaluation of vertical greenery systems for building walls. Build Environ 45:663–672
47. Cheng CY, Cheung KKS, Chu LM (2010) Thermal performance of a vegetated cladding system on façade walls. Build Environ 45:1779–1787
48. Perini K, Ottelé M, Fraaij ALA, Haas EM, Raiteri R (2011) Vertical greening systems and the effect on air flow and temperature on the building envelope. Build Environ 46:2287–2294
49. Wong NH, Tan AYK, Tan PY, Wong NCh (2009) Energy simulation of vertical greenery systems. Energy Build 41:1401–1408
50. Yuan J, Emura K, Farnham C (2017) Is urban urban albedo or urban green covering more effiective for urban microclimate improvement? A simulation for Osaka. Sustain Cities Soc 32:78–86
51. Stec WJ, van Paassen AHC, Maziarz A (2005) The modeling of the double skin facade with plants. Energy Build 37:419–427
52. Larsen SF, Filippin C, Lesino G (2014) Thermal simulation of a double skin façade with plants. Energy Proced 57:1763–1772
53. Lobaccaro G, Acero JA (2015) Comparative analysis of green actions to improve outdoor thermal comfort within typical urban street canyons. Urban Climate 14:251–267

54. Feng H, Hewage K (2014) Energy saving performance of green vegetation on LEED certified buildings. Energy Build 75:281–289
55. Gros A, Bozonnet E, Inard Ch, Musy M (2016) A new performance indicator to assess building and district cooling strategies. Process Engineering 169:117–124
56. Pons O, Nadal A, Sanye-Mengual E, Llorach-Massana P, Cuervad E, Sanjuan-Delmas D, Muñoze C, Rovira MR (2015) Roofs of the future: rooftop greenhouses to improve building metabolism. Process Eng 123:441–448
57. Jaffal I, Ouldboukhitine S-E, Belarbi R (2012) A comprehensive study of the impact of green roofs on building energy performance. Renew Energy 43:157–164
58. Gross G (2012) Effects of different vegetation on the temperature in an urban building environment. Micro-scale numerical experiments. Meteorologische Zeitschrift 21(4):399–412
59. Koyama T, Yoshinaga M, Maeda K, Yamauchi A (2014) Room temperature reductions in relation to growth traits of kudzu vine (Pueraria lobata): experimental quantification. Ecol Eng 70:217–226
60. Ballarini I, Corrado V (2012) Analysis of the building energy balance to investigate the effect of thermal insulation in summer conditions. Energy Build 52:168–180
61. Coma J, Pérez G, Solé C, Castell A, Cabeza LF (2014) New green facades as passive systems for energy savings on buildings. Energy Proced 57:1851–1859
62. Ip K, Lam M, Miller A (2010) Shading performance of a vertical deciduous climbing plant canopy. Build Environ 45:81–88
63. Castleton HF, Stovin V, Beck SBM, Davison JB (2010) Green roofs; building energy savings and the potential for retrofit. Energy Build 42:1582–1591
64. Susca T, Gaffin SR, Dell'Osso GR (2011) Positive effect of vegetation. Urban heat island and green roofs. Environ Pollut 159:2119–2126
65. Ascione A, Bianco N, de Rossi F, Turni G, Vanoli GP (2013) Green roofs in European climates. Does it have effective solutions for energy savings in air conditioning? Appl Energy 104:845–859
66. Perini K, Rosasco P (2013) Costebenefit analysis for green façades and living wall systems. Build Environ 70:110–121
67. Ottelé M, Perini K, Fraaij ALA, Haasa EM, Raiteri R (2011) Comparative life cycle analysis for green façades and living wall systems. Energy Build 43:3419–3429
68. Bianchini F, Hewage K (2012) How "green" are the green roofs? Lifecycle analysis of green roof materials. Build Environ 48:57–65
69. Sozer H (2010) Improving energy efficiency through the design of the building envelope. Build Environ 45:2581–2593
70. Djedjig R, Bozonnet E, Belarbi R (2015) Experimental study of the urban microclimate potential of green roofs and green walls in street canyons. Int J Low-Carbon Technol 10:34–44
71. Roehr D, Laurenz J (2008) Living skins: environmental benefits of green envelopes in the city context. Trans Ecol Environ 113:149–158
72. Hoelscher M-T, Nehls T, Jänicke B, Wessolek G (2016) Quantifying cooling effects of facade greening: Shading, transpiration and insulation. Energy Build 114:283–290
73. Chanampa M, Rivas PV, Ojembarrena JA, Olivieri F (2010) Systems of vegetal façade and green roofs used as a sustainable option in architecture. Des Princ Pract: Int J 4:1–10
74. Williams NSG, Hughes RE, Jones NM, Bradbury DA, Rayner JP (2010) The performance of native and exotic species for extensive green roofs in Melbourne, Australia. Acta Hort 881:689–696
75. Onmura S, Matsumoto M, Hokoi S (2001) Study on evaporation cooling effect of roof lawn gardens. Energy Build 33:653–666
76. Takakura T, Kitade S, Goto E (2000) Cooling effect of greenery cover over a building. Energy Build 31:1–6
77. Köhler M (2009) Energetic aspects of green roofs. In: Appl R, Ansel W (eds) Green roofs—bringing nature back to town. Proceedings: International Green Roof Congress. International Green Roof Association, Berlin, 181 p
78. Sala M (1998) Advanced bioclimatic architecture for buildings. Renew Energy 15:271–276

79. Ascione F (2017) Energy conservation and renewable technologies for buildings to address the impact of climate change and minimize the use of cooling. Solar Energy, in press
80. Sadineni SB, Madala S, Boehm RF (2011) Passive building energy savings: a review of building envelope components. Renew Sustain Energy Rev 15:3617–3631
81. Oberndorfer E, Lundholm J, Bass B, Coffman RR, Doshi H, Dunnett N, Gaffin S, Köhler (2007) Green roofs and urban ecosystems: ecological structures. Funct, Serv BioScience 57(10):823–833
82. White EV, Gatersleben B (2011) Greenery on residential buildings: Does it affect preferences and perceptions of beauty? J Environ Psychol 31:89–98
83. Tong Z, Whitlow TH, Landers A, Flanner B (2016) A case study of air quality above an urban roof top vegetable farm. Environ Pollut 208:256–260
84. Buchin O, Hoelscher M-T, Meier F, Nehls T, Ziegler F (2016) Evaluation of the health-risk reduction potential of countermeasures to urban heat islands. Energy Build 114:27–37
85. Schuster Ch, Burkart K, Lakes T (2014) Heat mortality in Berlin—spatial variability at the neighborhood scale. Urban Climate 10:134–147
86. Tzoulas K, Korpela K, Venn S, Yli-Pelkonen V, Kazmierczak A, Niemela J, James P (2007) Promoting ecosystems and human health in urban areas using Green Infrastructure: a literature review. Landsc Urban Plan 81:167–178
87. Kundrata et al (2017) Studie adaptačních opatření na zmírňování vlivu klimatických změn pro město Brno
88. CzechGlobe a kolektiv (2016) Zásady pro rozvoj adaptací na změnu klimatu ve městě Brně: s využitím ekosystémově založených přístupů. Východiska pro zpracování Strategie pro Brno 2050
89. Sorvig K, Thompson JW (2018) Sustainable landscape construction: a guide to green building outdoors. Island Press
90. Brandenburg Ch et al (2015) Urban Heat Islands—Strategieplan Wien
91. Adaptation futures (2016) [Online]. Available from: http://www.adaptationfutures2016.org/ Accessed on 12 June 2018
92. Adaptation futures 2018 [Online]. Available from: https://adaptationfutures2018.capetown/ Accessed on 12 June 2018
93. Otevřená zahrada—vzdělávací a poradenské centrum. [Online]. Available from: http://www.otevrenazahrada.cz/energie Accessed on 12 June 2018
94. Svobodová K (2018) Stanovení ochlazovací funkce zelené střechy v areálu Nadace Partnerství v Brně. Bachelor thesis MENDELU
95. Beinhauer P (2008) Standard-Detail-System (Systémy štandardných detailov: konštrukčné detaily pre stavebné zámery s popisom častí stavebných konštrukcií a cenami). Eurostav, Bratislava. ISBN 978-80-89228-11-9

Chapter 4
Infiltration of Rainwater in Urban Areas

D. Duchan and J. Říha

Abstract Contrary to traditional practice when rainwater is collected and conveyed from urban areas by combined, or storm sewers, present approach to the integrated stormwater control consists in the accumulation and infiltration of rainwater at the place of its origin. For design of the rainwater infiltration facility, the geological survey must be carried out, the method of infiltration has to be proposed and hydraulic calculation should be carried out. The design is subject to numerous uncertainties, arising from different conditions the infiltration tests, in design parameters and in operating an infiltration facility. In this chapter, the review of the most used facilities for infiltration is presented together with a brief description of geotechnical investigation and hydraulic approach recommended in the Czech Republic. An analysis of factors influencing the process of infiltration is carried out, related uncertainties influencing the design of the storage volume of the infiltration facility are discussed.

Keywords Surface water infiltration · Rainwater · Infiltration facility · Coefficient of infiltration

4.1 Introduction

The management with rainwater at urbanized territories has been traditionally carried out using a system of urban drainage consisting of the system of gullies and the network of combined or/and storm sewer conduits. These practices result in overdimensioning of sewer mains, to occasional and temporary overloading of sewers and to the necessity to release the rainwater to receiving streams via the storm overflow chambers. The latter brings hygienic problems in the streams and deterioration of stream water quality. Moreover, this traditional approach leads to significant reducing of natural infiltration of stormwater and thus reducing of groundwater replenishment. This effect is more stressing in arid areas with the lack of natural groundwater

D. Duchan (✉) · J. Říha
Faculty of Civil Engineering, Institute of Water Structures,
Brno University of Technology, Veveří 95, 602 00 Brno, Czech Republic
e-mail: duchan.d@fce.vutbr.cz

M. Zelenakova et al. (eds.), *Management of Water Quality and Quantity*,
Springer Water, https://doi.org/10.1007/978-3-030-18359-2_4

resources, but plays a significant role also in the Czech Republic, namely during the dry spells frequently experienced in the last decades.

Recent alternative approaches to the integrated control of rainwater prefer the accumulation and infiltration of rainwater at the place of its origin [1–8]. At present, this approach is preferred, implemented and applied using directives and regulations. During the last decades, guidelines and standards have been published in the developed countries for the design of infiltration facilities [9–14]. The guidelines outline the key processes and requirements for environmental, sustainable and water-sensitive urban planning and disposal of stormwater. They include water policy as well as planning and construction provisions. Key issues are regulatory considerations, planning and design procedures, construction and maintenance of infiltration facilities.

In the Czech Republic (CR), the requirements for rainwater infiltration are listed in the building code [15], in the Water Act [16] with supplementary decrees and in the Czech technical standards ČSN 75 9010 [17] and TNV 75 9011 [18] (see Sect. 4.2). These regulations follow international practice and reflect experiences from abroad [11, 19]. The specification on the site investigation, geological and hydrogeological survey, hydraulic calculations including numerical modeling, selection of reliable and appropriate infiltration facility and its operation and maintenance are relevant issues, which are part of the rainwater management in urban areas.

In this chapter, the present approach to stormwater management in urban areas in the Czech Republic is introduced. The types and design of infiltration facility are described, the factors influencing the reliability of the design are discussed together with the uncertainties of the individual parameters entering the calculation.

4.2 Legislative Framework in the Czech Republic

As mentioned above, principle rules for rainwater management are anchored in the Water Act [16]. According to the Water Act [16], the rainwater management should include utmost approach the natural water regime. Not only rainfall waters but also treated wastewater should be infiltrated to the geological strata. Such water must not contain dangerous or unhealthy compounds. Polluted waters may exceptionally be infiltrated based on the statement and agreement of certified professional personnel.

The building code [15] prescribes retaining and/or infiltrating surface water originated from rainfall for all new or reconstructed civil structures. Infiltration of rainwater is also specified by the decrees [20, 21] as the fundamentals for dealing with rainwater.

The standard ČSN 75 9010 [17] describes the extent and methods of geological and hydrogeological survey, the result of which is the determination of local geological conditions and the ability of the geological strata to infiltrate surface water. The principal characteristics to be determined are the type of the soil (soil classification), its porosity, hydraulic conductivity, the coefficient of infiltration and groundwater level and its natural variation. The standard also gives guidance on designing the volume of rainfall water, infiltration water and the corresponding volume of infiltration facility. These data are necessary for reliable hydraulic and technical design of stormwater retention/infiltration facility.

Information about water quality arriving from drained areas, typical polluting agents and expected pollution sources are described in the standard [18], Field [22] and Hlavínek et al. [3]. Precipitation waters from paved areas and roofs that may be discharged into an infiltration facility are divided into two categories according to standard [17] according to the presumed or established concentration of pollutants and the possible threat to groundwater when infiltrated into the ground:

- *acceptable stormwater,* the quality of which does not pose a risk on soil contamination and of groundwater quality; water comes from grasslands, natural and cultural landscapes with the potential to drain rainwater into drainage systems, roofs with area <200 m^2, terraces, communications for pedestrians and cyclists, entrances to individual garages and terrace houses,
- the stormwater which is *conditionally acceptable*, the quality of which may be degraded by specific pollution; water from roofs with the area $\geq$200 m^2, roofs made of inert materials and metallic elements (copper, zinc, tin) [19], roads for motor vehicles, public car parks, communications of industrial and agricultural sites and airport landing areas.

The design of infiltration facility for conditionally acceptable stormwater must achieve the greatest possible effect of pre-treatment of water and of self-cleaning processes through the soil layers. If these stormwaters already changed their quality during the flow on the surface, it is recommended to treat them as wastewaters and in such cases special permission based on the agreement of competent certified person according to Decree 432/2001 Coll. [23] is needed. Stormwater from potentially polluted areas is not recommended to be infiltrated. Infiltration may be permitted only in exceptional cases, after effective pre-treating and permitting by the Water Management Authority.

4.3 Conceptual Considerations

Conceptual consideration contains planning of the management with stormwater, necessary survey and the design of the facility. The scope of the documents and their details is proportional to the degree of documentation. In the Czech Republic, following types of documentation [15, 24] correspond with particular preparation phases:

- Documentation for zoning decision—this is needed for approval about fitting or changing territorial plan and for consequent obtaining the land permit. Geological survey and preliminary hydraulic design are a part of this documentation.
- Documentation for a building permit—this is needed for starting the construction. This documentation is elaborated in more details than the previous one.

- Documentation for manufacturing the structure—this is needed for the contractor to elaborate details of the structure and for the construction procedures. This documentation is usually elaborated by the contractor and includes stability assessment, organization plan of the construction including the schedule, etc.
- Documentation of real state—after construction detailed documentation with all changes must be elaborated for the building approval purposes.

The documentation of infiltration facilities differs based on the type of project stage, type of the facility and its size. Underestimation of financial means, especially at the beginning of the solution (study, investment plan and documentation for zoning decision) may increase the final investment costs, problems in further documentation, resulting from the re-evaluation of the land acquisition capacity or the underestimation of the amount of rainwater for the infiltration. The technical documentation may be of such difference that it can hardly be processed by a single authorized body. The chief designer is required to invite competent specialists and professional consultants to participate in the design.

4.3.1 Planning

In the Czech Republic, planning procedures are prescribed in the first part of the building code [15]. The planning starts with urban plans where the general characteristics of particular areas are defined. The general concept of water management has to be outlined including the management with rainfall water [5]. The following step in the planning is the project for the land permit and corresponding official decision. This phase is important to establish site characteristics, targets and preferred type of management with water and to consider all constraints that may arise through the design, construction and operation.

As a rule, site assessment should be undertaken to determine the site and drainage characteristics. The site assessment should provide the information about location, type of development (residential, administrative, industrial, agricultural), description of the area and a land cover. The most important data should be collected about the proposed outfall, point of discharge, amount and quality of discharged water. Potential site constraints and environmental considerations should also be identified.

4.3.2 Conceptual and Detailed Design

The conceptual design contains more detailed data and information about the site including the knowledge about the climate and hydrology (precipitation, evaporation, discharges, probabilistic analysis of individual hydrologic characteristics), the results of the geological and hydrogeological survey, its reporting and evaluation. Conceptual considerations deal with the selection of appropriate type of the infiltration facility (Sect. 4.4) corresponding with the available site, considering all constraints and taking into account geological conditions at the site. In some cases, a combination of individual types may be an advantage.

The detailed design follows on the approved conceptual design. It includes the development of detailed design drawings, reports and other documentation according to the building code [15] and related decrees [24]. The design should be systematically negotiated by involved parties.

The design should take into account all principles, considerations and constraints defined during the planning period. At all preparation steps, necessary permits and agreements have to be acquired. These are permits to locate the structure, building permit and final building approval which are provided by corresponding local authorities. In the case of water-related structures, these are the departments of environmental and water affairs.

4.3.3 Input Data

The necessary data concern the following:

- Information about the locality.
- Hydrological and climate data.
- The data about the rainfall water.
- Results of a geological survey.

Locality data include:

- site location in terms of administrative districts and regions,
- data on the property to the land taken from the Land Registry [25],
- topographical data including maps, usually digital (1:50,000, 1:25,000, 1:10,000, 1:5000, digital terrain model, aerial and satellite imagery, map coverage maps, basic water management maps 1:50,000, other thematic maps provided by municipalities, river authorities, etc.),
- documentation of buildings and other structures in the area of interest, which includes:
  - the type and the level of foundation of the structures, the subsurface parts of the buildings and the way of their isolation,
  - ground plan of the drained area and the layout of the infiltration facility,
- land cover and character at the drained areas and catchment, i.e., drawings showing the layout of the drained area and the neighborhood in a scale of max 1:1000,
- data on the protective zones of water resources in the vicinity,
- data about engineering networks in the location.

Hydrological data include information about catchment area and local watercourses. This concerns the area of the river basin and the data on the runoff conditions. Data on associated watercourses and fluctuations of water conditions at selected characteristic flow conditions are important with regard to the possible influence of groundwater levels on surface watercourses (and vice versa) and the possible limitation of the capacity of the proposed facility.

Climate data include rainfall, temperature, evaporation and low-temperature data. The data are on the intensity of precipitation, precipitation totals for individual precipitation periods and rainfall periodicity [17]. To design the infiltration facility, the above data must be provided or at least verified by the Czech Hydrometeorological Institute (CHMI).

The data about the rainfall water contain information about the amount of the runoff coming mostly from processed hydrological data, rainfall-runoff computations and hydraulic design of the facility [26, 27]. Important is the knowledge about the quality of approaching water and about the sediment transport. This governs the necessity of constructing sand traps and other devices to improve the quality of water for the infiltration.

Geological and hydrogeological conditions of the site have a dominant influence on the reliability and economy of the infiltration facility. In many cases, they are a limiting factor in terms of feasibility and financial costs. From this perspective, it is important to carry out a qualified and adequate site survey. In the past, in the case of inadequately conducted and incomplete surveys, a number of infiltration facilities have had project changes, construction costs increased and difficulties during the performance occurred. The geological survey contains information on the occurrence and the properties of groundwater and on the permeability of rocks in the subsoil. The subject of the geological survey must be both the location of the infiltration facility and the adjacent area.

In the following sections, selected crucial steps in the design procedure are discussed in more details. The issues are the list of infiltration facility types (Sect. 4.4), the guidance on the geological survey for the infiltration purposes (Sect. 4.5) and the guidance for the hydraulic design (Sect. 4.6). The consideration about uncertainties affecting the design is discussed as well (Sect. 4.7).

4.4 Types of Infiltration Facilities

In practice, numerous types of infiltration facilities are available. According to standard [17, 18] infiltration, facilities may be divided based on spatial and structural arrangement into above-ground and subsurface installations. Under above-ground infiltration facilities, following can be counted as in standard [18]:

- surface infiltration,
- swale infiltration,
- trench type infiltration,
- infiltration basin.

Subsurface infiltration facilities according to standard [18] may be divided into:

- infiltration shaft,
- subsurface infiltration blocks,
- shaft and pipe infiltration.

Surface and subsurface infiltration facilities are often combined to achieve higher efficiency of an infiltration. An appropriate type of the facility is chosen by the consulting engineer based on the results of geological survey, space conditions and requirements and other specifics of the site like amount and type of water, protective zones for water supply, etc.

4.4.1 Surface Infiltration

In surface infiltration, stormwater is gravitationally discharged to the designated infiltration area with lower altitude. This may be any park area with grass or gravel cover positioned lower than adjacent areas such as pavements, roads and parking lots. The area is adapted for infiltration of rainwater. The surface may be equipped with grass, permeable concrete tiles, permeable soils (e.g., gravel) or combination of mentioned materials (Figs. 4.1 and 4.2). The soil below the surface cover the surface must be permeable. This type of infiltration is used when enough space for spilling rainwater is available.

4.4.2 Swale Infiltration

Swale type infiltration is applicable in areas where not enough surface for infiltration is available. The swale provides mostly infiltration function and only small retention effect. The swale infiltration consists of shallow terrain depressions with the depth of approximately 0.3 m with grass surface or permeable material (Fig. 4.3). Soil below the swale surface must be permeable (Fig. 4.4). The combination with the trench infiltration facility (Sect. 4.4.3) bringing water to the swale is also an option. Swales are often applied at highly urbanized areas such as commercial precincts or car parks and are located in the median strip of divided roads.

In some cases, bioretention swales or bioretention trenches are applied within the base of a swale [13]. They provide both infiltration function for stormwater and treatment through filtration, extended detention and some biological uptake. This system is particularly efficient at removing nitrogen and other soluble or fine particulate contaminants.

Fig. 4.1 Surface infiltration—Brno University of Technology Campus, CR

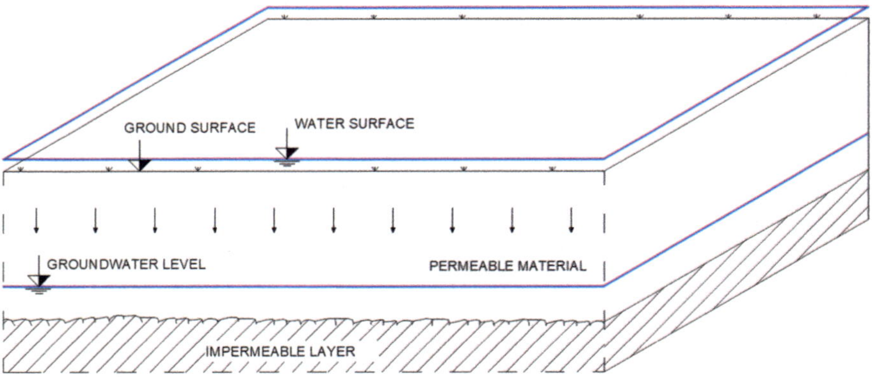

Fig. 4.2 Scheme of surface infiltration, adapted according to standard [18]

Fig. 4.3 Swale infiltration—Kohoutovice, Brno, CR

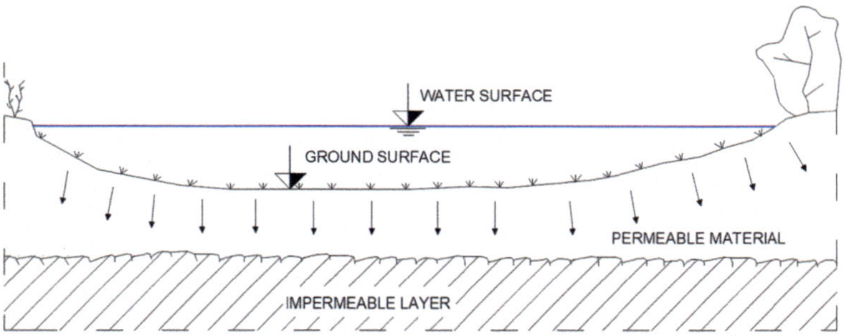

Fig. 4.4 Scheme of swale infiltration, adapted according to standard [18]

4.4.3 Trench Type Infiltration

Trench or furrow type infiltration facility is typically linear (Fig. 4.5 and 4.6). It may combine accumulation and infiltration function. Surface material should be naturally permeable (e.g., grass, gravel, sand) or artificial permeable material. Furrows are constructed along roads, sidewalks, railways for both drainage and infiltration purposes. The trench may have trapezoidal, bowl or triangular cross-section shape with water depth about 0.4 m and slopes 1:2–1:2.5 depending on the material used and space available. Trench furrows along the roads or railways are designed based on the rainwater volume as open channels [28]. According to the spatial configuration, they may also serve as conduits passing rainwater to other infiltration facilities.

Fig. 4.5 Trench (furrow) infiltration—Campus, Brno, CR

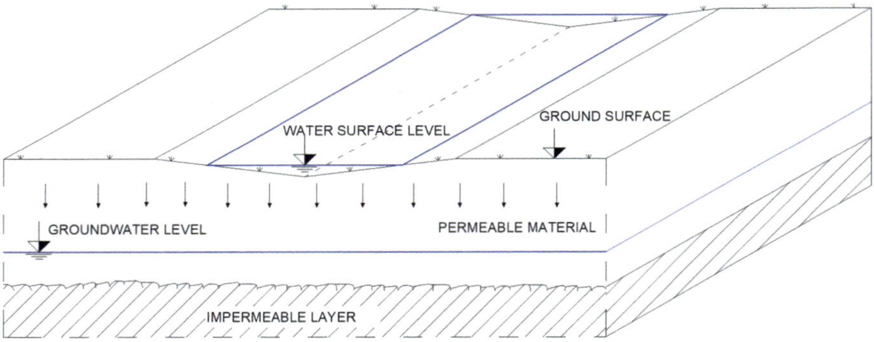

Fig. 4.6 Scheme of trench (furrow) infiltration, adapted according to standard [18]

4.4.4 Infiltration Basin

Infiltration basins (Fig. 4.7) serve as accumulation and infiltration facilities. These are mostly artificial reservoirs whose bottoms are filled with a permeable layer (Fig. 4.8). The inlet to the infiltration basin must be equipped with the sedimentation device like a sand trap or small sedimentation basin for settling of fine particles and suspended sediments which may cause clogging of permeable reservoir bottom. Another configuration of this type is reservoir where infiltration is realized only along permeable areas at the slopes of the reservoir or areas around the reservoir (Fig. 4.8). Such type of basin has only limited accumulation ability due to the permanent water volume at the lower portion of the basin. The advantage is only minor clogging of peripheral infiltration zones. An inflow of stormwater to the basin can be arranged via surface channels or subsurface pipes.

Fig. 4.7 Infiltration basin with a reservoir for sedimentation—Královo pole, Brno, CR

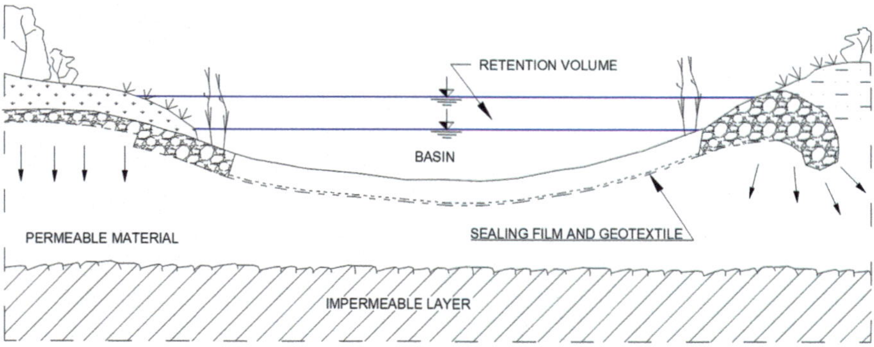

Fig. 4.8 Scheme of a basin with permeable slopes, adapted according to standard [18]

4.4.5 Infiltration Shaft

Infiltration shaft is a local facility similar to large diameter infiltration well (Fig. 4.9) with permeable bottom and/or walls. In front of the shaft, the device for capturing contaminants must be placed preventing clogging and sub-soil pollution. Groundwater table must be deep enough below the bottom of the shaft. The shafts are used where there is lack of space and if permeable subsoil underlies less permeable soil layer. It can also be used in combination with other infiltration facilities like infiltration pipes or trenches.

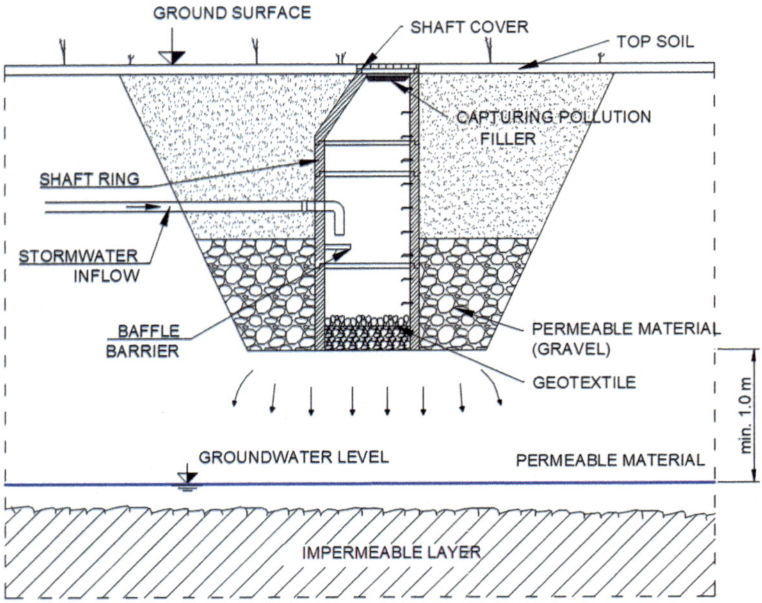

Fig. 4.9 Scheme of infiltration shaft in the permeable layer, adapted according to standard [18]

4.4.6 Subsurface Infiltration Blocks

Subsurface infiltration blocks (Fig. 4.10) are installed below the ground surface as a body made of permeable materials (gravel, rock) or plastic blocks. The latter can be filled with permeable soil. The plastic elements are flexible for mounting and enable creating blocks of various size and capacity. The shafts for the sedimentation of suspended solids and contaminants are recommended to precede an inlet to the block. The protection of the blocks is assured by placing the geotextile. Infiltration blocks are often combined with swale [18], where swale is on the surface, and infiltration blocks are located below the ground surface (Figs. 4.11 and 4.12).

4.4.7 Shaft and Pipe Infiltration

To the infiltration pipe, the stormwater is supplied via infiltration shaft which serves for the sedimentation of suspended solids contained in stormwater. The shaft is connected to the perforated pipe embedded in the filter material supporting the accumulation of water and assuring soil stability against its deformation (suffusion, clogging). Stormwater pre-treated in the shaft continues to the perforated pipe and infiltrates into surrounding subsoil (Fig. 4.13).

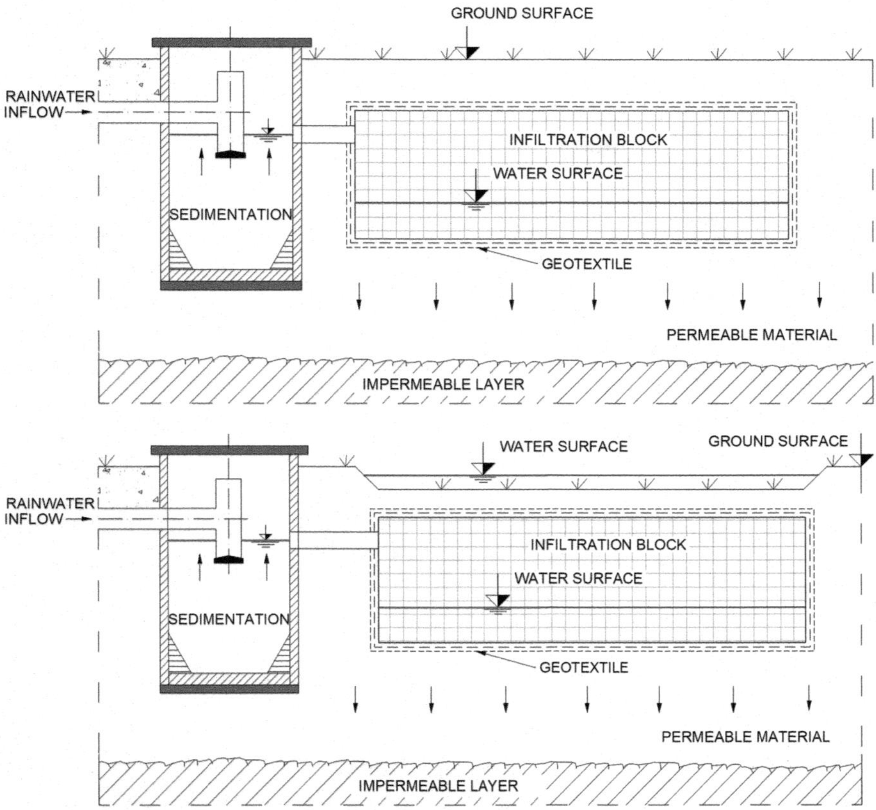

Fig. 4.10 Scheme of infiltration blocks, adapted according to standard [18]

4.5 Geotechnical Investigation

4.5.1 Generally

The basis for the high-quality technical design is well organized and carefully performed geological survey and geotechnical investigation. Its type and extent are generally governed by the layout plan of the drained area including buildings and on expected geological conditions. Based on the extent of the drained area and the stage of the documentation, the geological survey is carried out as [17]:

- preliminary, indicative survey,
- detailed survey,
- supplementary survey,
- risk analysis.

Fig. 4.11 Combination of swale and infiltration blocks in the subsoil—Campus, Brno, CR

Fig. 4.12 Infiltration blocks in subsoil during construction—Campus, Brno, CR [60]

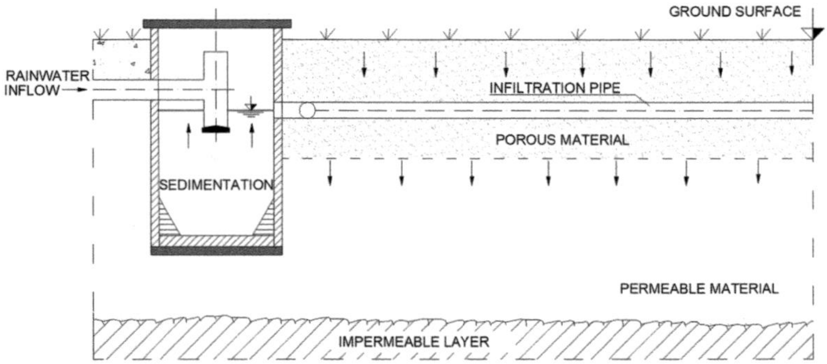

Fig. 4.13 Scheme of infiltration pipe, adapted according to standard [18]

The indicative (preliminary) survey is based on archival data (Geofond [29]) or other surveys and studies conducted in the past in the area of interest. The survey is conducted primarily in the case of drained areas smaller than 200 m^2 and for simple geological conditions featured by monotonous geologic profile, soils with similar geotechnical properties, unconfined aquifer, etc.

The detailed survey is realized for the detailed recognition of the site conditions. This is necessary to be performed at large drained areas larger than 200 m^2 or areas with complex geological structure. The detailed survey is also carried out in cases when an indicative survey cannot give enough information (no data in databases).

The aim of the supplementary survey is to obtain additional geotechnical and hydrogeological data or if the project documentation has changed. Risk analysis is performed only exceptionally in case of possible deterioration of groundwater quality due to infiltration, namely in case of important groundwater resources.

For detailed and supplementary survey, it is necessary to elaborate the project of the survey. Types and number of infiltration tests have to be specified including the location, depth of testing, the method of execution and analysis and the number of samples for laboratory tests. Some guidance for the project elaboration may be found in standard [17] or in guidance [4].

Input data for the geological survey

For the development of the project of geological survey and geotechnical investigation, following data have to be summarized, processed and analyzed (see also Sect. 4.3.3):

- Topographical and geodetic data include maps of the appropriate scale corresponding to the stage of the project documentation. Geological and hydrogeological maps may be obtained from central geological and cartographical databases [25, 30]. The more detailed data are taken from land geodetic surveying.
- Data about the site and drained area, its surface and land cover, related structures and buildings are provided by the relevant documentation (Sect. 4.3). The data about present state and planned civil structures should contain all relevant altitudes

such as level of foundation, the ground plan of the drained area and expected location of the infiltration facility specified by the boundary of available plots.
- Expected type, the shape of infiltration facility and altitude of its bottom level are necessary for the proposal of the depth of trial boreholes or borrow pits carried out as a part of the detailed geological survey.

4.5.2 Soil Properties

During the geotechnical investigation, basic soil properties are provided including soil classification. The properties are granulometric composition, bulk density, relative compaction, porosity, moisture, water content, etc. Soil permeability is tested, and related characteristics describing infiltration capacity are evaluated:

- the coefficient of infiltration k_v,
- the hydraulic conductivity k.

In most of the cases, the hydraulic design is usually based on simplified procedures applying the coefficient of infiltration k_v (Sect. 4.6). Values of k_v obtained from infiltration tests reflect mostly local conditions only. This is obviously disadvantage as the infiltration facility may have different size and location. Moreover, variable hydrological and hydrogeological conditions at the site represented by changing groundwater level and water content are of great importance, may affect soil pores capacity and may change the instantaneous value of the coefficient of infiltration. This means that the coefficient of infiltration obtained from infiltration tests usually suffers from considerable uncertainty in true infiltration capacity during the performance period.

When designing large and complicated infiltration facilities namely if individual facilities are interfering and may affect surrounding buildings, it is necessary to carefully describe the path of the water infiltrated into soil and proceeding in an aquifer for all potential scenarios of hydrogeological conditions and regimes. Here, hydraulic conductivity k should be determined during the geotechnical investigation, and advanced techniques of groundwater flow modeling should be applied. More details can be found in references [31–34].

4.5.3 Types of Tests

As mentioned above, the main aim of the tests is to obtain permeability and infiltration characteristics of the soil at the place of the proposed infiltration facility. Such tests may be divided into laboratory and field tests.

For laboratory tests, samples are taken from different depths representing the arrangement of the infiltration facility. Based on those samples, the following can be determined in the geotechnical laboratory:

- Basic geotechnical properties (bulk density, porosity, water content, granulometric composition, etc.).
- Classification of soil based on granulometric composition.
- Hydraulic conductivity from permeameters and grain size distribution by empirical formulae [35].

Field tests are carried out in the depth corresponding to both saturated and unsaturated zone. Following hydrogeologic and hydraulic characteristics are determined:

- The coefficient of infiltration.
- Hydraulic conductivity.
- The groundwater level in cases of boreholes or borrow pits reaching groundwater level (phreatic surface).

The *coefficient of infiltration* k_v is determined using the results from an infiltration test in an unsaturated zone as specified in standard [17]. The test may be performed in a borehole, borrow pit or in an infiltration cylinder (Fig. 4.14). For tests carried out in gravel and sand soils, stabilized (constant) water level at the testing device is maintained, recommended minimal volume of infiltrated water for such test is 1 m^3. For loamy and clayey soils mixed with coarse grains, tests with changing water level are applied. Infiltration tests are never performed in purely fine-grained soils like silts and clays as they are not permeable and infiltration at such conditions is practically impossible.

An important factor in determining the coefficient of infiltration is also the duration of an infiltration test. The coefficient of infiltration according to ČSN 75 9010 [17] should be evaluated by the end of the test after 24 h of infiltration. In practice, however, the length of the test is usually shorter. Due to this fact, over-evaluated coefficient of infiltration is usually obtained by the test (see Sect. 4.7).

Determining the *hydraulic conductivity* in unsaturated zone above the groundwater level is rather complicated and needs field measurements of the time change of the water level or inflow to the testing device during infiltration test. Following devices are in use:

- Infiltrometers—measurements on the surface of the soil or in the borrow pit,
- Permeameters—measurements to determine hydraulic conductivity at different depths of the borehole profile.

More details about hydraulic conductivity determination in the field can be found in extensive references [35–46].

The most reliable values of hydraulic conductivity can be obtained using pumping and intake tests in the saturated zone. These tests can be time-consuming and costly, and the design of the infiltration facility is performed only rarely. The procedures for implementation and evaluation of those tests are described in detail in the available references [13, 36, 47–49]. Simple and less time-consuming methods for determining "saturated" hydraulic conductivity are slug or bail tests which provide approximate values of hydraulic conductivity. Because of the short duration of the tests, they only characterize the closest neighborhood around the borehole so that the skin effect and

Fig. 4.14 Infiltration test using an infiltration cylinder

other resistances may affect the results. The most common methods for evaluating these tests are described in references [49–52].

Groundwater level has also to be monitored during the geotechnical survey in existing boreholes or newly established boreholes, shafts realized on purpose during the geological survey. Information about groundwater level may also be taken from archival data, other surveys and studies. Valuable information about long-term variation in groundwater level may also be obtained from the state monitoring network provided by the Czech Hydrometeorological Institute. For the design of the infiltration facility, maximum groundwater level under the bottom of the infiltration facility must be considered.

4.5.4 Final Report on the Geological Survey

All results from surveys are summarized in the final report containing information about the suitability of the area for stormwater infiltration, documentation of tests performed and recommendations for the design of infiltration facility. According to standard [17], the final report must contain:

- summarized geological and hydrogeological conditions taken from archive data, past studies and other surveys,
- information about other infiltration facilities, sources of water and its pollution, watercourses and structures related to groundwater,
- the summary of performed tests and their results,
- assessment of infiltration coefficient, hydraulic conductivity and groundwater levels,
- description of the path of pollutants,
- assessment of the suitability of the infiltration from the view of protection of existing and planned water resources, general protection of groundwater, possible slope instabilities, threats on existing structures, etc.,
- assessment of the interaction of projected and existing infiltration facilities,
- hydraulic effects of infiltration from the facility designed, an assessment of the suitability of the infiltration from a geological point of view, recommendation on a suitable type of infiltration facility and the construction and location; generally, the main prerequisite for the infiltration of stormwater is the existence of such subsoil, which is permeable and can lead away the infiltrated water from the place of infiltration. Generally, hydraulic conductivity is required to be greater than $k = 5.10^{-6}$ m/s [18, 53, 54],
- references to the literature and data sources.

4.6 Hydraulic and Technical Design

For the hydraulic design of infiltration facilities, various techniques are used. The traditional analytical methods using numerous simplifying assumptions are gradually replaced by more sophisticated numerical methods using unsaturated flow approach. The methods of modeling infiltration and flow of water in the saturated and unsaturated zones have been described in numerous studies being a basis for compiling efficient software products [31–34]. For the use of numerical models, it is necessary to obtain relevant geological and hydrogeological information about the structure of the groundwater body, properties of porous materials in the zone of infiltration, their deposition and the groundwater regime. In case of the design for smaller or less important facilities, the extent of the geological survey is usually limited, the use of groundwater flow models is practically excluded for the reason of a lack of financial resources.

Hydraulic computations of an infiltration process are based on the local geological conditions, legal requirements related to the site and namely on the type and arrangement of the facility (Sect. 4.4.3). The design consists of the following steps:

- calculation of the amount of rainfall water to be infiltrated,
- determining the volume infiltrated water,

- determining of the storage volume of an infiltration facility,
- technical design and elaboration of the technical description and drawings.

The *volume of stormwater* is identified according to [18, 47] using procedures of the hydrology of small catchments applied in the design of sewer systems. The amount of rainfall water is calculated using simple intensity formula:

$$V_s = A \cdot \varphi \cdot i \cdot t_c, \tag{4.1}$$

where V_S is the volume of precipitation per time t_c, φ is the runoff coefficient, i is the design rainfall intensity and A is the effective area determined as:

$$A = A_{red} + A_{VZ}, \tag{4.2}$$

where A_{red} is the plan view of the drained area, A_{VZ} is the area of the infiltration facility (in case of more extensive surface installations). When expressing Eq. (4.1) using the total design precipitation height h_d (mm) with the duration t_c (min), one obtains

$$V_S = \frac{h_d}{1000} \cdot (A_{red} + A_{VZ}). \tag{4.3}$$

The calculation for *the volume of infiltrated water* and an infiltration facility is carried out using variables characterizing the infiltration capacity of soil—the coefficient of infiltration [17], hydraulic conductivity [10, 31] or infiltration rate [9]. The parameters of infiltration are generally determined on the basis of the results of infiltration tests (Sect. 4.5.4). It is also considered that such determined parameters sufficiently represent the conditions at a site. The characteristics are the permeability of a groundwater body, homogeneity and anisotropy of materials, the moisture content of the soil, the state of groundwater table and the depth to the impermeable basement. Practical experience shows that the design is subject to numerous uncertainties which may lead to the underestimation of the volume of the facility when not taken properly into the account.

The *storage volume* of an infiltration facility V_{VZ} is determined in standard [17] as:

$$V_{VZ} \geq \max_{t_c = 0, t_{c,\max}} (V_S - V_{VSAK}), \tag{4.4}$$

where V_{VSAK} is the infiltrated volume per time t_c and t_{max} is the maximum duration time of constant intensity design storm (e.g., 72 h according to ČSN 75 9010). After substituting Eq. (4.3) into (4.4) and expressing infiltration volume using infiltration coefficient, one obtains:

$$V_{VZ} \geq \max_{t_c = 0, t_{c,\max}} \left(\frac{h_d}{1000} \cdot (A_{red} + A_{VZ}) - \frac{1}{f} \cdot k_v \cdot A_{VSAK} \cdot t_c \cdot 60 \right) \tag{4.5}$$

where h_d is the total design precipitation (mm) with the duration t_c (min) and selected frequency, A_{VSAK} is the infiltration area (m^2), f is the coefficient of safety recommended $f \geq 2$ by standard [17].

The *technical design* is governed by the type of the infiltration facility (Sect. 4.4). In the following text, only general guidance on the technical design is given for surface and subsurface installations.

Surface facilities are close to the natural infiltration into the soils at which water usually proceeds through the surface vegetation, mostly through the grass cover. The surface should be resistant against the erosion. The infiltration area must be periodically maintained and cleaned from sediments. The facility is usually created by relatively flat infiltration fields, ponds, ditches or trenches. Water depth in the facility is normally low. The depression of the infiltration field at the flat surface should be about 0.3 m. The surface should be equipped by suitable permeable layers corresponding to the lower geological layers. In case of more sloped configuration, the infiltration areas (reaches) are divided into sections by small walls or dikes. If water is brought by concentrated pipes or ditches, the intake has to be equipped by the lining protecting inlet against erosion.

Subsurface installations require carefully pre-treated water as it is difficult to clean them during the maintenance procedures. The facility is usually created by the subsurface volume filled with gravel or man-made blocks (e.g., plastic) in some cases combined with drain pipes. Other options are subsurface tunnels, infiltration galleries and infiltration shafts or wells. Sometimes individual elements are combined creating thus complex infiltration facility.

Following principles should be kept at the design:

- When deciding about the type of infiltration facility, the most important issue is the quality of rainfall or other wastewater. The subsurface facility must be equipped by efficient pre-treatment of water, e.g., by the sediment trap or settling tank.
- It is necessary to take into account the results of the geological survey.
- The design must take into account the possibility of facility maintenance. The approach road is essential.
- During its operation, the facility must not cause any harm to the surrounding structures and plots by infiltrated water. This concerns waterlogging of the land and seeping infiltrated water into cellars and other subsurface stowages. It is recommended to equip the infiltration facility by the emergency spillway.
- It is recommended by standard [17] that the available volume V_{VZ} of the facility should be emptied at least after 72 h after the end of the storm.

4.7 Dealing with Uncertainties

In Eq. (4.5), all variables are subject to uncertainties. The shortage of the deterministic approach mentioned in Sect. 4.6 is that the coefficient of safety is related only to the infiltrated volume. Practical experience suggests that most designers choose the lowest recommended value of the coefficient $f = 2$, namely for economic reasons. Strictly deterministic safety factor approach is presently substituted by more sophisticated methods. In practice, the method of partial reliability factors is presently in use at numerous technical disciplines [55].

The coefficient of infiltration k_v is determined by an infiltration test with a recommended duration [17]:

$$k_v = \frac{Q_{\mathrm{ZK}}}{A_{\mathrm{ZK}}}, \tag{4.6}$$

where Q_{ZK} is the volume of infiltrated water during the infiltration test or the inflow of water into a trial object and A_{ZK} is the infiltration area. However, the coefficient of infiltration is a time-dependent variable during the infiltration test. The value of the coefficient may be determined by using the basic relationships of groundwater hydraulics [31], the infiltrated flow rate can be expressed by means of the equation [32]:

$$Q_{\mathrm{VSAK}}(t) = \int\limits_{A_{\mathrm{ZK}}(t)} \mathbf{q}(x, y, z, t) \cdot \mathrm{d}A, \tag{4.7}$$

where $\mathbf{q}$ is the vector of the specific flow rate that is defined for the saturated and unsaturated zones as:

$$\mathbf{q} = \mathbf{K}(\theta) \cdot \operatorname{grad} h(\theta), \tag{4.8}$$

where $\mathbf{K}$ is the tensor of the saturated/unsaturated hydraulic conductivity, h is the piezometric head and θ is the moisture content of soil. By assuming $Q_{\mathrm{VSAK}} = Q_{\mathrm{ZK}}$ and substituting Eqs. (4.7) and (4.8) into (4.6), one obtains:

$$k_v = \frac{1}{A_{\mathrm{ZK}}} \int\limits_{A_{\mathrm{ZK}}} \mathbf{K}(\theta) \cdot \operatorname{grad} h(\theta) \cdot \mathrm{d}A. \tag{4.9}$$

All variables in Eq. (4.9) are time (t) and position (x, y, z) dependent. The hydraulic gradient depends on the shape and properties of the infiltration area, i.e., it depends on the shape of the infiltration facility, on the geological composition of the strata, on the boundary and initial conditions, on the water level in the infiltration facility, the level of the impermeable base, the initial groundwater level and moisture content of soil. $\mathbf{K}(\theta)$ and grad $h(\theta)$ can be determined, for example, using a calibration of numerical model of flow in the unsaturated zone [32–34, 46, 56–58].

Uncertainties are brought into the solution by the changing initial moisture content of the soil, the position of the groundwater level and the position of the impermeable basement as well. Another factor influencing the course of infiltration is the size of the infiltration facility, its spatial arrangement and type (an infiltration furrow, an infiltration well, perforated piping). These factors are usually different for the designed infiltration facility and for the test at which the coefficient of infiltration was determined.

At a practical design, the guidance of the standard ČSN EN 1990 [55] may be applied. Here, the storage volume of the infiltration facility should be determined using the condition of the limit state. According to Duchan, Říha [59], this can be expressed in a more general form as follows:

$$\gamma_{VZ} \cdot V_{VZ} > \gamma_n \cdot \max_{t_c=0,t_{c,\max}} (\gamma_S \cdot V_S - \gamma_{VSAK} \cdot V_{VSAK}) \tag{4.10}$$

where V_{VZ} is the storage volume of the infiltration facility, V_{VSAK} is the volume of infiltrated water and V_S is the volume of precipitation water. In the condition, in a broader concept, the left side represents the "resistance" of the object and the right side "load." The maximum of the term in parentheses on the right side of Eq. (4.9) is determined with a whole range of the duration times t_c of the design rain. The following partial coefficients have been introduced into Eq. (4.10):

- γ_{VZ} for geometric uncertainties of the volume of the flood-control storage,
- γ_n of the significance of the facility,
- γ_S expresses the reliability of the volume of precipitation water,
- γ_{VSAK} expressing uncertainty in the volume of infiltrated water.

The geological survey and field tests are usually not extensive, mainly in case of smaller installations in which the budget is limited. In practice, numerical models are used for designing infiltration facilities less often than simplified procedures applying the coefficient of infiltration k_v. The coefficient is determined using an infiltration test that usually reflects only instantaneous local conditions. When using the coefficient of infiltration, it is considered that it expresses characteristics such as hydraulic conductivity, inhomogeneity and anisotropy of the groundwater body. An important factor in determining the coefficient of infiltration is the duration of an infiltration test. The coefficient of infiltration according to standard [17] should be evaluated at the end of the test after 24 h; in practice, however, the length of the test is usually shorter. Due to this fact, the over-evaluated coefficient of infiltration is usually obtained (Fig. 4.15).

The more detailed analysis of related uncertainties and guidance on the quantification of abovementioned partial reliability coefficients is above the scope of this chapter. More details including nomographs for determining the values of individual coefficients for four types of soils and variable groundwater level may be found in the guidance [4].

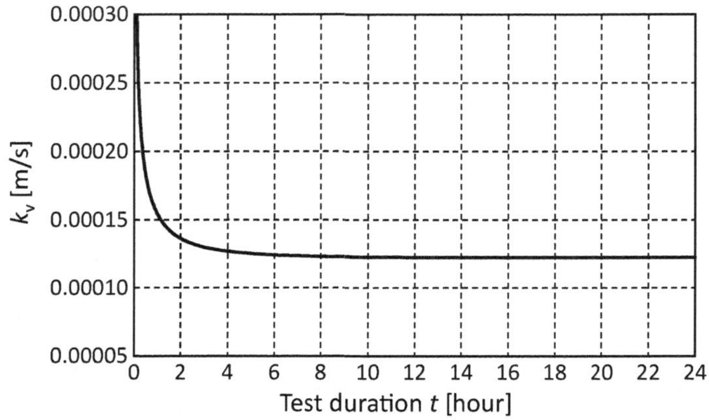

Fig. 4.15 Dependence of k_v on infiltration test duration [4]

4.8 Conclusions

The reduction of surface runoff belongs to the topical tasks of urban water management. The efficient measures of stormwater management are water retention and infiltration at the place of its origin, i.e., just within the built-up areas [19]. Efficient arrangements may be installed at or close to individual estates such as private dwellings, blocks of flats, industrial and agricultural facilities. The effect on the technical drainage systems such as sewerage and open channels may be considered if the retention and infiltration capacities are efficiently planned, designed and performed.

From the view of soil jeopardizing and potential deteriorating artificial surface, water infiltration is a contentious topic worldwide. Variable composition and quality of stormwater in contradiction to generally considered high value of groundwater eventuate in numerous uncertainties and frequently also contradictory opinions. The soil protection is also accentuated during the last decades in this context.

At the design, wide range of related issues has to be taken into account. These are uncertainties in geological structure of the soil profile, limitations of available areas for infiltration in urbanized territories, interference with existing engineering networks, concerns about affecting existing buildings by increased groundwater level due to infiltration and others. Thus infiltration of stormwater belongs to the most challenging problems of urban engineering. Several promising techniques related to stormwater infiltration, such as the use of new adsorption materials, are presently developed and tested [5].

4.9 Recommendation

The planning, design and performance of rainwater infiltration systems in urban areas have to be part of integrated urban drainage-wastewater systems. As such, the location of retention and infiltration facilities has to be planned and harmonized within the overall urban plan. Following recommendations may be formulated:

- Careful and professional hydrological, geotechnical and hydrogeological survey and analysis have to be performed before the design to determine soil profiles and infiltration rates. Hydrogeological investigations are required in cases when there is a likelihood of groundwater discharge or high seasonal water tables. Experience shows that saving money for the survey usually results in increased expenses for the design, construction and performance.
- At the survey and the design of infiltration device surrounding facilities and sub-surface structures must be taken into account. The infiltration facility must not be designed separately but in a wider context of other similar facilities and elements influencing groundwater regime, phreatic surface and flow direction at the entire area.
- As the overall planning, surveying and designing process is complex and multi-disciplinary, respected specialists and experts must participate in the solution.
- The infiltration facilities such as trenches, basins, wetlands and shafts must meet safety requirements provided, e.g., by signage, fences, locked shaft covers, etc.

References

1. Grischek T et al (1996) Urban groundwater in Dresden. Hydrol J 4(1):48–63
2. Watkins DC (1997) International practice for the disposal of urban runoff using infiltration drainage system. In: Groundwater in the urban environment, vol 1: problems, processes and management: proceedings of the XXVII IAH congress on groundwater in urban environment. Balkema, Nottingham. Rotterdam, pp 205–210
3. Hlavínek P et al (2007) Rainwater management in urban area. Ardec, Brno, p 164
4. Říha J et al (2015) Guidance: determining parameters for the design of infiltration facilities, 1st edn., p 120
5. Krejčí V et al (2000) Drainage of urbanized areas-conceptual approach. Noel, Brno, p 560
6. Pazwash H (2016) Urban storm water management, 2nd edn. CRC Press, p 684
7. Ferguson BK (2017) Stormwater infiltration. CRC Press, p 288
8. Guo JCY (2017) Urban flood mitigation and stormwater management. CRC Press, p 589
9. Bloomberg MR, Strickland CH (2012) Guidelines for the design and construction of stormwater management systems. Department of Environmental Protection and Department of Buildings, New York, p 137
10. DWA (2005) Planning, construction, and operation of facilities and infiltration of precipitation water. Arbeitsblat DWA-A 138. Hennef, Deutschland, p 60
11. DWA (2007) Recommendations for dealing with rainwater. Merkblatt DWA-M 153. Hennef, Deutschland, p 8
12. PWD (2014) Stormwater management guidance manual version 2.1. SMGM 2.1. Philadelphia Water Department, Philadelphia, USA, p 381

13. Water Sensitive Urban Design Guidelines (2013) South Eastern Council, Melbourne Water, Melbourne, Victoria, p 44
14. DESW (2013) Eastern Washington low impact development guidance manual. Department of Ecology State of Washington, p 249
15. The Act No. 183/2006 Coll. (2006) Act on town and country planning and building code (The Building Code). Czech Republic
16. The Act No. 254/2001 Coll. (2001) Act on Water (The Water Act). Czech Republic
17. ČSN 75 9010 (2012) Stormwater soakaways. Institut for standardization and testing. Czech Office for Standards, Metrology and Testing, Prague, p 44 (in Czech)
18. TNV 75 9011 (2013) Sustainable stormwater management. Ministry of Agriculture in Czech Republic, Prague, p 65 (in Czech)
19. ATV-DVWK, A 138 (2001) ATV-Abwassertechnische Vereinigung: Plannung, Bau, und Betrieb von Anlagen und Versickerung von Niederschlagen. Arbeitsblat. German Association for Water Management (in Deutch)
20. Implementation Decree No. 268/2009, Coll. (2009) Decree on technical requirements for civil structures. Czech Republic
21. Implementation Decree No. 501/2006, Coll. (2006) Decree on general land use requirements. Czech Republic
22. Field R (2018) Integrated stormwater management. CRC Press, p 399
23. Decree 432/2001, Coll. (2001) Decree of the Ministry of Agriculture on the documents for the request for a decision or statement and on the permits, permissions and statements of the water authority. Czech Republic
24. The Announcement No. 499/2006, Coll. (2006) About the documentation of civil structures. Czech republic
25. Land registry of the Czech Republic. Retrieved from www.geoportal.cuzk.cz. Accessed on Aug 2018
26. MACR (2011) Report on water management in the Czech Republic. Ministry of Agriculture of the Czech Republic, p 100
27. Butler D, Davies JW (2011) Urban drainage, 3rd edn. Spon Press, p 652
28. TP-83 (2014) Drainage of roads, technical conditions. Ministry of Transport, Department of Roads, Czech Republic, p 60 (in Czech)
29. Archive Geofond, Czech geological survey services. Retrieved from www.geology.cz. Accessed on Aug 2018
30. State administration of land surveying and cadaster. ČÚZK. Retrieved from www.cuzk.cz. Accessed on Aug 2018
31. Bear J, Verujiit A (1992) Modeling groundwater flow and pollution. D. Reidel Publishing Company, Holland, p 414
32. Lu N, Likos WJ (2004) Unsaturated soil mechanics. Wiley, New Jersey, p 584
33. Šejna M, Šimůnek J (2007) HYDRUS (2D/3D): Graphical user interface for the HYDRUS software package simulating two- and three-dimensional movement of water, heat, and multiple solutes in variably-saturated media (online)
34. Šimůnek J, van Genuchten MTh, Šejna M (2006) The HYDRUS Software package for simulating the two- and three-dimensional movement of water, heat, and multiple solutes in variably-saturated media. Praha, p 241
35. Benetín J (1958) Movement of water in the soil. Vyd. SAV, Bratislava, p 216 (in Slovak)
36. Jetel J (1982) Determination of hydraulic parameters of soils by hydrodynamic tests in boreholes (in Czech)
37. Girinskij NK (1953) Opredelenije koefficienta filtraciji peskov i suspesej po dannym naliva v šurfy. In: Trudy VEGINGEO Voprosy gidrogeologii i inženěrnoj geologii. Gosgeoltěchizdat, Moskva, pp 28–60 (in Russian)
38. Galabov M (1966) Nomogrami za chidrogeložki izčislenija. Izd. Technika, Sofija, p 144
39. Reynolds WD, Elrick DE, Topp GC (1983) A re-examination of the constant head well permeameter method for measuring saturated hydraulic conductivity above the water table. Soil Sci 136(4):250–268

40. Elrick DE, Reynolds WD (1992) Methods for analyzing constant-head well permeameter. Soil Sci Soc Am J 56:320–323, Canada
41. Elrick DE, Reynolds WD, Tan KA (1989) Hydraulic conductivity measurements in the unsaturated zone using improved well analysis. Ground Water Monit Rev 9:184–193
42. Bouwer H (2002) Artificial recharge of groundwater: hydrogeology and engineering. Hydrogeol J 10(1):121–142. https://doi.org/10.1007/s1040-01-0182-4
43. ČSN EN ISO 22282-5 (2012) Geotechnical investigation and testing-Geohydraulic testing. Part 2, water permeability tests in a borehole using open systems. Institut for standardization and testing. Czech Office for Standards, Metrology and Testing, Prague, p 32
44. Earth manual part 2 (1990) 3rd edn. Denver, Colorado, p 1278
45. Bouwer H, Rice RC (1976) A slug test method for determining hydraulic conductivity of unconfined aquifers with completely or partially penetrating wells. Water Resour Res 12(3):423–428
46. van Genuchten MTh, Leij FJ, Yates SR (1991) The RETC code for quantifying the hydraulic functions of unsaturated soils, version 1.0. EPA report 600/2-91/065, U.S. Salinity Laboratory, USDA, ARS, Riverside, California, p 93
47. Bareš V, Kabelková I, Stránský D (2013) TNV 75 9011 precipitation water management, Part 3: dimensioning of objects and facilities. Water Manage 11(2013):383–386
48. Vukovič M, Soro A (1992) Hydraulics of water wells. Theory and application. Water Resources Publications, LLC, Chelsea, Michigan, p 354
49. ČSN EN ISO 22282-2 (2012) Geotechnical investigation and testing—geohydraulic testing—-Part 2: water permeability tests in a borehole using open system. Czech Office for Standards, Metrology and Testing, Prague
50. Hvorslev MJ (1951) Time lag and soil permeability in ground-water observations. U.S. Army, Corps of Engineers, Waterways Experiment Station, Vicksburg MS, Bull. 36, p 53
51. Bouwer H (1974) Renovating municipal wastewater by high rate infiltration for groundwater research. J Am Water Works Assoc 159–162
52. Cooper HH, Bredehoeft JD, Papadopulos SS (1967) Response of a finite-diameter well to an instantaneous charge of water. Water Resour Res 3(1):263–269
53. Custodio E (1986) Recarga artificial de acuíferos. Ministerio de Obras Públicas y Urbanismo. Boletín de Informaciones y Estudios, no 45 (Febrero 45), p 134 (in Spanish)
54. MUNLV (2001) Ministerium für Umwelt und Naturschutz, Landwirtschaft und Verbraucherschutz des Landes Nordrhein-Westfalen (MUNLV NRW), 2001. Naturnahe Regenwasserbewirtschaftung. Zukunftsfähige Wasserwirtschaft in Industrie- und Gewerbegebieten, p 118
55. ČSN EN 1997 (1997) Eurocode 7. Geotechnical design, p 138
56. Harbaugh AW, Banta ER, Hill MC, McDonald MG (2000) MODFLOW-2000, the U.S. Geological Survey modular ground-water model—user guide to modularization concepts and the ground-water flow process. Open-file report 00-92. U.S. Geological Survey, p 127
57. Lin H-CJ, Richards DR, Talbot CA (2011) A three-dimensional finite element computer model for simulating density-dependent flow and transport in variably saturated media, version 3.0, p 150
58. Šimůnek J, Šejna M, Saito H, Sakai M, van Genuchten MTh (2009) The hydrus-1D software package for simulating the movement of water, heat, and multiple solutes in variably saturated media, version 4.08, HYDRUS software series 3. Department of Environmental Sciences, University of California Riverside, Riverside, California, USA, p 332
59. Duchan D, Říha J (2017) Stormwater infiltration facilities-philosophy, design. Int J Water Resour Arid Environ 6(2017):33–40
60. Dubar s.r.o. Retrieved from www.dubar.cz. Accessed on Aug 2018

Part III
Water Quality and Reuse

Chapter 5
Stream Water Quality Modelling Techniques

Jaromír Říha

Abstract Water quality in open channels plays a significant role in water resources planning and management. Water quality models are important decision support tools for water pollution control, the study of the aquatic ecosystems and the assessment of the effects of point and non-point (diffuse pollution) sources. Modelling of stream water quality is broadly used to assess current conditions and impacts of proposed measures in water quality management. The modelling is based on knowledge about parameters describing transport processes like hydrodynamic dispersion, advection and decay rates. According to the purpose, the scale of the studied area and the detail of solution various models of different levels are used. In this chapter, practical examples of screening, overall and detailed water quality studies applying modelling techniques are presented.

Keywords Watercourses · Water quality · Stream water quality modelling · Pollution sources

5.1 Introduction

In the worldwide context, and in the Czech Republic, the protection of waters is traditionally a complex activity consisting in the protection of both quantity and quality of surface and subsurface water. Water quality studies are integral part of research and tutorials focused on water management and environmental issues [1–7] and others. In the Czech Republic, the water quality is dealt with according to the Czech law represented, namely, by the Water Act [8] and by the European regulations, namely the Water Framework Directive [9]. The Directive [9] aims at maintaining and improving the aquatic environment and primarily the quality of the waters concerned. In the Czech Republic, the responsible authorities are the Ministry of Agriculture and the Ministry of Environment who annually report the state of water management to the

J. Říha (✉)
Faculty of Civil Engineering, Institute of Water Structures,
Brno University of Technology, Veveří 95, 602 00 Brno, Czech Republic
e-mail: riha.j@fce.vutbr.cz

M. Zelenakova et al. (eds.), *Management of Water Quality and Quantity*,
Springer Water, https://doi.org/10.1007/978-3-030-18359-2_5

government of the country. For the water quality assessment in the Czech Republic national decrees [10, 11] supplement Water Act [8] and incorporate European regulations into the Czech legislation system taking into account local conditions.

For the assessment of the long-term conditions of water quality in watercourses, periodic monitoring is carried out at 263 profiles located at significant streams in the Czech Republic (Fig. 5.1). The frequency of the sampling is annually 12 or 24 samples. Normally, 78 water quality indicators are observed, however not at all sampling profiles. The monitoring programme [12] was elaborated by the Czech Hydrometeorological Institute and was issued by the responsible ministries providing guidance on the methods and the extent of the monitoring. As a part of individual research projects, on-purpose monitoring is frequently carried out at locations of interest and during the limited period. This monitoring is sometimes organized as tracer experiments [1].

In the European context, the European Water Framework [9] formulates objectives concerning stream water quality. In the Czech Republic, the immission and emission standards for the release of both municipal and industrial effluents to the surface streams are specified in the [10, 11]. They are expressed in terms of maximum permissible concentration limits for individual water quality indicators like biochemical oxygen demand (BOD) and chemical oxygen demand (COD), nitrogen compounds and many others. In practice, the majority of permanent effluent discharges, for example, from waste water treatment plants (WWTP) are subject to periodic water quality checks carried out by authorized bodies. However, the limit values are not prescribed for instantaneous releases of sewage water from storm water overflows, which is mostly due to the difficulty of measuring released wastew-

Fig. 5.1 The location of monitoring profiles in the Czech Republic [16]

ater during heavy rainfall [2]. For this type of effluents, permissible limits have to be specified individually based on the character and quantity of the released wastewater, on the water flow and quality in the receiving stream and its environmental value. Cost–benefit analysis is required to indicate the effectiveness of the measures proposed. During the last decades, numerous studies have dealt with both municipal and industrial effluents to the surface waters [3] and others. Particular attention has been paid to water quality monitoring, measurement and contamination tracking in relation to drainage system [4, 13, 14]. Here, various pollution indicators and polluting agents have been studied. These were dissolved oxygen, BOD_5 and COD, nitrogen and phosphorus compounds, temperature, faecal contamination and many others.

The critical part of the stream water quality management is mathematical models, which may serve for the interpolation of monitoring data and are an essential decision tool. They enable optimization of proposed measures for stream water quality improvements like investments into the sewer systems or intensification of wastewater treatment plants (WWTP). Stream water quality models are used to be part of general water management plans of inhabited and outdoor areas as a base for the elaboration of urban plans namely of the municipalities. The results of coupled hydraulic and water quality modelling of sewer–watercourse may indicate the need for arrangements at the sewer network like surge chambers and storm water overflows separating part of wastewater during heavy rainfalls.

Due to one-dimensional (1D) character of open channels, the water quality models in streams are essentially proposed as 1D. According to the nature of the simulated problem, the water quality models may be steady state or transient in combination with steady-state or transient water flow.

5.2 Classification of Water Quality Models

Classification of stream water quality models may be carried out according to various factors. This may be according to the model level (global, basic, detailed), domain character (urban area, water courses related to river basin, models including surface and diffusion pollution) and time dependence of variables (steady state, transient).

5.2.1 Models According to Their Level

Global models, sometimes called as screening models, provide fast and general information about the state of water quality in large catchments. At the modelling fixed constants and parameters are used in the simplified governing equations. Such models usually do not reflect the interaction between individual environmental components, water quality indicators and pollution agents. Global models are used for the assessment of the relative effect of crucial measures at the catchment, forecasting of relative changes in water quality, evaluation of the impact of hydrological characteristics on water quality or for identification of dominating processes and variables in the catchment.

Basic models contain more detailed description of the most important processes governing stream water quality. The results provide a rough quantitative view on the concentration of individual indicators. Basic models enable evaluation of limit (maximum, minimum) concentrations at individual channel reaches, the approximative extent of potential accidents in river pollution and consequence of both accidental and permanent spills. Based on results of basic modelling, the areas for more detailed solution may be pinpointed.

Detailed models concern local water quality conditions, namely in particular areas and watercourses with significant pollution origin and spills where water quality may be permanently or temporarily harmed. Simulation of individual pollution indicators is possible, and interaction between the selected substances may be taken into account. All types of pollution sources like surface, linear and point are included either invariable or changing during the time.

Borders between single model types are not sharp; particulars of the models are governed by various factors like size of the river basin, number, and significance of pollution sources, the density of monitoring network, etc.

In this chapter, applications of three mentioned models are presented. The general location of individual models is shown in Fig. 5.2.

Fig. 5.2 Approximate location of individual models. Blue—global model, green—basic balance model, red—detailed model

5.2.2 Models According to the Regime of Pollution Release

Dynamic models concern situations when the pollution is suddenly released at river cross section. The pollution causes "wave" which propagates along the stream. Such pollution escape can be originated by controlled and uncontrolled releases from the wastewater treatment plant (WWTP), car crash or lack of discipline in the industrial, municipal or agricultural waste management. When solving the sudden release of the pollution to the stream, water quality management aims to the most effective protection and a decrease of injuries of human life, human health, property and real estates. The purpose of the modelling is to forecast time-spatial distribution of the pollution concentration along the watercourse downstream of the spillage point. The mathematical model is therefore based on the unsteady transport-dispersion equation. Due to relatively short time, conservative solids are usually assumed, and the decay of substances is not taken into account. The numerical model can be used as so-called "alarm model" compiled in advance for the given stream, calibrated and verified using the field injection test data obtained from tracing experiments.

Balance models are used for proposing solutions for the long-term pollution state. Also, they help to permanently taking care of the stream water quality improvement. This can be achieved by testing different scenarios such as reducing pollution sources by investments to remedial measures with the aim to reduce the pollution emission at the sources (WWTP intensification, landfill management, system of penalties, etc.). The main goal of the modelling is to evaluate scenarios tending to the overall stream water quality improvement. The target water quality is determined by the national and international water quality standards and directives, environmental demands (water organisms, plants) and by the manner of water use (drinking water). The quasi-steady-state balance stream water quality models serve for the evaluation of present water quality conditions at various flow and pollution scenarios and for the simulation of selected variants of water quality improvement (building or intensification of WWTP, erosion control, etc.) and optimization of the improvement based on cost–benefit analysis.

In general, two approaches mentioned above may be used as models of all three conceptual levels mentioned in the Sect 5.2.1. However, the dynamic approach is common at more detailed models (Sect. 5.7) with boundary conditions and other state variables changing with time.

5.3 Input Data for the Modelling

Input data concern both watercourses network and corresponding river basin. For the analysis, integrated system of water courses is characteristic, and the subject of modelling is significant streams regarding the length, discharge conditions or pollution sources.

In general punctuality of input data should follow the aims of modelling and required accuracy. Before the modelling, the data should be completed, summarized, evaluated eventually corrected, completed and presented in tables, graphs and thematic maps. The structure of the data may be as follows:

1. General characteristics of the catchment like geographical data (location, altitude, management, land cover), morphology, cadastral and population data, protected areas with special water regime, river network topology, etc.
2. River data contain geometrical characteristics of the streams, water bodies (reservoirs, lakes), hydraulic structures, water management rules and arrangements (water diversions).
3. Hydrological and climatic data concerning precipitation, temperatures, discharges in streams and water quality data.
4. Pollution sources are classified according to their type (municipal, agricultural, industrial) and according to spatial character (point, linear, surface and diffuse sources).
5. Data from water quality monitoring serve for model calibration and verification. It concerns data from permanent long-term monitoring and from tracer experiments.

Sometimes, it is desirable to aggregate basic data into groups according to their acquisition, a method of processing or method of their incorporation into the model. It is important to relate all data to the unique global coordinate system (e.g. using GIS) or to the river stationing which may, however, differ in time with the gradual morphological processes and technical interventions like shortening of streams by cutting meanders.

The data acquisition represents considerable financial means and needs sufficient time reserve. In case of newly established monitoring for long-term pollution balance modelling, its duration exceeds several years to obtain seasonal water quality variations for selected indicators.

The data are acquired from diverse sources. The first information may be taken from the maps, in the Czech Republic comprehensive water management maps in scale 1:50,000, and also other river and catchment data are freely available from web-oriented hydroecological information system [15]. Hydrological and climatic data and data from monitoring provides the Czech Hydrometeorological Institute [16]. More detailed river data and information about hydraulic structures are managed by river authorities. The pollution data should be provided by competent representatives of relevant sources like water companies managing WWTPs, industrial or agricultural facilities. It is recommended to check and verify reported data using population (population equivalent) data and other traditional and empirical knowledge about pollution sources and their treatment [17, 18]. The worst identifiable pollution amounts are spills from storm water overflows. These may be estimated only indirectly using calibrated sewer hydraulic and pollution models. In all cases, site investigation is necessary.

Data analysis usually contains basic statistical analysis, regression methods, time series analysis and comparison of monitoring data with water quality limits. It is also necessary to adjust data to appropriate form suitable for the numerical model. For the model calibration and verification, temporally homogeneous data should be provided. The homogeneity refers to the river topology and geometry, hydrology and also water quality where all data from measurements, site investigation, hydrological and water quality monitoring should correspond to the same period. This may be difficult especially when processing long-term water quality monitoring data when during the period changes may occur in the catchment, streams and pollution sources.

5.4 Mathematical Formulation

The transport processes in open channels are traditionally described by one-dimensional (1D) convection-diffusion Eq. (5.1) expressing mass conservation law of the studied matter [19–21]. In such models, the density of polluted water is assumed to be constant and similar to "clean" water density, the substance is well mixed over the cross section, and only longitudinal hydrodynamic dispersion is taken into account. The basic Eq. (5.1) expresses the time change of the mass (expressed via concentration c) as the function of advection, dispersion, dilution, constituent reactions and interactions, sources and sinks [19]:

$$\frac{\partial c}{\partial t} + \frac{\partial (uc)}{\partial x} - \frac{\partial}{\partial x}(D_{\mathrm{L}}\frac{\partial c}{\partial x}) = R + \frac{S}{A}, \quad u = \frac{Q}{A}, \tag{5.1}$$

where c is concentration, u is cross-sectional mean velocity, A is flow area, Q is discharge, D_{L} is longitudinal hydrodynamic dispersion coefficient, R expresses constitutional changes of matter (e.g. retardation, sedimentation and degradation), S is source/sink per reach length, t is time, and x is spatial coordinate along the centre line of the channel.

The boundary conditions are defined as the known concentration at the upstream end $x = 0$:

$$c(0, t) = c_0(t) \tag{5.2}$$

and zero concentration gradient in the downstream end $x = x_{\mathrm{L}}$,

$$\frac{\partial c(x_{\mathrm{L}}, t)}{\partial x} = 0 \tag{5.3}$$

An initial condition is defined as known concentration along the assessed reach at time $t_0 = 0$.

$$c(x, t_0) = c_I(x) \tag{5.4}$$

In Eq. (5.1), the key parameters are flow velocity u and coefficient of longitudinal dispersion D_L, which both generally change in time and along the stream axis. Basically, the flow velocity $u(x, t)$ is determined by hydrodynamic modelling [7]. In practical applications, the dispersion coefficient D_L is frequently assumed to be constant in time and along the typical river reaches. The longitudinal dispersion coefficient D_L is related to the hydraulic and geometric characteristics of a given open channel and also to fluid properties [1, 22]. Its value may be determined using empirical formulae published by various authors. The best fit verified by tracing experiments [1] provide following Eq. (5.5) [20] and Eq. (5.6) [23]:

$$D_L = 0.011 \frac{C w^2}{h\sqrt{g}} .u, \tag{5.5}$$

$$D_L = \frac{2.0h\sqrt{g}}{C} \left(\frac{w}{h}\right)^{3/2} .u \tag{5.6}$$

For the solution of Eq. (5.1) with boundary and initial conditions, appropriate numerical methods are used. These are finite difference method, finite elements method, a method of characteristics, random walk procedures and others.

5.5 Global Models

Global models serve for obtaining preliminary, general and fast information about water quality in larger catchments and river networks and for the rough estimate of effects of intended remedial measures within stream water quality management. The global models are characterized by only general data without more detailed data about stream geometry, hydrology of the catchment and pollution sources. Usually, the studies are "short term" with relatively low budget. The algorithm is simple with beforehand set model parameters. The assessment has more or less qualitative nature rather than quantitative one. As an example, the study [6] may be mentioned at which global model comprised about 1000 km of water courses in the Dyje river basin. The main goal of the study was to assess an impact of technical improvements on sewers and wastewater treatment facilities financed from European sources on the water quality of the Dyje river which at its lower part is a boundary stream with Austria.

The structure of necessary data generally corresponds to the Sect. 5.3. However, only rough and generalized data are provided about watercourses and water bodies mostly using water management maps 1:50,000 [15] and general water management plans. Quasi-steady-state hydrological and hydraulical conditions are assumed, the solved scenarios usually deal with average annual discharges and also low discharges with exceedance 270, 330 or 355 days. Long-term (usually not exceeding 5 years) averages of pollution sources and monitoring results are used.

One of the goals of global river pollution studies is a qualified estimate of total present and a future mass load of individual polluting agents. This is focused on total averages and on unfavourable conditions at low discharges during dry spells.

As the survey into all pollution sources over large catchments may be time demanding, quite expensive and hardly realized, many times only sources which are subject to improvements are credibly taken into account in the analysis. The model is set up via present water quality identified by long-term monitoring by introducing "fictive" pollution sources and constitutive changes following monitored water quality. By changing real pollution sources, qualitative rate of improvement may be modelled along the system of the concerned streams. Both mass flow and concentrations are simulated for individual water quality indicators.

It can be seen that numerous simplifying assumptions are taken into account at global models. These come mostly from the lack of more detailed and reliable data. For given scenario constant mass flow not changing with time is anticipated; change of concentration is given only by changing the discharge in streams (average, various exceedance). Constitutive changes may be approximated by the kinetics of the first-order [24] with constants estimated using an analogy with similar catchments and river systems.

From previous studies [25], it comes that at the steady-state approach the effect of hydrodynamic dispersion may be neglected. Thus, Eq. (5.1) may be rearranged using $Q = u.A$ and $R = - K.c$ as follows:

$$\frac{\mathrm{d}(Q \cdot c)}{\mathrm{d}x} - S + A \cdot K \cdot c = 0 \tag{5.7}$$

with boundary condition $c(0) = c_0$ expressing known concentration at the uppermost profiles of the domain. K is a first-order rate constant which in natural streams varies between 0 and 15 (1/day) for different open channels and flow conditions, type of water quality indicator and intensity of the processes in the streams.

The mass flow (kg/s) is defined as follows:

$$L = c.Q = A.u.Q \tag{5.8}$$

For the numerical solution, simple explicit difference scheme was applied (Fig. 5.3).

Denoting L_j, L_k mass flow in nodes j, k, the sum of all pollution sources along the reach with the length Δx is denoted $L_i = S.\Delta x$ where S is a pollution source along the river reach. Then, Eq. (5.7) may be rewritten in terms of differences:

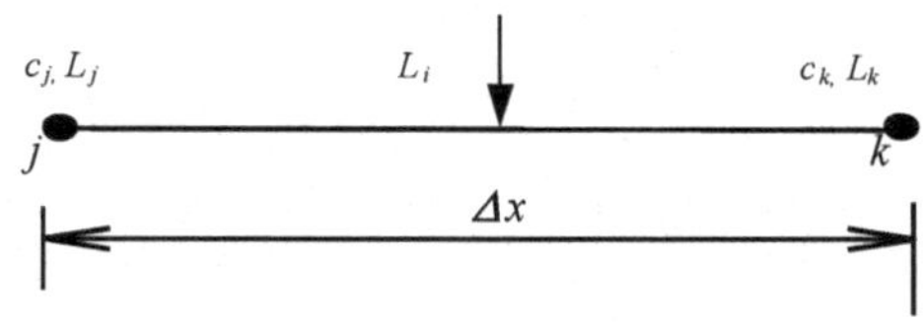

Fig. 5.3 The scheme of the computational river reach

$$L_k - L_j - L_i + A.K.c.\Delta x = 0 \tag{5.9}$$

Constitutive changes in Eq. (5.9) may be expressed using detention time Δt along the reach Δx:

$$A.K.c.\Delta x = K.\Delta t.L_j, \tag{5.10}$$

and after some manipulation:

$$L_k = L_j + L_i - L_j.K.\Delta t = 0. \tag{5.11}$$

The pollution "input" L_i along the reach length Δx is expressed for the present state (subscript S) and improved state after realizing anticipated measures (subscript O). The present mass flow L_{Sk} in the node k may be expressed via average discharge Q_{Sk} and concentration c_{Sk} taken from river monitoring:

$$L_{Sk} = Q_{Sk} \cdot c_{Sk} \tag{5.12}$$

Mass flow L_{Sk} thus represents terms in Eq. (5.11), i.e. mass inflow from upper catchment L_{Sj}, pollution sources along the reach L_{Si} and the effect of constitutional changes along the reach between nodes j and k. The mass flow for the scenario with implemented measures L_{Ok} may be expressed via present mass flow, expected impact of all improvements upstream of given reach and difference in constitutional change along the reach due to the change of total mass load:

$$L_{Ok} = L_{Sk} + K \cdot \Delta t \cdot (L_{Sj} - L_{Oj}) - \sum_{p=1}^{n} (L_{Si}^{p} - L_{Oi}^{p}), \tag{5.13}$$

where n is number of pollution sources upstream of given reach (node k). The effect of weir pools and reservoirs may be estimated using data from monitoring and based on direct proportion with present state based on mass flow entering and leaving the reservoir:

$$L_{OP} = \frac{L_{SP}}{\sum_{p=1}^{n} L_{SN}^{p}} \cdot \sum_{p=1}^{n} L_{ON}^{p} \tag{5.14}$$

where L_{OP} and L_{SP} are mass flows downstream of the reservoir after and before adopting remedial measures on pollution sources, $\sum_{p=1}^{n} L_{ON}^{p}$ and $\sum_{p=1}^{n} L_{SN}^{p}$ are total mass inflows to the reservoir by n tributaries after and before adopting remedial measures.

The algorithm is straightforward to be compiled using a spreadsheet where individual streams are configured at separate sheets. The calculation proceeds from the upstream river reaches and follows the river network topology. At river junctions,

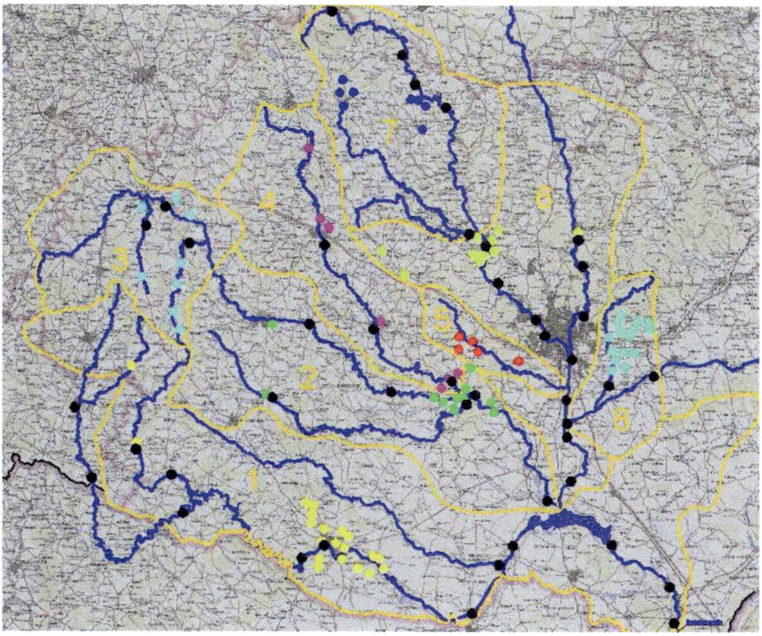

Fig. 5.4 The scheme of the Dyje river network. Black—monitoring profiles, colours—clusters of improvements

the mixing law is applied (Fig. 5.7). The present averaged mass flow is derived from available discharge and concentration measurements in the monitoring network (see Figs. 5.1 and 5.4). The scenarios of pollution reduction from individual sources are introduced to the model.

In the study [6], the effect of improvements on sewer and wastewater treatment facilities aggregated into eight "clusters" according to sub-catchments was modelled. The objective of the simulations of 11 scenarios was to assess the effect on water quality in the Dyje river at the boundary profile in the city of Břeclav close to the Czech-Austrian boundary (Fig. 5.2) for four water quality indicators, namely BOD, COD, ammonia nitrogen (N-NH_4) and phosphorus (P). In Fig. 5.4, black dots indicate the profiles with hydrological and water quality monitoring, and coloured dots show improvements on pollution sources.

Figure 5.5 shows the water quality map according to [26] for the BOD indicator. The light and dark blue depict streams with very good and good water quality; further classes are marked by green, yellow and red colour indicating the worst stream water quality. Two lines along the streams show water quality classes before and after expected improvements. Another analysis was done via longitudinal sections of concentration and mass flow of individual water quality indicators. In Fig. 5.6,

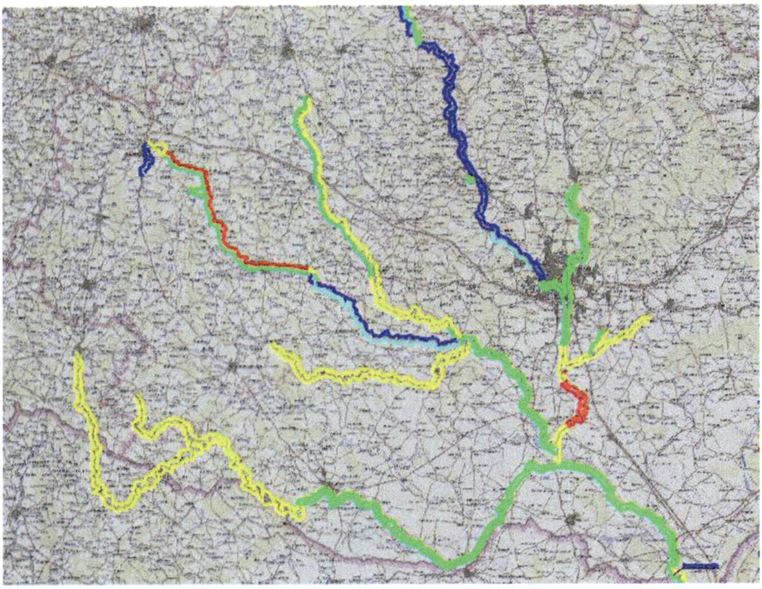

Fig. 5.5 Water quality map for scenario "all improvements" (scenario 11) showing improvement of water quality for BOD along individual streams

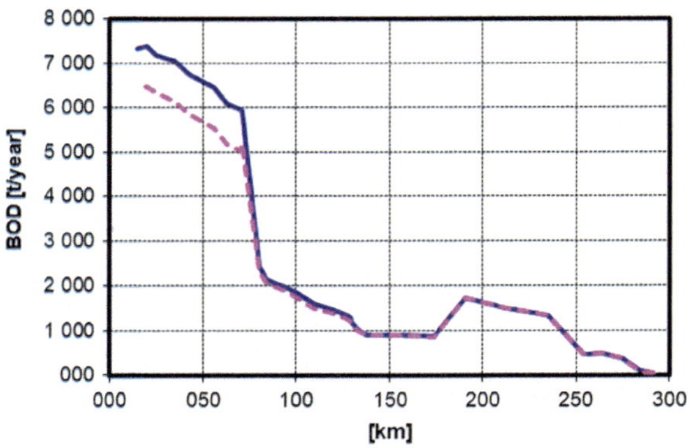

Fig. 5.6 Longitudinal section along the Dyje river for BOD mass flow. Solid dark blue—present state, dashed violet—scenario 11 "all improvements"

the longitudinal section of the total mass flow in (t/year) along the principal stream the Dyje river is shown for present state and for all realized improvements (scenario 11).

The results of the modelling showed that measures at all 8 groups of pollution sources at scenario 11 resulted in water quality improvement at the Dyje river boundary profile by 12% for BOD, 5.5% for COD_{CR}, 10% for N-NH_4 and about 10% for P. The water quality class usually improves only at the upper part of the catchment where the improvements represent higher portion of total load spilled to the stream.

5.6 Basic Models

Basic models serve as the long-term balance solutions in middle-sized catchments in more details than global screening models. Due to the inadequate monitoring frequency (usually once in a month) and no information about the time course of pollution sources, it is not possible to reliably model short-term fluctuations of concentrations of individual pollutants. Similar to global models discharges, velocities, pollution sources and water quality characteristics are averaged over time. Thus, the model is formulated as quasi-steady state with varying punctuality of input data. However, more realistic values of all model parameters are input and resulting water quality variables obtained, so the model provides not only qualitative but also quantitative information about the pollution sources and their impact on stream water quality.

The problem has to be treated as the combination of hydrodynamic river analysis and transport-dispersion solution. The mathematical model for the individual problem solution must issue from the catchment, stream and pollution data available and from the anticipated accuracy of results. In case of basic models, the hydrodynamics and also pollution transport is usually approximated by steady-state approach. The mathematical model of water quality consists of 1D advection-dispersion mass balance Eq. (5.1) expressed for given pollution parameter and corresponding boundary conditions. When solving the long-term pollution balance problem, following additional simplifications have to be assumed:

- the river flow is quasi-steady, the pollution modelling is carried out for selected average and low discharge scenarios,
- the description of the modelled domain is performed with regard to the time delay of water in streams, reservoirs, weir pools and ponds,
- the pollution transport is assumed to be steady and time-independent due to poor data about time variations of the pollution sources; the sampling is performed in relatively long-time intervals (once in the month); seasonal concentration variations are analysed using the mass flow approach and appropriately "smoothed",
- it can be proved [25] that the longitudinal dispersion can be neglected at the steady-state approximation; the error in results does not usually exceed 5%. This error is negligible when compared with sampling error and inaccuracy in determining the pollution sources.

Applying mentioned above simplifications on Eq. (5.1), i.e. $\partial c(x, t)/\partial t = 0$, omitting the second member on the right side and assuming steady-state conditions $R(x, t) = R(x)$, $c(x, t) = c(x)$, $S(x, t) = S(x)$, $A(x, t) = A(x)$, substituting $Q = A.u$ and after some manipulations it reads:

$$\frac{1}{A}\frac{\mathrm{d}(Qc)}{\mathrm{d}x} = R + \frac{S}{A}, \tag{5.15}$$

Only the headwater boundary condition (5.2) is applied, at the steady-state model, the condition in the node $x = L$ is not specified and is determined during the computation (free parameter).

For the solution of Eq. (5.15), the explicit finite difference scheme may be used. When denoting the sum of all pollution sources, S_i at the river reach i with the length Δx, Eq. (5.1) can be written as follows (Fig. 5.3):

$$\frac{Q_k \cdot c_k - Q_j \cdot c_j}{\Delta x} = R_i \cdot A_i + S_i. \tag{5.16}$$

An example of numerical solution deals with the modelling of BOD. In accordance with [24] the first-order kinetic model may be rewritten:

$$R^{\mathrm{BOD}} = -K_{\mathrm{BOD}} \cdot c^{\mathrm{BOD}}. \tag{5.17}$$

After substituting Eq. (5.17) to (5.16), the BOD concentration in the node k is expressed:

$$c_k^{\mathrm{BOD}} = \frac{c_j^{\mathrm{BOD}} \cdot Q_j + \Delta x.S_{\mathrm{BOD}}}{Q_k + A_i \cdot K_{\mathrm{BOD}} \cdot \Delta x}, \tag{5.18}$$

where $K_{\mathrm{BOD}}(T)$ is the rate coefficient of BOD change due to sedimentation, scouring, biochemical oxidation, c^{BOD} is BOD concentration, T is the temperature, R^{BOD} is the rate of BOD change, S_{BOD} is the BOD pollution source along the reach k. The mass flow L_j and L_k in nodes j and k is expressed as follows:

$$L_k^{\mathrm{BOD}} = \frac{L_j^{\mathrm{BOD}} + \Delta x.S_{\mathrm{BOD}}}{Q_k + A_i \cdot K_{\mathrm{BOD}} \cdot \Delta x} \cdot Q_k \tag{5.19}$$

Substituting the average cross-section A_i on the reach i

$$A_i = \frac{Q_k \cdot v_j + Q_j \cdot v_k}{2 \cdot v_k \cdot v_j} \tag{5.20}$$

the Eq. (5.19) is expressed as follows:

$$L_k^{\mathrm{BOD}} = \frac{L_j^{\mathrm{BOD}} + \Delta x.S_{\mathrm{BOD}}}{Q_k + \frac{Q_k \cdot v_j + Q_j \cdot v_k}{2 \cdot v_k \cdot v_j} \cdot K_{\mathrm{BOD}} \cdot \Delta x} \cdot Q_k \tag{5.21}$$

At the stream junction, the mixing equation is applied (Fig. 5.7):

$$c_k^{\mathrm{BOD}} = \frac{c_j^{\mathrm{BOD}} \cdot Q_j + c_p^{\mathrm{BOD}} \cdot Q_p + \Delta x.S_{\mathrm{BOD}}}{Q_k + A_i \cdot K_{\mathrm{BOD}} \cdot \Delta x} \tag{5.22}$$

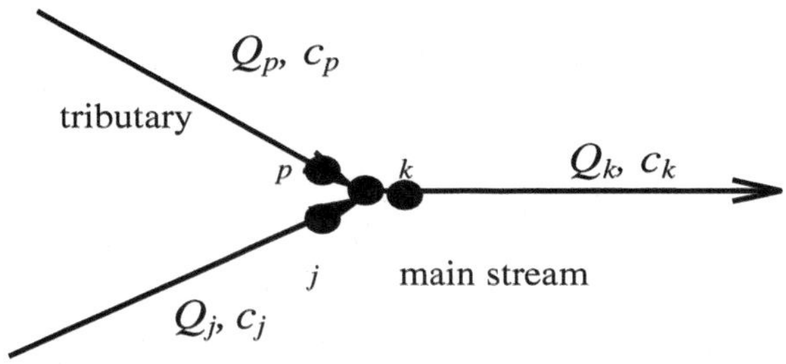

Fig. 5.7 The stream junction notation

The influence of the temperature on the biological processes can be expressed by the change of temperature dependent rate coefficients K in Eq (5.17) according to the Arrhenius formula [24]:

$$K(T) = K_{20} \cdot \Theta^{(T-20)} \tag{5.23}$$

where $K(T)$ is temperature dependent rate coefficient, T is temperature, K_{20} is rate coefficient at temperature 20 °C, Θ is empirical temperature coefficient differing for typical reactions [24]. Temperature field has to be input in the individual river network nodes. It is derived from measured or modelled temperatures for given scenario, e.g. season.

An algorithm is based on explicit procedure when the mass flow for particular pollution parameter (analogically to BOD) is modelled. The calculation is carried out from headwater nodes of individual stream branches in the downstream direction. Input data are dealing with stream network topology, geometric and hydraulic properties, pollution sources and calibration data. The results are saved in appropriate formats to be transferred to graphical applications (GIS), various types of graphical outputs are available—pollution sources, calibration results, mass flow, and concentration.

Practical application using basic model was carried out at four catchments of Czech rivers the Jihlava, Želetavka, Oslava, and Svratka (Fig. 5.2). The total length of modelled streams at all rivers was about 600 km. Following examples are referred to the study of Jihlava river catchment [27].

For the calibration, the extensive set of observation data has to be collected and analysed. The data from sampling are usually obtained in the form of concentration of given water quality indicator. For the processing, this should be completed with the discharge data and finally, the mass flow values for each profile and sampling time has to be calculated. Assuming that the stream mass flow is generally higher during the wet periods (rainwash, street pollution, sewer separators), it is advisable to classify mass flow results according to stream discharges and arrange them into two calibration sets for "wet period" characterized by approximately annual average discharge Q_a and "dry period" characterized e.g. by Q_{355} days discharge or rather higher.

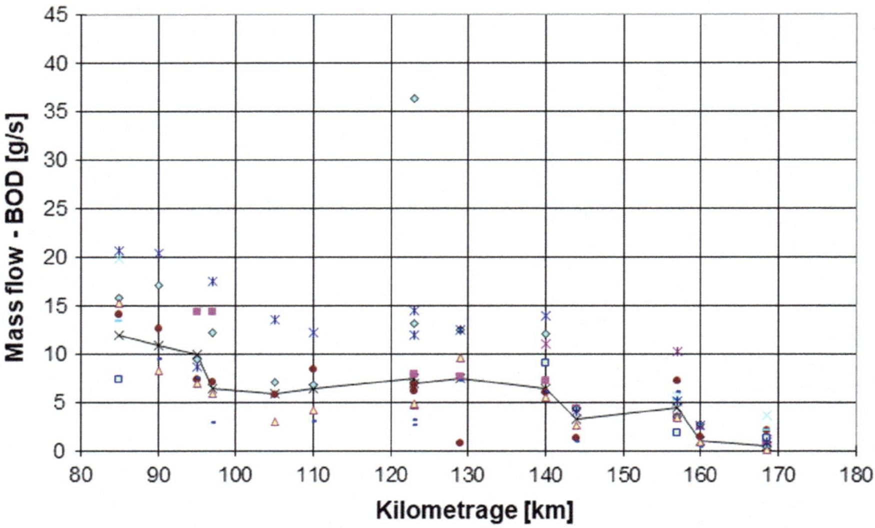

Fig. 5.8 The Jihlava river—sampling data analysis—BOD mass flow, dry spell

Example of such analysis at the Jihlava river for the dry period is shown in Fig. 5.8. In such a way the numerical model was calibrated for two scenarios mentioned using analysed and smoothed sampling data. The calibration proceeded from headwater nodes of individual stream branches in the downstream direction. During the calibration, coefficient K_{BOD} in Eq. (5.21) was determined along the streams by the trial-and-error method. Characteristic values of coefficients were set for modelled flow and pollution conditions.

Verification of the model was performed for the set of independent observation data obtained from various NGO's. Results of model calibration are shown in Fig. 5.9 for the dry period, and BOD mass flow in the Jihlava river. The simulations of selected scenarios were based on calibrated and verified model. Quite an extensive number of variants dealing with improvements of various pollution sources was evaluated and analysed.

The graphical outputs enabled the assessment of individual pollution sources, their significance and impact on water quality along water courses. The outputs were expressed in the form of mass flow and concentration of individual pollution parameters in tabular and graphical form. An example of the results is shown in Fig. 5.10 where the proportion of inputs from municipal, industrial and agricultural pollution sources in terms of COD_{CR} can be seen along the Želetavka river.

In Fig. 5.11 another presentation of results is shown where individual water quality indicators at selected profile are compared with their division according the pollution sources in the catchment upstream.

The balance models may serve as powerful and efficient decision support tool when allocating financial sources at stream water quality improvement. The results of modelling may indicate the most significant pollution sources considerably affecting

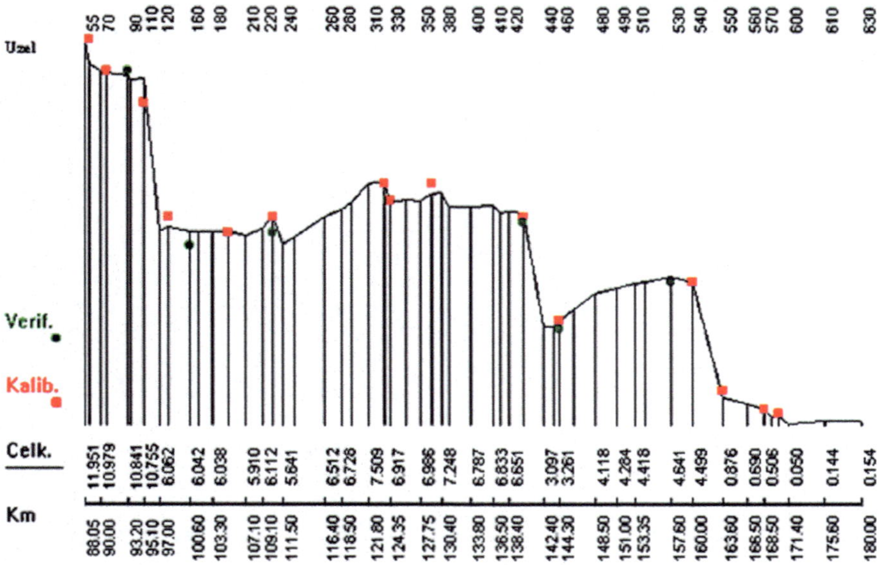

Fig. 5.9 The Jihlava river model calibration—BOD mass flow in (g/s), dry period. Results of model ———,calibration data ·,verification data •

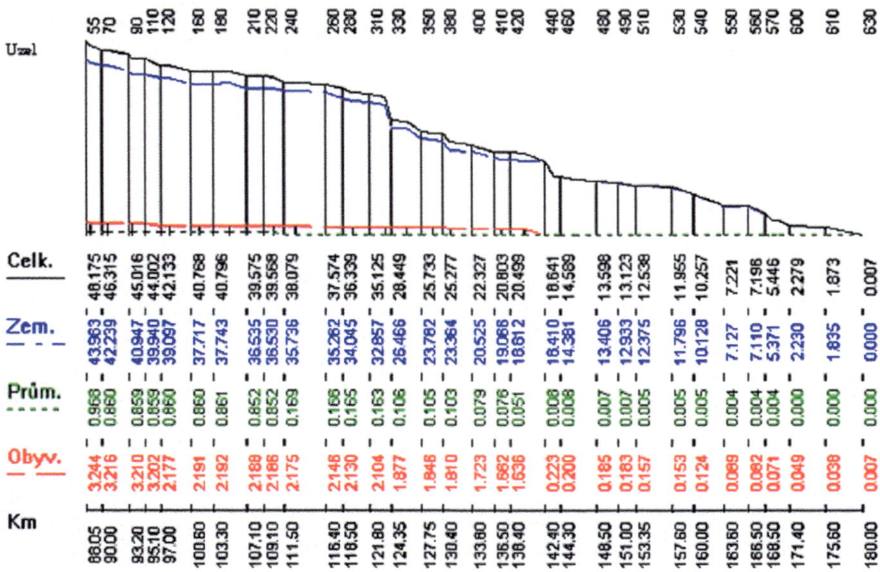

Fig. 5.10 The Želetavka river—COD mass flow (g/s) with the division to agricultural, industrial and municipal pollution origin

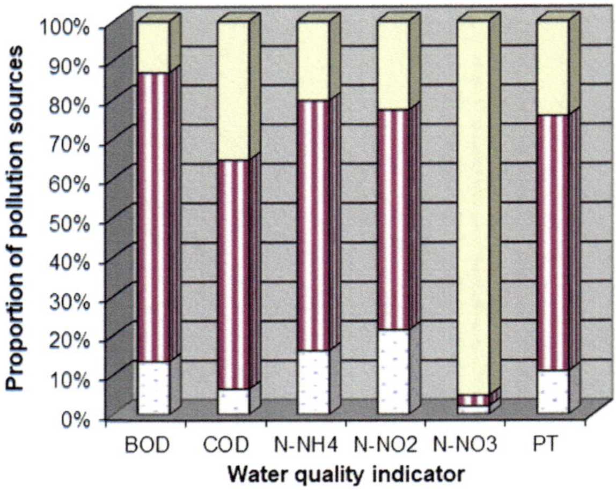

Fig. 5.11 Comparison of the percentual contribution of the pollution from individual sources for individual water quality indicators at the Jihlava river profile at the entrance to the Dalšice reservoir. White dotted—industrial, red vertical strips—municipal, yellow—agricultural and diffusion

water quality. For example, it can be seen from Fig. 5.11 that BOD, N-NH_4 and phosphorus pollution come mostly from the municipalities, on contrary nitrates N-NO_3 come dominantly from agriculture.

Experience shows that more sophisticated transient models may not give more accurate results. The most crucial is to collect all relevant information, namely those about all important pollution sources, to be included into the model and relevant data from water quality sampling.

5.7 Detailed Models

Detailed models deal with local water quality solutions. They focus mostly on the limited part of the catchment or river reach. The objectives of the solution may be the impact of the interaction of sewer network on stream water quality via sewer overflows, detailed modelling of the progression of spills due to accidents, development of emergency plans and others.

The models are usually conceived as transient, time series of the discharge and pollution sources have to be provided as input data. When the duration of events is short, and the transport of the solids is fast, constitutive changes may often be neglected and the pollutants are considered to be conservative. On the other hand concentration gradients are considerable so dispersion together with advection are the prevailing and the most important processes influencing the time and spatial course of concentration.

For the solution, full shape of Eq. (5.1) is used. To realistically describe the dispersion processes, the dispersion coefficient has to be reliably estimated. For this purpose tracing experiments are executed and evaluated [1]. Another, less reliable method is to derive the dispersion coefficient using empirical formulae, e.g. by Eqs. (5.5) and (5.6).

Table 5.1 Summary of the streams in the Vyškov city

River	Reach length (km)	Average width of the river bed (m)	Number of sewer overflows
Velká Haná	0.7	5	3
Malá Haná	0.4	4	1
Drnůvka	2.0	3	13
Roštěnický potok	1.0	4	5
Haná	5.0	9	31

One example is the use of the simulation models for the optimization of the management of sewer systems and sewer overflows releasing sewage waters to the receivers during extensive storm events. The municipalities have been traditionally provided with combined sewer systems. Newly built urban areas have been appended to the existing sewer mains, causing their frequent overloading. To release water and avoid sewer overloading sewer overflows (SO) are proposed along the sewer mains usually following local rivers [28]. The appropriate tool for the assessment of the impacts of sewer overflow effluent discharges into surface streams is the pollution transport modelling [5]. In practice, both single models of sewer and open channel networks and coupled models including both systems may be used [29]. It is true that the single stream water quality models often enable easier, faster and more transparent data handling.

The input data consist in the detailed topology and geometry of the streams, including local arrangements like water diversions, intakes, headraces, etc. The stream topology has to be completed by the locations of pollution sources. These are frequently storm sewer overflows, both present and proposed to be built. Hydrological conditions in streams have to be described both for the no-rain period and for the design storm event. Usually, the storm event and its impact on sewer system is the subject of the modelling of sewer network response in terms both water amount and its quality. In this case outflow hydrographs from sewer overflows and water quality resulting from the sewer model are inputs to the open channel water quality model.

This kind of model is demonstrated in the following example where the impact of effluents from sewer overflows in the city of Vyškov and their superposition is modelled by 1D transient water quality model. The purpose of the modelling was to assess the changes over time in the concentration of six water quality indicators, namely BOD, COD, Ammonia nitrogen (N-NH_4), total Nitrogen (N), total Phosphorus (P), and suspended solids (SS) in the principal rivers in Vyškov. Assessed rivers represent small- and middle-sized streams. The study was the part of General Water Management Plan as a part of the urban plan of the city of Vyškov in the South Moravian Region of the Czech Republic (Fig. 5.2). The scheme of local streams and sewer overflows along the receivers are marked in Fig. 5.12. The summary of local streams is shown in Table 5.1.

In Vyškov, the revision and improvements of existing sewer overflows have been planned together with the eventual design of new ones. The rehabilitation of sewerage

Fig. 5.12 River network in the Vyškov city with marked sewer overflows: red—existing remaining, blue—existing, blank—proposed

also involves the design of storm water retention tanks, which attenuate the peak discharges in the sewers and so decrease the released amount of polluted water to receiving streams via storm water overflows. In urban areas, the common problem is spatial constraints which limit wider application of large storm tanks. Certain problems are also maintenance and keeping tanks clean from sediments after each storm event. An assessment of above mentioned arrangements for water quality improvement was also an objective of the modelling.

In Figs. 5.13 and 5.14 modelling results for BOD concentrations in (mg/l) during the design storm event in the Vyškov city are plotted along the Haná river (see Fig. 5.12). The graphs depict the envelopes of maximum concentrations reached at given channel profile during the transient propagation of pollution gradually spilled from sewer outflows and the superposition of pollution "cloud" along the Haná river during the storm event. In the figures, both present state and state corresponding to proposed arrangements at the sewer network are shown.

Figure 5.13 shows the conditions along the Haná river during the average discharge Q_a, Fig. 5.14 the conditions at the low discharge Q_{270} with the exceedance of 270 days.

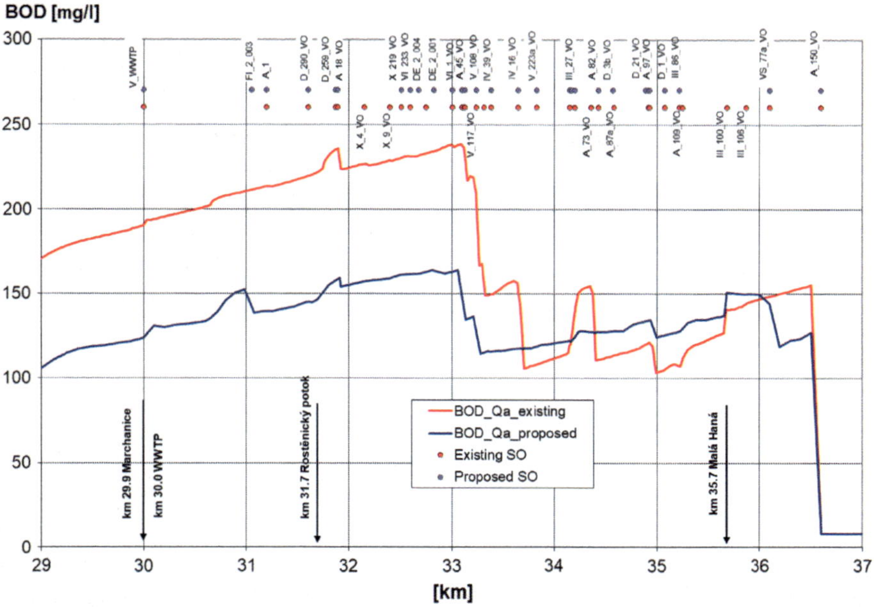

Fig. 5.13 BOD concentrations along the Haná River—average discharge Q_a. Marked sewer overflows: red—existing, blue—proposed including remaining

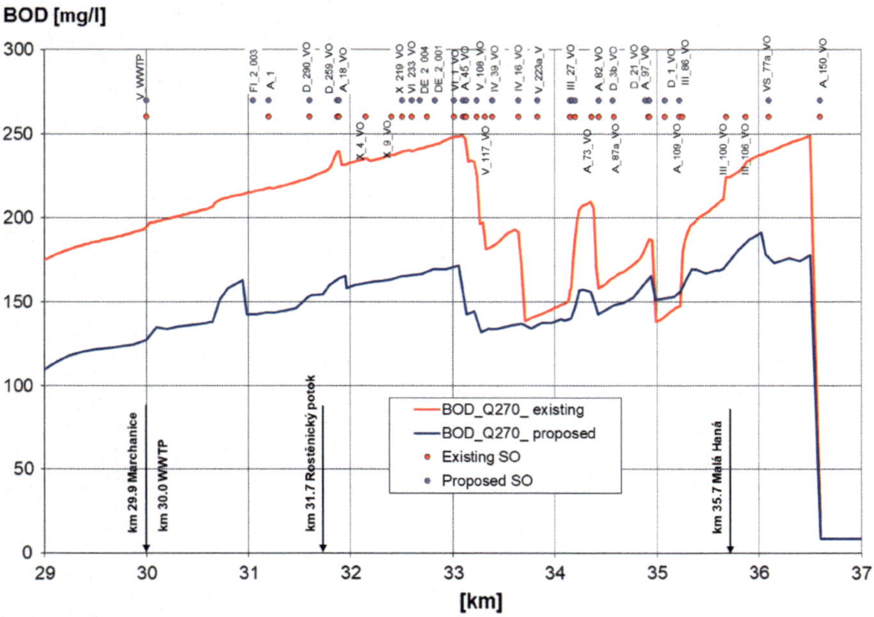

Fig. 5.14 BOD concentrations along the Haná River—average discharge Q_{270}. Marked sewer overflows: red—existing, blue—proposed including remaining

Table 5.2 Maximum modelled concentrations at Q_a—perspective, comparison with immission standards

Water quality indicator	The Haná	The Rostěnický potok	The Drnůvka	Imission standard
BOD	164	157	191	5
COD_{CR}	440	423	508	26
$N\text{–}NH_4$	15	14	17	0.23
N_{TOTAL}	21	20	24	6
P_{TOTAL}	4.1	3.6	4.6	0.15
Suspended solids	264	253	310	20

It can be seen that the most influential is the very upstream sewer overflow which increases significantly (more than 15 times) the concentration in the Haná river, another significant worsening of water quality in the Haná may be observed close to the town centre where numerous sewer overflows with significant effluents increase BOD up to the 250 mg/l which practically comply with sewage concentration in the sewer. It can be seen that due to relatively low annual average discharges Q_a = 0.44–0.64 m^3/s when compared to the outflow from sewer separators (in total about 7.3 m^3/s) the BOD concentration in the Haná river is not so much influenced by original river discharge in the stream before the storm event. This is true namely downstream of the town centre. However, it can also be seen that the effect of technical arrangements on the sewer network upstream of the town centre is more significant in case of small discharges in the Haná river. This is due to minimal discharge Q_{270} = 65–84 l/s.

Downstream of the town centre where the most of improvements concentrated the maximum effect reaches about 35% of the original concentration. All the same even in case of improvements anticipated the effluents from sewer overflow would harm water quality in surface streams for a limited time during the storm which is about 1.5–2 h. As an example, modelled maximum concentrations of individual water quality indicators along individual watercourses in Vyškov for the Q_a scenario and proposed improvements are shown in Table 5.2. In the last column, the immission water quality standards according to [10] are mentioned for simulated water quality indicators. The comparison shows that even if considerable investments into water quality improvement were introduced the temporary exceeding of acceptable concentrations specified for long-term stream water quality was significant.

Experience shows that imission standards should not be employed in case of little incidents like e.g. spills from sewer overflow during extreme storm events with low periodicity. The acceptable concentrations have to be designated at the negotiation of all involved bodies like river agencies, water companies, municipalities, modellers and independent consultants. At the decision the nature of the streams, endangered habitat and also possibilities for technical arrangements on the sewer system and at the urban area have to be taken into account.

5.8 Conclusions

At present water quality, numerical modelling plays a significant role in stream water quality management, in the assessment of present water quality in water courses and predicting impact of individual pollution sources. Modelling techniques are recently frequently used for the optimization of remedial measures adopted on pollution sources at different levels of the river basin extent and input data accuracy. For this purpose models of various classes may be used, like global screening models, basic balance models and detailed transient flow and contaminant transport models answering particular questions about local water quality issues.

In this chapter, the principles of individual model types are mentioned. Their use is demonstrated on case studies carried out by the author during last three decades on the territory of the Czech Republic. Particular problems, results achieved and questions arose at the modelling are briefly discussed together with comments on the of interpretation of typical graphical and tabular outputs.

5.9 Recommendations

When modelling stream water quality in water courses the appropriate model has to be proposed based on conceptual considerations containing set of simplifications and preliminary assumptions. The reasons of these assumptions should be carefully discussed and justified.

The crucial issue are comprehensive, complete, reliable and homogeneous data on catchment and channel characteristics, pollution sources and also water quality in streams obtained by monitoring. For the data assembling sufficient time and adequate financial sources have to be reserved.

The reliability of the model may be significantly improved by its calibration and verification. When the stream water quality data are missing sensitivity study is strongly recommended. Here the data about the parameters influencing water quality should be taken from the literature sources and from the previous studies.

For decision makers the results of the modelling should be properly summarized, depicted and interpreted.

Generally, the last step of the modelling process should be feedback on the compliance between predicted values of water quality indicators by modelling and the real values obtained by monitoring after adopting proposed measures. Unfortunately, this step is mostly omitted. The impact usually manifests itself after numerous years after the initial proposals and modelling. Moreover, there are usually no extra financial sources for additional on-purpose monitoring and for such comparisons and evaluation of possible shortcomings of the forecasts.

References

1. Julínek T, Říha J (2017) Longitudinal dispersion in an open channel determined from a tracer study. Environ Earth Sci 76(17):592, springer
2. Harmel RD, Slade RM, Haney RL (2010) Impact of sampling techniques on measured stormwater quality data for small streams. J Environ Qual 39(5):1734–1742
3. Berndtsson JC (2014) Storm water quality of first flush urban runoff in relation to different traffic characteristics. Urban Water J 11(4):284–296
4. Miskewitz RJ, Uchrin ChG (2013) In-stream dissolved oxygen impacts and sediment oxygen emand resulting from combined sewer overflow discharges. J Environ Eng 139(10):1307–1313
5. Morales V, Mier J, Garcia M (2017) Innovative modeling framework for combined sewer overflows prediction. Urban Water J 14(1):1–15
6. Hlavínek P, Říha J (2001) Stream water quality model as a tool for the assessment of impact of improvements of pollution sources in the River Dyje Basin. In: 2nd International conference Interurba II, Lisabon, Poster, CD
7. Crabtree B, Earp W, Whalley P (1996) A demonstration of the benefits of integrated wastewater planning for controlling transient pollution. Water Sci Technol 33(2):209–218
8. The Act No. 254/2001 Coll. Act on Water (The Water Act)
9. Directive 2000/60/EC of the European Parliament and of the Council of 23 October 2000 establishing a framework for community action in the field of water policy
10. Decree No. 401/2015 Coll. on indicators and values of permissible pollution of surface water and wastewater, details of the permit to discharge wastewater into surface water and into sewerage systems and in sensitive areas (CZ legislation)
11. Decree No. 416/2010 Coll. on indicators and values of permissible pollution of wastewater and the details of the permit to discharge wastewater into groundwater (CZ legislation)
12. Framing monitoring programme (2013) Czech Hydrometeorological Institute, 28 p (in Czech). Retrieved from http://www.env.cz/cz/ramcovy_program_monitoringu. Last accessed on Jan 2018
13. Lee S, Lee J, Kim M (2012) The influence of storm-water sewer overflows on stream wq and source tracking of fecal contamin. KSCE J Civil Eng 16(1):39–44
14. Pálinkášová Z, Šoltész A (2012) Hydrologic and hydraulic evaluation of drainage system in eastern Slovak Lowland. Pollack Periodica 7(3):91–98
15. WRI (2018) Hydroecological information system. Retrieved from http://heis.vuv.cz
16. CHMI (2018) Network of monitoring of surface water. Retrieved from http://hydro.chmi.cz/hydro
17. Imhoff K (1953) Taschenbuch der Stadtentwasserung. VEB Verlag Technik, Berlin, 286 p
18. Novotny V, Olem H (1994) Water quality—prevention, identification and management of diffuse pollution. Van Nostrand Reinhold, New York, 1054 p
19. Fischer HB (1967) Analytical predictions of longitudinal dispersion coefficient in naturel streams. In: Proceedings of the twelfth international association for highway research congress, vol 4, part 1
20. Fischer HB et al (1979) Mixing in inland and coastal waters. Academic Press, New York
21. Knopman DS, Voss CI (1987) Behaviour of sensitivities in the one-dimensional advection-dispersion equation: implications for parameter estimation and sampling design. Water Resour Res 23:253–272
22. Seo IW, Cheong TS (1998) Predicting longitudinal dispersion coefficient in natural streams. J Hydraul Eng 124(1):25–32
23. Iwasa Y, Aya S (1991) Predicting longitudinal dispersion coefficient in open-channel flows. Lee J, Cheung Y (eds), Environmental hydraulics: proceeding of the international symposium on environmental hydraulics, Hong Kong, China, pp 505–510
24. Chapra SC (1997) Surface Water quality modeling. McGraw Hill, USA, p 844
25. Říha J, Daněček J, Glac F (1997) An influence of dispersion on the concentration in open channels. J Hydrol Hydromech 45(1–2):165–179

26. ČSN 75 7221 (2017) Water quality. Classification of surface water quality
27. Říha J (2000) Simple numerical model for river network water quality decision support. In: The 4th international conference on hydroscience and engineering, Seoul, Korea, p 24
28. Thorndahl SL, Schaarup-Jensen K, Rasmussen MR (2015) On hydraulic and pollution effects of converting combined sewer catchments to separate sewer catchments. Urban Water J 12(2):120–130
29. Arheimer B, Olsson J (2001) Integration and coupling of hydrological models with WQ models: applications in Europe. Swedish Meteorological and Hydrological Institute, Sweden

Chapter 6
Pharmaceuticals in the Urban Water Cycle

Adéla Žižlavská and Petr Hlavínek

Abstract Pharmaceuticals are an integral part of modern medicine. We can hardly imagine the functioning of modern society without them. Thanks to medicines we can eradicate serious pandemic illnesses, help people overcome serious psychological or psychiatric problems, prolong life by lowering high levels of cholesterol and the occurrence of diabetes, offer comforting benefits like pain relief, and provide hormonal drugs to prevent unwanted conception. Pharmacology underwent a boom in the second half of the nineteenth century. Drugs continued to be developed in the twentieth century. It was only in 1965 that scientists noticed substances were finding their way into the environment, with a number of studies identifying pharmaceuticals in surface water. It was only in the 1990s that scientists began to draw attention to the impact of pharmaceuticals on fish and biota in watercourses. At the turn of the millennium, drug concentrations were found in drinking water and groundwater. Currently, we know that most persistent drugs are already distributed throughout the water cycle. We know they affect the environment but we are not clear as to what extent this impacts us as a human beings—we can only speculate about such impacts and hope that rising numbers of cancers and increasing infertility are not the consequences of this. This chapter aims to summarize the existing knowledge about entry pathways, occurrence, degradation, and behavior of drugs in the water cycle.

Keywords Pharmaceuticals · Water · Wastewater treatment · Environment

6.1 Introduction

The current way of using pharmaceuticals divides modern society into two different views. While one part is glorified for their amazing healing effects of eradicating various illnesses or prolong human life, another part of the human society screaming with horror that the tax on comfort is too high, and we are cutting down under ourselves.

A. Žižlavská (✉) · P. Hlavínek
Faculty of Civil Engineering, AdMaS Centre,
Brno University of Technology, Brno, Czech Republic
e-mail: gotzingerova.a@fce.vutbr.cz

M. Zelenakova et al. (eds.), *Management of Water Quality and Quantity*,
Springer Water, https://doi.org/10.1007/978-3-030-18359-2_6

Without drugs, we would not be able to withstand plagues, prevent low mortality in new-born babies, or deal with different hereditary diseases. Contrarily, it is clear from different studies that the use of pharmaceutical substances has changed the natural environment around us, for example, pharmaceutical substances act as a hormonal form of contraception in rainbow trout or crayfish populations, or result in aggressive or suicidal behavior in marine animals exposed to antidepressants [1].

To be able to build a comprehensive picture of the extent to which drugs influence today's environment we need to summarize the development of medicines and how and when they started to become a dangerous phenomenon. The name pharmaceutical originated from the Latin word "pharma"—the medical branch of medicine is called "pharmacy." This discipline includes several other sub-disciplines, such as pharmacology, pharmaceutical chemistry, and pharmacognosy. The first comprehensive references to the use of drugs date back to the sixteenth century in the book of Canadian medicine written by the Persian scholar Avicenna. This book represented the basic teaching text used by Christian and Muslim medical apprentices. Subsequent centuries showed no significant development in the field of pharmacy. It was only in the nineteenth century that emphasis was put on modern methods in medicine and the development of new drugs. The first university institute for pharmacology was founded in 1847 by the German pharmacologist Rudolf Buchheim. From the outset, pharmacologists devoted themselves primarily to natural substances, especially medicinal herbs and natural drugs, or attempted to exploit the effects of various animal or plant poisons [2].

The development of modern laboratory techniques enabled many new drugs to be developed, for which many such discoveries or discoverers have received the Nobel Prize, such as Alexander Fleming or Selman Abraham Waksman for discovering antibiotics. Other important milestones in this field include the discovery of purine metabolism (Gertrude Elion and George H. Hitchings), beta blockers, cimetidine [3], and statins [4]. Another pioneer in the development of chemical analogies of known active substances was Sir David Jack at the medical products company Allen and Hanbury, who went on to advocate the development of an inhaled selective beta-2-adrenergic agonist for the treatment of asthma [5]. This was followed by the development of antivirals, hormonal contraceptives, an organ transplantation immunosuppressant, and medicines used in antitumor therapy.

The cloning of human proteins has made it easier to target treatment processes associated with specific diseases [6]. Today's pharmacologists use genetics, molecular biology, chemistry, and other advanced sciences to convert information about molecular motifs and mechanisms to treat diseases. Through such advances it is possible to develop methods of preventive care, diagnostics, and even create personalized medication.

From the 1960s onward there occurred a boom in the use of new pharmaceuticals, such as hormonal contraceptives, diverse types of antidepressants, and cancer treatment substances. The great enthusiasm that these drugs evoked meant that their impact on the environment was overlooked. However, in 1965 Theo Colborn's book *Silent Spring*, drew attention to issues with insecticides and pesticides and Stumn-Zollinger and Fair pointed out that steroid hormones, detected at the inflow to wastewater treatment plants (WWTPs), were also detected in sewage discharge. Subsequent studies on the occurrence of drugs in wastewater up to 1994 revealed a link between the occurrence of a synthetic hormone ethinylestradiol and impacts on fish health. This started new research into emerging pollutants in the environment. At present, it is possible to say quite openly, based on numerous studies from all over the world (e.g., from Spain, France, and the Netherlands), that drugs have spread freely through the entire water cycle and are even found in mains drinking water.

It is alarming how long it has taken to initiate research into the occurrence of drugs in the environment—monitoring of cytotoxic agents used for the treatment of oncological diseases only started in 2011 [7], even though at the time they were known to have ecotoxic, mutagenic, and cancer-causing effects according to the World Health Organization (WHO).

At the same time, the issue of endocrine disruption and the increased need for drug control and commodities are emerging through the Endocrine Disruptor Screening Program (EDSP) in the United States—methods for identifying endocrine disruptors (EDCs) are being developed within the Organisation for Economic Co-operation and Development (OECD).

Most drugs are present in the water cycle in the form of microcontaminants, often at only very low detection levels (with concentrations in the nanogram and microgram ranges). Moreover, due to the changing chemical, biological, and physical conditions within the water cycle, there may be different syntheses that produce conjugates of these substances or different co-metabolites of the original substances. Therefore, it is very difficult to complete accurate detection of substances via common analyses methods like HPLC (High-Performance Liquid Chromatography). Therefore, in some cases sample testing, based on the overall estrogenic or androgenic effect on organisms, represents a better approach.

The issue of drug removal from the water cycle is problematic. We know that conventional methods of treatment, without specific approaches, are not very effective and are capable of degrading only those substances that are readily degradable, such as some types of analgesics or substances with high adsorption abilities. It is important to realize that the term pharmaceutical covers substances with very different physicochemical properties. This means that there is no single technology that is able to remove all drugs from the water cycle. Another issue is that drugs can degrade only in a few places throughout their journey from excretion to entering surface water, via water treatment plants or straight effluent from hospitals. However, drugs in the water cycle and those in animal manure, which is spread on fields, allows substances to seep into the groundwater or be flushed into surface streams by surface run-off. Such a situation can only be changed by government legislation. It is alarming that some substances that are already forbidden for human use, like sulfonamide antibiotics, are still served to livestock and thus, through biomagnification, can return into the human metabolism.

6.2 Effects, Behaviors, and Characteristics of Pharmaceuticals

The occurrence of medicinal products in water and their effects on organisms depend on both the physicochemical properties of the individual drugs and on the conditions that exist in their pathways into the water cycle.

In the case of the physicochemical properties of medicinal products, of importance here is whether a chemical is:

- Hydrophilic (i.e., is water soluble).
- Hydrophobic (i.e., is insoluble in water).
- Lipophilic (i.e., has the ability to dissolve in fat).
- Lipophobic (i.e., is insoluble in fat).

Various other factors and coefficients are used to describe the characteristics of such substances. These can partially be used to predict to behavior within the aquatic environment (i.e., whether a substance will, for example, be sorbed on solid particles or activated sludge). These factors are briefly outlined in the following text.

Biodegradation coefficient (constant) $\mathbf{K_{biol}}$. This expresses the kinetic rate of biodegradation of a substance in liters per gram of suspended solids per day. Compounds can be classified according to their K_{biol} as: very highly ($K_{biol} > 5$ L g SS^{-1} d^{-1}); highly ($1 < K_{biol} < 5$ L g SS^{-1} d^{-1}); slightly ($0.5 < K_{biol} < 1$ L g SS^{-1} d^{-1}); and barely ($K_{biol} < 0.5$ L g SS^{-1} d^{-1}) biologic degradable [8].

Adsorption coefficient $\mathbf{K_d}$. This coefficient is sometimes also denoted as the solid–water distribution coefficient (K_d, in liters per kilogram), defined as the ratio between the concentration of a substance in a solid matrix to its concentration in aqueous phase.

Partition coefficient n-octanol/water $\mathbf{K_{ow}}$. This is defined as the percentage of a chemical concentration in *n*-octanol and the concentration of the substance in water (all in equilibrium, at a specified temperature).

Bioconcentration factor (BCF). This characterizes the equilibrium ratio between concentration in an organism and concentration in the surrounding environment—most often water. It is mainly used for adsorption and incorporation into biomass [9].

Individual coefficients to a certain extent depend, for example, on the bioconcentration factor and the K_{ow} coefficient, since the higher the logarithmic value of K_{ow}, the higher the bioconcentration coefficient. In the case of degradation by conventional convection processes, the individual parameter ratios can identify the behavior of the substance throughout the cleaning cycle, as in the case of the solid–water distribution coefficient K_d and the biodegradation constant K_{biol}, such that [10]:

1. Compounds with a high biodegradation constant K_{biol} and a low sludge–water K_d distribution coefficient, like ibuprofen, should be sufficiently transformed independently of sludge residence time or water residence time.

2. Compounds with low K_{biol} and high K_d, like musk substances, are retained in aerated areas of activation by sorption and are significantly transformed with sufficient sludge degradation.
3. Compounds with low K_{biol} and mean K_d values, such as estrone and 17β-estradiol, are slightly transformed independently of sludge residence time and with a slight dependence on the residence time of water.
4. Low K_{biol} and low K_d compounds, like as carbamazepine, are neither biotransformed nor removed from sewage irrespective of sludge and water residence time.

In addition to the coefficients mentioned above, the degradation rate I also influences other important parameters such as time of idle delay, temperature, and pH [8, 10].

Basic information regarding the presence of pharmaceuticals in the aquatic environment is linked to their paths of entry into the system. One unquestionable means of entry, representing the main gateway for the introduction of drugs into the environment, is from wastewater treatment plants (WWTPs). Such plants represent an imaginary boundary between modern human society and the natural environment; interactions occur between substances originating from the degradation of original substances, via physical, chemical, and biological reactions that form different metabolites, co-metabolites of conjugates, or deconjugates. The influence of such actors can also be felt on biological processes, for example, the effects of antibiotics on diverse types of anaerobic bacteria in WWTPs [11].

The main means of detection of drugs in wastewater is pharmacokinetic in approach—determining the metabolism of a drug in an organism, that is, how many parent substances are excluded by the body in percentage terms and how much of a particular metabolite is produced. For most drugs, this information is collected by the European Medicine Agency (EMEA).

The negative effects of a drug always depend on four basic conditions:

1. Exposure time.
2. Organism (species).
3. Concentration of contaminant.
4. An individual's stage of development.

Many of the underlying negative effects are provided in the following text. Some of these effects overlap for certain substances, for example, reprotoxicity in some cases corresponds with endocrine disruption—one substance may have multiple classifications in terms of its impact on an organism.

Bioaccumulation, bioconcentration, and biomagnification

Bioaccumulation is defined as either simple substance uptake, substance accumulation over time, or substance retention [12, 13]. Bioaccumulation is usually calculated as the ratio of the concentration of the compound of interest in the biota of a sample (plant, animals) to its concentration in the environment (e.g., soil or water) [14]. There is currently little empirical data to determine the exact value of bioaccumulation. One criteria for bioaccumulation is the octanol/water partition coefficient (K_{ow})

> 5, as shown in Annex D of the Stockholm Convention [15]. Alternatively, it can be defined according to the OECD, where log $K_{ow} > 3$ means that the substance will tend to accumulate [16].

Bioconcentration represents the accumulation of a chemical in or on an organism. It is determined by the value of bioconcentration factor.

Biomagnification is the transfer of xenobiotic substances within the food chain where the consumer demonstrates higher concentrations compared with concentrations found in the source. There is an increase in intrinsic concentrations (in fat) of treated substance in organisms at subsequent trophic levels in the food chain. Mechanisms triggering increased concentrations of contaminants at higher trophic levels should be considered when assessing whether a chemical is biodegradable [17].

Endocrine disruption

Endocrine disruption is the ability of substances to disrupt the endocrine system. Substances affecting the endocrine system are called endocrine disruptors (EDCs; derived from the English "disrupt" meaning to break or discontinue). The WHO define an EDCs as an "exogenous substance or mixture that changes the function of the endocrine system and consequently causes adverse effects on the health of the organism, its offspring or subpopulation" [18].

The demonstrable effects of pharmaceuticals on the endocrine system:

1. Effects on cytochrome P450 (CYP). An enzyme group that significantly contributes to the metabolism of xenobiotics (drugs) in the body, for example, diazepam, ibuprofen, codeine, and others. This set of enzymes is found in almost all animal and plant species including bacteria and fungi [19].
2. Effects on non-P450 aldehyde phenol oxidation enzyme systems. Such systems are used to catalyze the degradation of aldehydes, the most famous of which is aldehyde hydrogenase (ALDH), which affects embryonic development, oxidative stress, and cancer development [20].
3. Effects on responses to oxidative stress (lack of oxygen in the body) [21, 22].
4. A deregulation of free radicals (ROS) in the body (superoxide (O_2^-), hydrogen peroxide (H_2O_2))—cause tissue damage.
5. Effects on circadian biorhythms (the internal clock that is dependent on the production of melatonin hormone).
6. The synthesis of steroid hormones [23].

Toxicity

Toxicity is the ability of synthetic and natural substances to cause the living organisms with which they come into contact acute or chronic poisoning. Toxicity is expressed in terms of LD_{50} (the value attributed to LD_{50} for a substance is the dose required to kill half the members of a tested population after a specified test duration). Studies on aquatic organisms have revealed the toxicity of ciprofloxacin to green algae [24] and the toxicity of fluoroquinolone antibiotics (ciprofloxacin, lomefloxacin, ofloxacin, levofloxacin, enrofloxacin, and flumequine) to five aquatic organisms: cyanobacteria, *Microcystis aeruginosa*; capercaillie, *Lemma minor*; green algae, *Pseudokirchneriella subcapitata*; crustaceans, *Daphnia magna*; and *Pimephales promelas* [25].

Carcinogenicity
The exposure of carcinogenic compounds to organisms leads to unwanted cell proliferation leading to cancers. For example, there is sufficient evidence that EE2 has carcinogenicity in humans (used as a postmenopausal estrogen therapy and often combined with estrogen–progesterone) and evidence exists in animals for estrogen and estrone carcinogenicity [26].

Reprotoxicity
Reprotoxicity is the ability of a substance to interfere with the reproductive and immune systems of animals and humans. For example, in the case of pharmaceuticals, fluoxetine induces the early reproduction of glochidia (the microscopic larval stage of some freshwater mussels and aquatic bivalve mollusks) [27].

Mutagenicity (genotoxicity)
Mutagenicity causes irreversible genetic changes to either a single gene, a gene block, or the entire chromosome. The mutagenic effect is dependent both on the dose of genotoxic substance and time of exposure. There may be apparent physical changes in organisms that are transmitted into consequent subpopulations. The mutagenicity of individual substances can be tested either in vitro (i.e., tested directly on a given chemical) or via an injection (individuals are exposed to a substance). Typical representatives of genotoxic agents are diclofenac and 17α-ethinylestradiol, with mutagenic changes most commonly impacting algae, fish, and some amphibians [28–30].

Teratogenic effect
The teratogenic effect is the ability of chemicals to cause physiological disorders. Generally, we designate these as teratogens. Main teratogenic drugs are cytostatics, tetracycline antibiotics, and antiepileptics.

Analysis, classification, and occurrence of pharmaceuticals
Different pharmaceuticals have different negative effects on living organisms. One group may affect the endocrine system while another may cause allergies, mutagenic changes, influence self-preservation instincts, or increase resistance of dangerous bacteria and parasites to healing and treatment processes (Table 6.1).

The following sections of this chapter outline some drug groups of non-steroidal origin.sections of this chapter outline some drug groups of non-steroidal origin.

6.3 Analgesics

This group represents a type of painkiller ("an," without, "algia," pain). Currently, analgesics can be delivered orally, locally as gels and ointments, or in an injectable form. Less potent analgesics used in most countries are not prescriptive (i.e., they can be obtained over the counter). They are consumed in large numbers, for example,

Table 6.1 Pharmaceuticals detected in wastewater [10, 31–35]

Steroidal pharmaceuticals	Estrogens	17-estradiol (E2)—component of menopause medicines; Estriol—(E3)—hormone creams for menopause, multiple sclerosis; 17α-ethinylestradiol (EE)—contraception	Estradiol is the most potent natural estrogen. EE activates estrogen receptors (estrogen activation)—acting as endocrine disruptors
	Progestrons	Norethisterone (or 17α-ethynyl-19-nortestosterone) combined oral contraceptives, menstrual treatment, menopause or postponing menopause, preventing uterine bleeding, and preventing premature labor. Progesterone is used to support pregnancy in assisted reproduction (ART)	These are substances that, in combination with estrogen, are estrogenic—acting as endocrine disrupters
	Anti-estrogens	Hormonal treatment of breast cancer	Prevents estrogen from binding to its receptor
	Androgens and glukokortikoids	Testosterone—anabolic steroids, growth stimulation, and osteoporosis treatment Beklomethasone—a combination of various drugs used, for example, in the treatment of asthma and skin eczema Hydrocortisone—treatment of allergic reactions, psoriasis, and eczemas	Endocrine disruption
	Phytoestrogens	Alleviate the symptoms of menopause and menstrual pain	Exhibit estrogenic activity

(continued)

Table 6.1 (continued)

	Veterinary growth hormones	Pregnancy terminations in cattle and support for growth of livestock and animals for slaughter	Endocrine disruption
Non-steroidal pharmaceuticals	Agents for treatment of blood and blood-forming organs	Acetylsalicylic acid—reducing blood clotting Pentoxyphylline improves blood flow of peptic vessels	Pentoxifylline is also a receptor antagonist. Acetylsalicylic acid increases the activity of hepatic enzymes
	Agents for the treatment of the heart and circulatory system	Statins—lowering blood cholesterol	Endocrine disruption
	Antibiotics	Treatment of infectious conditions	Antibiotic resistant bacteria
	Analgesics	A group of pain-relieving medicines (ibuprofen, carbamazepine, and clofibric acid)	Increased inhibition of an organism; bioaccumulative effects
	Antidepressants	Monoamine reuptake inhibitors (fluoxetine and citalopram)	Can affect the influence of signalling by dopamine and noradrenaline; can initiate mutagenesis
	Agents for the treatment of allergies and asthma	Glucocorticoids, antihistamines, and sympathomimetics	Endocrine disruption
	X-ray media	Magnetic resonance, radiology, during surgery	Highly persistent, carcinogenic
	Cytostatics	Treatment of cancer	Show carcinogenicity, mutagenicity, or embryotoxic parameters
	Antiepileptics	Treatment of epilepsy	Strongly persistent
	Beta blockers	Beta blockers reduce the effect of stress hormones. Treatment of the cardiovascular system, anxiety disorders, and migraine	Strongly persistent

in the Czech Republic in 2017 alone, acetylsalicylic acid represented the third most often purchased medicine—529,920 packs were sold compared with 248,881 packs of diclofenac [36].

The most common reasons for use of analgesics are [37]:

1. Pain relief associated with wounds, post-operative procedures, or chronic pain (joint pain, etc.).
2. Reduction of pain in the muscles due to viral or infectious diseases (fever reduction, etc.).
3. Control of allergic reactions when taken in conjunction with antihistamines.

Analgesics can be grouped accordingly [37]:

1. Non-opioid analgesics that suppress inflammation and block prostaglandin production (e.g., salicylates—acetylsalicylic acid medicines).
2. Opioid analgesics that act on opioid receptors in the central nervous system and other tissues (e.g., morphine, codeine, and bezitramide).

The history of analgesic use goes back to the search for pain relief linked to human society in the Neolithic period. The first used analgesics were herbal extracts and herbs from various plants, such as poppies, gooseberries, hops, and mandragora, as well as alcohol. The most popular among them was the dried juice from immature poppies known as opium. For a long while this substance was the most widely used pain-relieving drug. It is known to have been used since the Middle Ages up to the nineteenth century, when its use within Western society exceeded tolerable limits leading to pressure to develop new opioid analgesics that did not cause addiction. The first to do so was pharmacist Friedrich Wilhelm Adam Sertürner who discovered the alkaloid called morphine. However, the use of morphine became as controversial as opium in terms of patient dependence. Next came the process of morphine acetylation in 1874 by the English chemist Charles R. Alder Wright (1844–1894) who produced a derivative called heroin, which at the time was greeted with great pomp as a medicinal product for morphine-dependent patients. Since then the use and manufacture of heroin has been prohibited in most countries. At the same time heroin was being synthesized in the laboratories of the well-known pharmaceutical company Bayer, acetylsalicylic acid synthesis was also underway, with its commercial name aspirin. Synthesis of some analogies of this acid led to the discovery of other drugs such as diclofenac or diflunisal. As far as paracetamol is concerned, it has been in use since 1878, however, based on tests with contaminated preparations it was considered harmful and ineffective. It was only in 1946, after new tests, that this conclusion was reassessed. At that time paracetamol was identified as the main metabolite of phenacetin and became the most widely used antipyretic and analgesic in patients with fevers [38, 39]. Ibuprofen was discovered in the 1950s. It was originally designed for the treatment of rheumatoid arthritis, but its pain-relieving effects proved to be more effective than aspirin [97].

Historically, the pathways of non-opioid analgesics into the environment began in the 1950s. For some analgesics, like aspirin, this was identified as not being a problem since such substances degraded in normal biological processes. However, in the case of diclofenac it did represent a problem because if its high persistence and negative impact on rainbow trout and some species of predatory birds [40]. Table 6.2 details the types of non-opioid analgesics and provides their effects on the environment and their occurrence in the water cycle.

6.4 Antibiotics

Antibiotics represent a type of drug used to kill or suppress the growth of microorganisms. They are most commonly prescribed for the treatment of bacterial diseases but are also effective against some fungal species and parasitic protozoa [60]. This group includes up to 6000 substances that have antibiotic effects, however, currently only around 70 are used for treatments [61]. Taking antibiotics dates back to ancient China and Egypt where moldy soy milk and moldy bread were applied to open wounds [62]. Officially, however, the discovery of antibiotics is attributed to Alexander Fleming, who in 1929 described the effect of *Penicillium* fungus on bacteria, an observation for which he received the Nobel Prize. However, the success of this work was based on the work already outlined by his predecessors, such as Luis Pasteur, Rudolf Emmerich, Charles-Joseph Bouchard, and Ernest Duchesne.

According to mechanisms of action antibiotics can be divided into two groups:

1. Bactericidal—kills bacteria.
2. Bacteriostatic—stops bacterial proliferation.

Depending on the mechanism of action of an antibiotic on bacterial cells, antibiotics are divided into several groups (Fig. 6.1; Table 6.3):

- Antibiotics that inhibit the synthesis of lipids and other cell wall materials, such as beta-lactam antibiotics, monobactams, carbapenems, glycopeptides, and bacitracin.
- Antibiotics that disrupt the cytoplasmic membrane, such as polyenes, polymyxins, azoles, amphotericin B, and ionophores.
- Antibiotics that inhibit nucleic acid synthesis by inhibiting (1) DNA gyrase, such as aminoglycosides, novobiocin, and quinolones; or (2) RNA polymerase, like ansamycins.
- Antibiotics that interfere with bacterial protein synthesis by (1) binding to ribosomal subunit 30S, like tetracyclines; (2) binding to subunit 30S and 50S, like aminoglycosides; (3) binding to subunit 50S, such as macrolides, linkosamides, amfenicols, diterpens, and aminocyclitols; or (4) by preventing the incorporation of amino acids into proteins, like amphenicol.
- Antibiotics that inhibit the synthesis of folic acid, such as sulfonamides and trimedoprim.

One of the most significant threats associated with antibiotic penetration into the environment is so-called bacterial resistance. This is the ability of bacteria to resist antibiotic treatment. In some countries, due to bacterial resistance, carbapenem

Table 6.2 Non-opioid analgesics [36, 41]

Antipyretics				
Title	Effects and description	Occurrence and degradation	Negative effects	Degradation
Acetylsalicylic acid	Has antiplatelet effects, is given preventively in infarct and cerebrovascular events; as an analgesic and an antipyretic it is not used very much. It is contraindicated in children with viral diseases (influenza and varicella)	In the aquatic environment its metabolite is present. Salicylic acid is widely distributed in various environmental matrices. However, a 2002 Berlin study by Heberer showed a concentration in run-off of 0.04 μg/l [33]. Similarly, a 2013 study in Greece measured the outflow from one hospital at 1000 ng/l; the discharge from municipal WWTPs was not detected	Salicylic acid is also found in some plant species, representing an essentially natural plant hormone. Concerns arise mainly from high concentrations in water found in WWTPs	Good degradability at conventional WWTPs

(continued)

Table 6.2 (continued)

Antipyretics				
Title	Effects and description	Occurrence and degradation	Negative effects	Degradation
Paracetamol (USA, acetaminophen)	Effective in about 50% of patients. It is the safest analgesic at therapeutic doses and can be given during pregnancy. However, with intoxication (10–20 tablets), heavy toxic liver damage occurs	4-acetaminophen is one of the most common drugs in various environmental matrices, i.e., sediments, soil, or drinking water [42]. Recently, the River Tyne (United Kingdom) was found to have levels higher than 65 μg/l [43, 44]; groundwater, used as a drinking water supply, had concentrations measured at 211 and 1.89 μg/l [45]	None detected	Paracetamol can degrade only selected bacteria strains such as *Pseudomonas*, *Bacillus*, *Acinetobacter*, and *Sphingomonas*
Non-steroidal antirevmatics				
Ibuprofen	Used for mild and moderate pain in various locations (especially joints, muscles, teeth, and the head), menstrual pain, elevated temperatures, and inflammation	Ibuprofen is given a great deal of attention, especially because of its high concentrations in WWTPs It is one of the most heavily consumed drugs (per kilogram per capita), producing a relatively large number of metabolites via various degradation processes	Ibuprofen has been shown to significantly affect the growth of several bacterial and fungal species [46]. In combination with other analgesics, it is possible to treat bacterial diversity at the biological level of WWTPs and to affect their functioning	In general, ibuprofen exhibits a high level of removal in conventional WWTPs (about 90%) [46] The removal of ibuprofen depends on the time spent in WWTPs. A long rainy season causes a significant reduction in the elimination of ibuprofen and ketoprofen

(continued)

Table 6.2 (continued)

Antipyretics				
Title	Effects and description	Occurrence and degradation	Negative effects	Degradation
Ketoprofen	Used as an analgesic and anti-inflammatory agent for the treatment of pain and inflammation, especially in muscles and bones [36]	Up to 80% of the bile and urine are eliminated in unchanged forms [36]. It is part of almost all environmental matrices and the water cycle [47]	Has a negative effect on the growth and development of carp embryos and larvae [48]. Like diclofenac, it has a negative impact on wild vulture populations	Although ketoprofen exhibits a wide range of variability when removing it from sewage treatment [49] an assumption is made that removal levels are approximately 50–70% in WWTPs [50]
Diclofenac	Applied to reduce inflammation and treat pain. The use of medication for animals is controversial due to its toxicity, which can quickly kill predatory birds that have eaten dead animals. In many countries, it has been banned for veterinary use	Since 2013, increased attention has been paid to the occurrence of diclofenac in surface water [51]. It occurs both in sewage sludge and wastewater. Concentrations in surface water range from only a few ng/l to 15 μg/l [52]. Some 65–70% is excreted in urine and 20–30% in faeces [53]. Therefore, the legislative target for diclofenac approval in the aquatic environment is recommended at 100 ng/l in Europe [54, 55]	Diclofenac caused a sharp decline in the vulture population on the Indian subcontinent—a drop of 95% by 2003 and a decrease of 99.9% by 2008. The mechanism was renal failure [40]. Diclofenac also causes harm to species of freshwater fish like rainbow trout [56]. It is linked to chronic toxicity in invertebrates and a delay in embryo hatching [52]. It affects the diversity of nitrifying bacteria [57]	Degradation occurs via co-metabolic processes and metabolic processes involving soil microorganisms [58, 59]

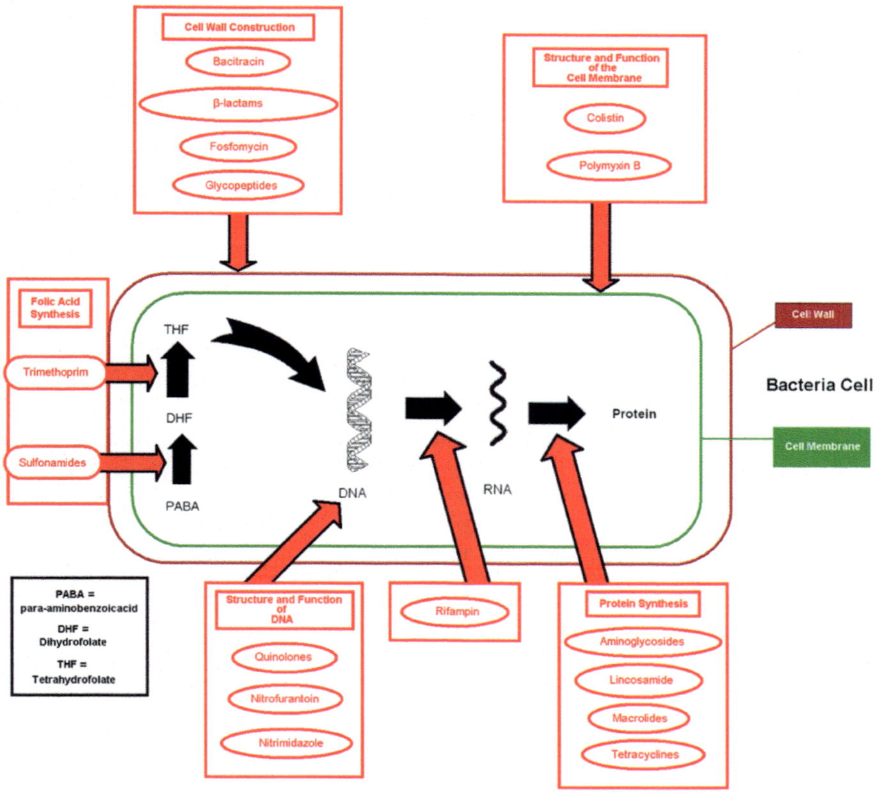

Fig. 6.1 Antibiotic action mechanisms [63]

antibiotics no longer work in more than half the cases treating *Klebsiella pneumonia*. The resistance of *Escherichia coli* to one of the most commonly used drugs for the treatment of urinary tract infections (fluoroquinolone antibiotics) is very widespread. In many parts of the world, there are countries where this treatment is now ineffective in greater than half the patients treated. In addition, treatment as ineffective (third-generation cephalosporin antibiotics) has been confirmed in at least 10 countries (Australia, Austria, Canada, France, USA) and 350,000 cases due to resistant bacteria [64]. Antibiotics may only be partially metabolized in humans and animals after administration—with metabolic rates estimated to be about 30% [65].

Table 6.3 Selected types of antibiotics and their use, negative effects, and occurrence in environmental matrices [67–69]

Aminoglycosides (streptomycin, neomycin, and gentamycin)	Treatment of infections caused by gram-negative bacteria, such as *Escherichia coli*, *Klebsiella*, or *Pseudomonas aeruginosa*, has been shown to counteract *Leptospira* and some *Staphylococcus*. They have much stronger side effects than other antibiotics and have only a very narrow spectrum of activity	Bacteria having kanamycin/neomycin resistance have been found in crude drinking water samples in Louisiana [70]
Amphenicols	Widescreen antibiotics effective against *Salmonella* infections and *Haemophilus influenzae* are also used to treat meningococcal or pneumococcal infections of the CNS or as an alternative to tetracyclines in severe rickettsia infections	Bacteria having resistance to chloramphenicol have been found in raw water samples for drinking purposes in Louisiana [70]
Cephalosporins	First-, second-, third-, fourth-, and fifth-generation preparations are distinguished. They are very effective antibiotics and are relatively often prescribed for the treatment of urinary tract inflammation, skin inflammation, or otorhinolaryngological infections. For example, in Germany, an average of 56.7 packs of cephalosporin were prescribed each day in 1997	These are top-priority antibiotics that are on the list of critically important antimicrobials for human medicine published by the WHO; should not be administered to farm animals [71]
Glycopeptide antibiotics	Multi-resistant gram-positive bacterial infections—a backup human antibiotic	Bacteria having vancomycin resistance have been found in crude drinking water samples in Louisiana [70]

(continued)

Table 6.3 (continued)

Carbapenems	A drug of choice only for life-threatening nosocomial infections caused by some strains of *Enterobacter* spp., *Serratia* spp., *Citrobacter freundii*, or *Campylobacter fetus*; acute necrotizing pancreatitis and bacterial meningitis	According to the WHO, due to resistant strains of *Klebsiella pneumoniae*—common intestinal bacteria that can cause life-threatening infections—the possibility of treatment with carbapenem antibiotics in over half the patients is totally ineffective [68]
Linkosamides	Treatment of bone and soft tissue infections. They are effective against *Staphylococci* and *Bacteroides* spp. May also be used in the treatment of parasitic diseases such as toxoplasmosis, babesiosis, or malaria	Bacteria having lincomycin resistance have been discovered in samples of crude drinking water in Louisiana [70]
Macrolides—erythromycins	Treatment of infections caused by *Streptococcus*, campylobacter infections, listeriosis, diphtheria, bacterial pneumonia, and atypical pneumonia (*Mycoplasma* chlamydia), and intestinal diseases caused by *Shigella* and *Salmonella*	Bacteria with resistance to lincomycin were detected in samples of crude drinking water in Louisiana [70] Following a Europe-wide campaign to reduce the use of macrolide antibiotics in Finland, a sustained decline of 16.5–8.6% in erythromycin resistance was observed in group A streptococcus [72]
Narrow-spectrum penicillin antibiotics	Treatment of streptococcal, pneumococcal, and meningococcal infections and infections caused by sensitive *Staphylococcus*, therapies for syphilis, gonorrhoea, diphtheria, spasms, angina, erysipelas, rheumatic fever, leptospirosis, actinomycosis, etc. It is the medicine of choice for Lyme borreliosis	Bacteria having oxacillin resistance were detected in samples of crude drinking water in Louisiana [70]

(continued)

Table 6.3 (continued)

Broad-spectrum penicillin antibiotics	Treatment of haemophiliac and enterococcal infections, uncomplicated urinary tract inflammation, biliary tract inflammation, bronchitis. Listeriosis therapy	Their removal via conventional WWTPs ranges between 60 and 90% [73, 74]
First-generation quinolones	Treatment of urinary tract infections including gout; only effective against gram-negative bacteria	Belongs to one of the substances on the list of critically important antimicrobials for human medicine issued by the WHO; should not be administered to farm animals [71]
Fluoroquinolones	Effective against most gram-positive and gram-negative bacteria (including *Klebsiella* spp., *Proteus* spp., and *Pseudomonas aeruginosa*), also affect chlamydia, mycoplasmas, and *Mycobacterium tuberculosis*. They are effective in the treatment of upper and lower respiratory infections, gout, prostatitis, anthrax, and legionary illness. They are expensive and reserve antibiotics with limited indications	Ciprofloxacin most commonly found in the environment However, a very low degree of biodegradation has a high level of adsorption [75]
Streptogramins	A reserve antibiotic that is used only in life-threatening infections caused by multi-resistant strains of *Enterococcus* that are resistant to vancomycin	Streptogramins were originally developed to replace inactive antibiotics for gram-positive bacteria. Unfortunately, streptogramin-resistant bacteria are now found, for example, in the Beijing River and the Xijiang River in the Pearl River Delta [76]

(continued)

Table 6.3 (continued)

Sulfonamides	Discovered even before penicillin. However, this broad spectrum of antibiotics has too many side effects and therefore are increasingly rarely prescribed. Rare uses include skin treatments	The most common antibiotic in both wastewater and surface water, occurring in river sediments, primarily in samples from Asian countries [77] Demonstrates an effect on *Daphnia* A combination of tetracycline and sulfamethoxazole enhances the transfer of genes associated with bacterial resistance within WWTPs [11]
Tetracyclines	Have a broad antimicrobial spectrum; they act on most gram-positive and gram-negative bacteria, including some anaerobes, rickettsia, chlamydia, mycoplasma, and some protozoa such as amoebae	Bacteria having tetracycline resistance have been found in crude drinking water samples in Louisiana [70] Like sulfonamides, they are one of the most frequently occurring antibiotics detected both in wastewater and sludge or river sediments

Antibiotics penetrate the water cycle in several ways:

1. From urban WWTPs.
2. In slurries from livestock applied to agricultural surfaces from which substances are flushed into rivers or infiltrate the soil and enter groundwater supplies.
3. By using antibiotics in aquaculture.

The WHO has consistently highlighted the dangerous activity of using preventive antibiotics administered in small doses to promote livestock growth. In 2006, the European Union banned preventive use of antibiotics for such purposes in livestock. Nevertheless, in other countries around the world (especially China and India) this continues. Since 2006, such a use of antibiotics has been consistently linked to increased antibiotic resistance.

Specific sites linked to bacterial resistance are WWTPs, due to high densities of bacteria species present in such places—there is a so-called horizontal transfer of genes between environmental bacteria and human pathogens. This horizontal transmission of genes between species is additionally evoked by the selective influence of heavy metals and biocides [66]. It is also interesting to note that hospital wastewater accounts for only 1% of the total amount of antibiotics in municipal water, demonstrating that the main contributor of antibiotics in wastewater is the general population [65].

Within Europe and the rest of the world, there are several projects and panels dealing with the issue of antibiotic bacterial resistance, including:

1. The Panel on Additives and Products or Substances used in Animal Feed (FEEDAP). This organization runs surveillance of antibiotic use in farm animals in the EU.
2. The European Center for Disease Prevention (ECDC). This organization considers the control of antimicrobial resistance and consumption of antibiotics in humans.
3. The European Antimicrobial Resistance Network (EARS-Net). This organization collects data on antibiotic resistance from national laboratories.
4. The European Committee for Antimicrobial Sensitivity Testing (EUCAST).
5. The Global Antimicrobial Resistance Tracking System (GLASS)—part of the WHO—supports a standardized approach to collect, analyze, and share data on antimicrobial resistance at the global level to inform decision-making and to manage local, national, and regional measures.
6. The Global Antibiotic Research & Development Partnership (GARDP) was established as a joint initiative by the WHO and the Drugs for Neglected Diseases initiative (DNDi). GARDP runs research and development through public–private partnerships. The goal of the 2023 partnership is to develop and deliver up to four new treatments by improving existing antibiotics and accelerating the introduction of new antibiotics to the market.
7. The Interdependent Coordination Group on Antimicrobial Resistance (IACG) consists of representatives of the main UN agencies and more sectoral agencies along with a similar number of individual experts.

6.5 Cytostatics

The history of cytostatic development began in 1940 when American scientists tried to use the therapeutic effects of chemical warfare applications. They tested whether it was possible to specifically use nitrogen mustards for the treatment of lymphomas (a lymphatic tumor) [78]. After the end of World War II, there was an increase in the development of cancer treatments. First, antifolates like methotrexate, shown to be successful in the treatment of breast cancer in women and the treatment of bone marrow cancer, were tested. Following on from that came the discovery of 6-mercaptopurine, which functioned as a highly effective anti-leukemic preparation. Drugs that are effective in the fight against cancer include alkaloids from Madagascar's periwinkle (*Catharanthus roseus*), originally intended for the treatment of diabetes. After 1955, the urgency to identify treatments for cancer increased, and the era of second-generation chemotherapeutic drugs arrived. After 1965, combination chemotherapy approaches started to be used, where several therapeutic agents, such as methotrexate, 6-merkaptopurine, vincristine (vinca alkaloids), and prednisone, were used in treatments—such treatments were shown to be very effective. At present, almost all successful cancer chemotherapy regimens use this paradigm of multiple drugs administered simultaneously—also known as polychemistry. Other success-

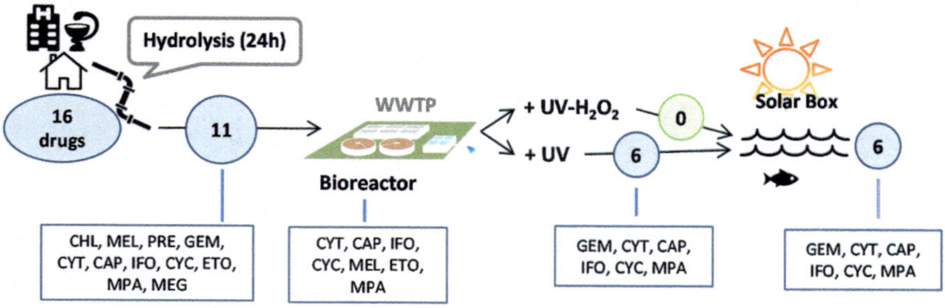

Fig. 6.2 "Life cycle" of cytostatic drugs in the environment, including number of drugs studied, degradation processes studied, and number of remaining substances after each treatment [82]

ful drugs used in chemotherapy include taxanes, camptothecins, platinum agents, nitrosoureas, anthracyclines, and epipodophyllotoxins; hormone-based drugs include tamoxifen, a medicine used to treat breast cancer. Another group of drugs used are metabolic pathway cell division inhibitors, such as Bcr-Abl tyrosine kinase inhibitors and monoclonal antibodies [3, 79].

Bearing in mind the historic overview outlined above, cytostatics first entered the water cycle in the 1950s. Cytostatic drugs have emerged as new water contaminants due to their widespread use as cancer treatments. Cytostatic drugs are thus genotoxic, mutagenic, cyanogenic, teratogenic, and fetotoxic. [80]. The gateways for cytostatics into the aquatic environment are WWTPs, hospital effluent, household wastewater, chemical discharges, and drug disposal. Conventional cleaning processes are not particularly effective at removing cytostatics. Again, this is a group of substances with different hydrophilic and hydrophobic properties. The use of anaerobic membrane reactors represents a very effective approach to degrading cytostatics [55] as does the advanced oxidation processes (AOP) using UV–H_2O_2 or ozone [81], verified in a study where laboratory testing of 24 types of cytostatic drugs considered their persistence against hydrolytic, biological, and photochemical decomposition (Fig. 6.2; Table 6.4).

6.6 Pharmaceuticals for Mental Illnesses

This group includes a relatively wide range of drugs with diverse origins and mechanisms of action. We can divide them according to two aspects of sense affecting in psychopharmacology (Table 6.5). Alternatively, this group of pharmaceuticals can be subdivided into three psychopharmaceutical groups based on chemical composition [87].

1. Neuroleptics include phenothiazines (chlorpromazine), xanthene and thioxanthene, various tricyclic structures, tetracyclic structures, pentacyclic structures (reserpine), butyrophenones (haloperidol), diphenylbutylpiperidines, piper-

Table 6.4 Types of cytostatics detected in the water cycle [80, 83]

Title	Therapy	Contamination and behaviour	Negative effects	Degradation
Mycophenolic acid (MPA)	Allogeneic prophylactic treatment against renal, cardiac, and hepatic organ rejection	60% of the drug is excreted in urine as mycophenolic acid glucuronide while 3% remains unchanged [36] in WWTPs. The glucuronide metabolite deconjugates and reforms its original compound, infitrate into the drinking water sources, MPA was detected in all river samples taken from the Llobregat River in Spain at concentrations ranging from 17.0 to 56.2 ng/l [82]	Endocrine disruption disrupts the DNA of animals	Estimated removal rates are 41% predicted pKa values range from 3.57 to 4.61, and octanol/water partition coefficients range from 2.8 to 4.2. However, in the treatment of potable water by sedimentation and sand filtration, complete elimination was achieved [82]
Cyclophosphamide	Treatment of leukemia and malignant lymphomas [36]	From non-detectable to 43 ng/l in hospital effluent; from 8 to 26 ng/l in sewage; from 7 to 25 ng/l in output from WWTPs; and from 0 to 20 ng/l in river water [84]	Carcinogenic	Low biological degradation [82]
Tamoxifen	Treatment of ovarian, uterine, and prostate cancers	In natural water found at levels up to 200 ng/l [85]	Carcinogenic	Absorbs anaerobic sludge [82]

(continued)

Table 6.4 (continued)

Title	Therapy	Contamination and behaviour	Negative effects	Degradation
Methotrexate	Treatment of bone marrow cancer, skin cancer, leukemia, and breast cancer	In hospital sewage levels ranged from non-detectable to 19 ng/l, lower that input water at WWTPs with levels ranging from non-detectable to 23 ng/l	Carcinogenic	High removal by anaerobic biodegradation
Ciprofloxacin	In addition to cancer treatment, it is also used, for example, to treat urinary tract infections, and so it is likely to be detected in high concentrations in wastewater	Detected at very high levels in hospital sewage (from 3089 to 14,826 ng/l), wastewater (from 1172 to 1558 ng/l), and even in surface water upstream from the WWTPs (from 8 to 56 ng/l). Values from 7 to 103 ng/l in effluents exiting WWTPs [86]	Toxic; inhibits growth of freshwater algae [85]	Highly persistent [82]

Table 6.5 Psychopharmacological groupings according to their impacts on mental function [88]

Parameter	Sense affecting	Group of psychopharmacologic	Examples
Vigilance (vigilita)	Positive	Psychostimulants	Amphetamine, amfetaminil, ephedrine, phenmetrazine, Mazindol, Mezokarb, Pemolin, and methylphenidate
		Nootropics	Piracetam, Pyritinol, and Meklofenoxate
	Negative	Hypnotics	Barbital, Amobarbital, hexobarbital; Glutethimid, Metachalon, Klomethiazol; nitrazepam, flunitrazepam, Triazolam; Zopiclone, zolpidem
Affectivity	Positive	Antidepressants	Imipramine, Desipramine, Amitriptyline, Nortriptyline, Clomipramine, Maprotiline, citalopram, fluoxetine, Fluvoxamine, Mianserin, Buspirone, Moclobemide, Li^+
		Anxiolytics	Guaifenesin, Meprobamate; Diazepam, alprazolam, oxazepam; neuroleptics and antidepressants
	Negative	Dysphoricum	Reserpine, clonidine, methyldopa
Psychological integration	Positive	Neuroleptics	Thioridazine, Chlorpromazine, Chlorprothixene, Levopromazin, haloperidol, Perphenazine, clozapine, Amisulprid, risperidone
	Negative	Hallucinogens	Lysergic, cocaine, amphetamines, ketamine, hashish, marijuana, phencyclidine, and mescaline

azines, benzimidazolinones, bicyclic 5-HT2 and D2 antagonists, imidazolidinones, naphthenic derivatives, and benzamides.
2. Antidepressants include tricyclic (imipramine), tetracyclic (mianserin, maprotiline), bicyclic (citalopram), and monocyclic compounds.
3. Anxiolytics (tranquilizers and antifungal agents) represent a group comprising benzodiazepines (diazepam), heterocyclic diazepines, benzazepines and benzodiazepine receptor ligands, diphenylmethane derivatives, piperazine and piperidine carboxylates, quinazolinones, thiazine, triazoles, other heterocyclic and imidazole derivatives, propanediol derivatives, carbamate components, and anthracene derivatives.

The history of drugs being used to treat mental illness dates back to ancient Egypt when people realized that some individuals were mentally more or less resilient. In ancient cultures, these deviations were attributed to evil spirits or supernatural powers impacting on behavior. Treatment of such mentally ill people included Ayurveda (originating in ancient India), which uses various odors of herbs and aromatic ointments, and the separation of individuals from the rest of society. The development of modern psychopharmacs dates to the second half of the nineteenth century, when morphine, potassium bromide, chloralhydrate, hyoscine, paraldehyde, etc., were introduced as treatment processes. The next stage of development came in the first half of the twentieth century, introducing the therapeutic processes of nicotinic acid, penicillin, and thiamine, which led to the eradication of psychiatric disorders caused, for example, by syphilis or cerebral palsy psychoses. The second half of the twentieth century saw the introduction of so-called psychotropic drugs including chlorpromazine, an aliphatic phenothiazine, and reserpine, an indole derivative; two structurally and functionally distinct antidepressants, iproniazide, a hydrazine monoamine oxidase inhibitor (MAOI), and imipramine, an iminodibenzyl (dibenzoazepine) monoamine (re)uptake inhibitor (MAUI); anxiolytics, such as meprobamate and propanediol; and mood stabilizers like lithium carbonate, an inorganic salt. Current modern psychopharmacology continues to use "psychotropic drugs" and seeks to target as much as possible the healing effects of brain monoamines and their metabolites, which are involved in the transfer of neurons to the synaptic cleft [89].

Major depressive disorder (MDD) is the most common mood disorder in the United States with a lifetime prevalence of 14.4% [90]. Antidepssants of the Selective serotonin reuptake inhibitors (SSRIs) are the first step in a treatment plan, and as patients progress during treatment, new antidepressant drugs with different mechanism of action are introduced (e.g., bupropion, tricyclic antidepressants) [91].

The entrance of mental illness pharmaceuticals into the water cycle once again dates back to the turn of the second half of the twentieth century. Increased attention to their occurrence in the water cycle was at the turn of the millennium when the use of analytical technology, with lower detection levels, revealed concentrations of antidepressants and antipsychotics in wastewater and surface water [35, 92]. It has been verified that the effects of such drugs on living organisms are dangerous even at low concentrations. Adverse effects have been revealed on aquatic organism due their ability to disturb homeostasis in the central and peripheral nervous system in vertebrates, with invertebrates facing other hazardous impacts as a result of their proven synergy with other pharmaceutical substances [93]. Because antidepressants are one of the most prescribed medicines in the treatment of mental health they are among the most commonly detected of drugs. Most commonly, these detected drugs include fluvoxamine, fluoxetine, norfluoxetine, paroxetine, sertraline, citalopram, fluvoxamine, desmethylfluvoxamine, mirtazapine and desmethylmirtazepine, desmethylsertraline, venlafaxine, O-desmethylvenlafaxine, bupropion, amitriptyline, nortriptyline, carbamazepine (often referred to as an antiepileptic drug, also prescribed for the treatment of manic conditions or for the treatment of bipolar disorders), 10,11-dihydrocarbamazepine, and tramadol [94–96]. Conventional sewage treatment methods, when capturing antidepressants and antipsychotic agents, offer an average removal rate of 30% [94].

6.7 Conclusions

The topic of pharmaceuticals in the water cycle is a complex one. There is diversity in the physicochemical nature of pharmaceuticals and high degradation degrees can only be provided based on accurate analyses of sewage from respective WWTPs. Systemically, it will always be a necessary aim to develop removal technologies targeted precisely at specific medicinal substances and their concentrations. The question remains where to place such removal technologies—at sewage treatment plants or outflows from hospitals or perhaps decentralized placement for larger territorial units? The concentration of drugs in wastewater is variable and depends on many factors such as seasonal periods, urban sprawl, the dilution of water spills due to heavy rain events, and synergies with other pharmacological agents in wastewater. At present, the best approach is probably to try to get an accurate idea of the conjugation and metabolic processes taking place in sewage and wastewater treatment plants. In addition, we should further try to build more efficient degradation technologies or at least more accurately predict the concentrations of drugs in the water cycle. As has been mentioned above, WWTPs represent an boundary between our modern "chemically contaminated" society and the "pure" natural environment. Our best efforts should be concentrated on preserving our natural environment for future generations and preserving as many animal and plant species as possible while at the same time providing our descendants with adequate and functional curative pharmaceuticals.

Acknowledgements This chapter was produced in collaboration with project No. LO1408 "AdMaS UP—Advanced Materials, Structures and Technologies," supported by the Ministry of Education, Youth and Sports under the "National Sustainability Programme I."

References

1. Fong PP, Alex T, Ford AT (2014) The biological effects of antidepressants on the molluscs and crustaceans: a review. Aquat Toxicol, 4–13
2. Katzung BG (2015) Basic & clinical pharmacology. McGraw-Hill Professional
3. Black J (1988) Drugs from emasculated hormones: the principles of synoptic antagonism. Nobel Lecture
4. Endo AA (2010) historical perspective on the discovery of statins. Proc Jpn Acad, 484–496
5. Watts G (2012) Obituary: Sir David Jack. Lancet
6. Swinney DC, Anthony J (2011) How were new medicine discovered? Nat Rev. Drug Discov, 507–519
7. Mompelat S, Thomas O, Le Bot B (2011) Contamination levels of human pharmaceutical compounds in French surface and drinking water. J Environ Monit, 2929–2939
8. Suarez S, Lema JM, Omil F (2010) Removal of pharmaceutical and personal care products (PPCPs) under nitrifying and denitrifying conditions. Water Res, 3214–3224
9. Pitter P (2009) Hydrochemie. VŠCHT Praha, Prague
10. Suarez S et al (2008) How are pharmaceutical and personal care products (PPCPs) removed from urban wastewaters? Rev Environ Sci Biotechnol, 125–138
11. Gao P, Munir M, Xagoraraki I (2012) Correlation of tetracycline and sulfonamide antibiotics with corresponding resistance genes and resistant bacteria in a conventional municipal wastewater treatment plant. Sci Total Environ, 173–183
12. Meador J (2006) Rationale and procedures for using the tissue-residue approach for toxicity assessment and determination of tissue, water, and sediment quality guidelines for aquatic organisms. Hum Ecol Risk Assess, 1018–1073
13. Zenker A et al (2014) Bioaccumulation and biomagnification potential of pharmaceuticals with a focus to the aquatic environment. J Environ Manag, 378–387
14. Shenker M et al (2011) Uptake of carbamazepine by cucumber plants—a case study related to irrigation with reclaimed wastewater, 905–910
15. Gobas F et al (2009) Revisiting bioaccumulation criteria for POPs and PBT assessments. Integr Environ Assess Manag 5(4):624–637
16. OECD (2008) Guideline 315 (Online)
17. Gray JS (2002) Biomagnification in marine systems: the perspective of an ecologist. Mar Pollut Bull 45(1–12):46–52
18. WHO (2002) Global assessment of the state-of-the-science of endocrine disruptors. World Health Organisation (Online)
19. Harrigan JA et al (2004) DNA adduct formation in precision-cut rat liver and lung slices exposed to benzo a pyrene. Toxicol Sci 77(2):307–314
20. Nebert DW et al (2000) Role of the aromatic hydrocarbon receptor and Ah gene battery in the oxidative stress response, cell cycle control, and apoptosis. Biochem Pharmacol 59(1):65–85
21. Gong Y, Han XD (2006) Nonylphenol-induced oxidative stress and cytotoxicity in testicular Sertoli cells. Reprod Toxicol 22(4):623–630
22. Hahn ME (1998) The aryl hydrocarbon receptor: a comparative perspective. Comp Biochem Physiol C-Toxicol Pharmacol 121(1–3):23–53
23. Yao TP et al (1993) Functional ecdysone receptor is the product of ECR and ultraspiracle genes. Nature
24. Li WC (2014) Occurrence, sources, and fate of pharmaceuticals in aquatic environment and soil. Environ Pollut, 193–201
25. Bringolf RB et al (2010) Environmental occurrence and reproductive effects of the pharmaceutical fluoxetine in native freshwater mussels. Environ Toxicol Chem, 1311–1318
26. Highman et al (1980) Neoplastic and preneoplastic lesions induced in femal C3H mice by diets containing diethylstilbestrol or 17beta-estradiol. J Environ Pathol Toxicol, 81–95
27. Heltsley RM et al (2006) Prozac elicits spawning in native freshwater mussels. In: 23nd annual meeting of the American Chemical Society, San Francisco, 10–14

28. Mehinto AC, Hill EM, Tyler CR (2010) Uptake and biological effects of environmentally relevant concentrations of the nonsteroidal anti-inflammatory pharmaceutical diclofenac in rainbow trout (*Oncorhynchus mykiss*). Environ Sci Technol, 2176–2182
29. Lyssimachou A, Arukwe A (2007) Alteration of brain and interrenal StAR protein, P450scc, and Cyp11beta mRNA levels in atlantic salmon after nominal waterborne exposure to the synthetic pharmaceutical estrogen ethynylestradiol. J Toxicol Environ Health, 606–613
30. Vannini C et al (2011) Effects of a complex mixture of therapeutic drugs on unicellular algae Pseudokirchneriella subcapitata. Aquat Toxicol, 459–465
31. Caliman FA, Gavrilescu M (2009) Pharmaceuticals, personal care products and endocrine disrupting agents in the environment—a review. Clean-Soil Air Water 37(4–5):277–303
32. Adeel M, Song X, Wang Y et al (2017) Environmental impact of estrogens on human, animal and plant life: a critical review. Environ Int 99:107–119
33. Heberer T (2002) Tracking persistent pharmaceutical residues from municipal sewage to drinking water. J Hydrol, 175–189
34. Vrcek V (2017) Pharmacoecology—The Environmental Fate of Pharmaceuticals. Kem U Industriji-J Chem Chem Eng 66(3–4):135–144
35. Snyder (2008) Occurrence, treatment, and toxicological relevance of EDCs and pharmaceuticals in water. J Int Ozone Assoc 1(30):65–69
36. SÚKL (2018) Databáze léku. SÚKL - Státní ústav pro kontrolu léčiv (Online). http://www.sukl.cz/leciva
37. Lincová D, Farghali H et al (2007) Základní a aplikovaná farmakologie. GRADA, Praha
38. Brune K, Hinz B (2007) History of analgesics. (book auth.) In: Willis WD, Schmidt RF (eds) Encyclopedia of pain. Springer, Berlin, 899–904
39. Nešmerák K (2016) History of analgetics. Res Gate (Online). https://www.researchgate.net/publication/309670737_History_of_analgesics
40. Oaks JL et al (2004) Diclofenac residues as the cause of vulture population decline in Pakistan. Nature, 630–633
41. Lüllmann H et al (2004) Farmakologie a toxikologie. Grada, Praha
42. Żur J et al (2018) Paracetamol—toxicity and microbial utilization. Pseudomonas moorei KB4 as a case study for exploring degradation pathway. Chemosphere, 192–202
43. De Gusseme B et al (2011) Degradation of acetaminophen by *Delftia tsuruhatensis* and *Pseudomonas aeruginosa* in a membrane bioreactor. Water Res, 1829–1837
44. Antunes SC et al (2013) Biochemical effects of acetaminophen in aquatic species: edible clams *Venerupis decussata* and *Venerupis philippinarum*. Environ Sci Pollut Res, 6658–6666
45. Li J, Ye Q, Gan J (2014) Degradation and transformation products of acetaminophen in soil. Water Res, 44–52
46. Ferrando-Climent L et al (2012) Comprehensive study of ibuprofen and its metabolites in activated sludge batch experiments and aquatic environment. Sci Total Environ, 404–413
47. Wang J et al (2017) Targeted eco-pharmacovigilance for ketoprofen in the environment: need, strategy and challenge. Chemosphere, 450–462
48. Prášková E et al (2013) The effects of subchronic exposure to ketoprofen on early developmental stages of common carp. Acta Vet Brno, 343–347
49. Verlicchi P, Al Aukidy M, Zambello E (2012) Occurrence of pharmaceutical compounds in urban wastewater: removal, mass load and environmental risk after a secondary treatment-a review. Sci Total Environ, 123–155
50. Fatta-Kassinos D, Bester K, Kümmerer K (2010) Xenobiotics in the urban water cycle mass flows, environmental processes, mitigation and treatment strategies. environmental pollution
51. EU (2013) Směrnice 2013/39 Evropského parlamentu a Rady ze dne 12. srpna 2013 Směrnice pozměňující 2000/60/ES a 2008/105/ES, pokud jde o prioritní látky v oblasti vodní politiky. Úřední věstník Evropské unie (Online). https://publications.europa.eu/cs/publication-detail/-/publication/296e91b8-4610-11e3-ae03-01aa75ed71a1/language-cs
52. Bouju H et al (2016) Elucidation of biotransformation of diclofenac and 4′ hydroxydiclofenac during biological wastewater treatment. J Hazard Mater, 443–452

53. Vieno N, Sillanpää M (2014) Fate of diclofenac in municipal wastewater treatment plant—a review. Environ Int, 28–39
54. Velo G, Moretti U (2010) Ecopharmacovigilance for better health. Drug Saf, 963–968
55. Wang X, Zhang J, Chang VWC, She Q, Tang ChY (2018) Removal of cytostatic drugs from wastewater by an anaerobic osmotic membrane bioreactor. Chem Eng J 339:153–161
56. Schwaigera J et al (2004) Toxic effects of the non-steroidal anti-inflammatory drug diclofenac: part I: histopathological alterations and bioaccumulation in rainbow trout. Aquat Toxicol, 141–150
57. Kraigher B et al (2008) Influence of pharmaceutical residues on the structure of activated sludge bacterial communities in wastewater treatment bioreactors. Water Res, 4578–4588
58. Groening J et al (2007) Transformation of diclofenac by the indigenous microflora of river sediments and identification of a major intermediate. Chemosphere, 509–516
59. Facey SJ et al (2018) Rapid and complete degradation of diclofenac by native soil microorganisms. Environ Technol Innov, 55–61
60. Buerge IJ et al (2006) Occurrence and fate of the cytostatic drugs cyclophosphamide and ifosfamide in wastewater and surface waters. Environ Sci Techno, 7242–7250
61. O'Shea D (2005) Internet archive—Dublin City University (Online). https://web.archive.org/web/20120105081602/http://www.dcu.ie/~oshead/BE401/brewing/lecture4.pdf
62. Pazdera J (2015) Jak moc roste spotřeba antibiotik? OSEL (Online). http://www.osel.cz/8164-jak-moc-roste-spotreba-antibiotik.html
63. Benutzer F (2012) Antibiotikum. wikipedie de (Online). https://commons.wikimedia.org/wiki/Category:Antibiotics#/media/File:Antibiotics_action.png
64. WHO (2018) Antimicrobial resistance. World Health Organisation (Online). http://www.who.int/news-room/fact-sheets/detail/antimicrobial-resistance
65. Kummerer K (2004) Resistance in environment. J Antimicrob Chemother, 311–320
66. Karkman A et al (2018) Antibiotic-resistance genes in waste water. Trends Microbiol, 220–228
67. Merck (2003). Antibiotics. Online medical library—Merck (Online)
68. WHO (2018) Antimicrobial resistance. World Health Organization (Online). http://www.who.int/en/news-room/fact-sheets/detail/antimicrobial-resistance
69. Lange-Ernst ME (2005) Antibiotika: omyly a pravda: šance, rizika, alternativy. Olomouc: FONTÁNA
70. Bergeron S et al (2015) Presence of antibiotic resistant bacteria and antibiotic resistance genes in raw source water and treated drinking water. Int Biodeterior Biodegradation, 370–374
71. WHO (2017) Critically important antimicrobials for human medicine, 5th revision. WHO (Online). http://www.who.int/foodsafety/publications/antimicrobials-fifth/en/
72. Seppala H et al (1997) The effect of changes in the consumption of macrolide antibiotics on erythromycin resistance in group A Streptococci in Finland. N Engl J Med, 441–446
73. Papageorgiou M, Kosma Ch, Lambropoulou D (2016) Seasonal occurrence, removal, mass loading and environmental risk assessment of 55 pharmaceuticals and personal care products in a municipal wastewater treatment plant in Central Greece. Sci Total Environ, 547–569
74. Harrabi M et al (2018) Analysis of multiclass antibiotic residues in urban wastewater in Tunisia. Environ Nanotechnol Monit Manage, 163–170
75. Yang Y et al (2017) Application of aptamers in detection and chromatographic purification of antibiotics in different matrices. TrAC Trends Anal Chem, 1–22
76. Jianga H et al (2018) Exploring the differences of antibiotic resistance genes profiles between river surface water and sediments using metagenomic approach. Ecotoxicol Environ Saf, 64–69
77. Hu Y et al (2018) Antibiotics in surface water and sediments from Hanjiang River, Central China: occurrence, behavior and risk assessment. Ecotoxicol Environ Saf, 150–158
78. Gilman A (1963) The initial clinical trial of nitrogen mustard. Am J Surg, 574–578
79. Cancer Research (2012) Cancer drugs. Cancer research UK (Online). https://about-cancer.cancerresearchuk.org/about-cancer/cancer-in-general/treatment/cancer-drugs
80. Dubey K et al (2017) Slurry photocatalytic membrane reactor technology for removal of pharmaceutical compounds from wastewater: towards cytostatic drug elimination. Sci Total Environ

81. Franquet-Griell H et al (2016) Do cytostatic drugs reach drinking water? The case of mycophenolic acid. Environ Pollut, 532–536
82. Franquet-Griell H et al (2017) Biological and photochemical degradation of cytostatic drugs under laboratory conditions. J Hazard Mater, 319–328
83. Janssens R et al (2017) Slurry photocatalytic membrane reactor technology for removal of pharmaceutical compounds from wastewater: towards cytostatic drug elimination. Sci Total Environ, 612–626
84. Ferrando-Climent L, Rodriguez MS, Barceló D (2014) Incidence of anticancer drugs in an aquatic urban system: from hospital effluents through urban wastewater to natural environment. Environ Pollut, 216–223
85. Ferrando-Climent L, Rodriguez-Mozaz S, Barceló D (2014) Incidence of anticancer drugs in an aquatic urban system: from hospital effluents through urban wastewater to natural environment. Environ Pollut, 216–223
86. Buerge IJ, Buser HR, Poiger T, Müller MD (2006) Occurrence and fate of the cytostatic drugs cyclophosphamide and ifosfamide in wastewater and surface waters. Environ Sci Techno, 7242–7250
87. Muller B (1992) Psychotropics 92/93. Lundbeck, Netherlands
88. Švestka J (1995) Psychotropic drugs in clinical practice. Grada, Praha. ISBN 80-7169-169-0
89. Ban TA (2001) Pharmacotherapy of mental illness—a historical analysis. Prog Neuro-Psychopharmacol Biol Psychiatry, 709–727
90. Kessler RC et al (2012) Twelve-month and lifetime prevalence and lifetime morbid risk of anxiety and mood disorders in the United States. J Methods Psychiatr Res, 169–184
91. Hillhouse TM, Porter JH (2015) A brief history of the development of antidepressant drugs: From monoamines to glutamate. Exp Clin Psychopharmacol, 1–21
92. Ternes TA (1998) Occurrence of drugs in German sewage treatment plants and rivers. Water Res 32:3245–3260
93. Giebułtowicza J, Nałęcz-Jaweckib G (2014) Occurrence of antidepressant residues in the sewage-impacted Vistula and Utrata rivers and in tap water in Warsaw (Poland). Ecotoxicol Environ Saf 104:103–109
94. Lajeunesse A et al (2012) Distribution of antidepressant residues in wastewater and biosolids following different treatment processes by municipal wastewater treatment plants in Canada. Water Res 46(17):5600–5612
95. Vasskog T et al (2008) Occurrence of selective serotonin reuptake inhibitors in sewage and receiving waters at Spitsbergen and in Norway. J Chromatogr A 2:194–205
96. Wick A et al (2009) Fate of beta blockers and psycho-active drugs in conventional wastewater treatment. Water Res 43:1060–1074
97. Nesměrák K (2016) Histore analgetik. Res Gate (Online). https://www.researchgate.net/publication/309670737_History_of_analgesics

Chapter 7
Biological Audits in the System of Water Treatment Control

Jana Říhová Ambrožová

Abstract The whole water-supply system from the catchment to the consumer should be studied and monitored as one continuum with many interrelations. This principle is one of the key points of Water Safety Plans which are included in the new WHO Drinking Water Quality Guidelines. Microscopic analysis can help to understand the sources and consequences of eutrophication and the pollution of both treated and untreated water, thus affecting the work of the designers, construction engineers and operatives, water managers, water suppliers, health authorities, and decision-makers involved. These methods can provide rapid results and reliable data, which are useful in selecting appropriate preventive measures or ecologically sustainable remedial actions. Biological and hydrobiological audits are discussed step by step in the context of risk analyses (HACCP), monitoring of water treatment technologies and the Water Safety Plans used in the Czech Republic.

Keywords Drinking water · Water Safety Plans · Water supply · Water treatment control · Biological audits · Microscopic analysis · Indicators of water quality

7.1 Importance of Biological Analyses in Technological Practice

It is needed to pay increased attention to the system of production, transport, and accumulation of drinking water because the water quality may be affected at any point of the continuous water-supply system, both internally and externally. Pursuant to the requirement of the Water Safety Plans (WSP), it is necessary to perceive the distribution network as a continuum, i.e. from the source up to the consumer [1]. The objective of the documents is a trouble-free supply of drinking water to the population from the actual source (surface, ground). The recommendations for the

J. Říhová Ambrožová (✉)
Department of Water Technology and Environmental Engineering,
Faculty of Environmental Technology, Institute of Chemical Technology Prague,
Technická 5, 166 28 Prague 6, Czech Republic
e-mail: jana.ambrozova@vscht.cz

M. Zelenakova et al. (eds.), *Management of Water Quality and Quantity*,
Springer Water, https://doi.org/10.1007/978-3-030-18359-2_7

quality of drinking water and its evaluation, issued by the World Health Organization (WHO), are currently crucial documents in water-supply engineering, which are gradually applied in water-supply companies. The concept of the documents includes the system for providing sanitary drinking water. This especially involves health criteria, plans for safe drinking water provision, and independent regulation. The plans of safe supply with water (Water Safety Plans, WSP) are significantly supported by the evaluation of risks within the scope of the entire water supply system (from the source up to the consumer's tap). The identification and monitoring of the checkpoints most efficient for decreasing the detected risks, and the creation of effective inspection systems of management and operating plans, are very important for the evaluation of risk [2, 3].

In the water supply system, it is possible to apply the risk analysis (HACCP; Hazard Analysis and Critical Control Points) that should consider the entire supplied area including the raw water source, collecting structures (reservoirs), water treatment (technology), distribution network together with the structures located in it (pumping stations, accumulation, water supply tanks) and the consumer endpoints in the network. Upon finding the risk points, the appropriate plan for securing safe water supply is then elaborated (EN 15975-1, EN 15975-2). The plan serves not only for dealing with breakdowns but also corrective actions, precautions, regular inspection, and for the needs of continuous monitoring and verifying by audits. The monitoring programme and the comprehensive risk assessment should form a part of the operating code. Having this system at our disposal certainly decreases the risk of epidemics from drinking water, improves water quality, enhances the understanding of the operated system, directs the investments into the renewal of the distribution system in a targeted and more economical way and the consumer definitely enjoys greater confidence in the quality of drinking water [4].

Drinking water must not contain any pathogenic and conditionally pathogenic microorganisms. In the case of their appearance, these microorganisms are an indication of wastewater penetration (surface water, water with microbial contamination) into the water-supply pipeline, a defective connection of the water-supply system (leaks, breakdowns, technological water), insufficient technical securing of wells (backflow valves, connection with contaminated surface water), secondary contamination (by air, water), etc. From the point of view of hygiene, the presence of these bacteria in the water represents an urgent threat of the emergence of disease. Prevention of the creation and spread of infectious illnesses are determined in the Czech Republic by Law No. 258/2000 Coll., on the protection of public health. The hygienic requirements for drinking water (including packaged water), the scope and frequency of inspection of the observance of the hygienic limits of microbiological, biological, physical, and chemical indicators of drinking water are determined by Decree No. 252/2004 Coll. Safe drinking water will only be in the entire supply system if water source contamination is prevented, if the water is treated sufficiently and if secondary contamination is prevented during accumulation, distribution, and usage of drinking water [3, 5, 6]. Due to the time, the water is kept in the distribution network. An important aspect of water quality in drinking water distribution system is the residence time; microbial contamination may occur due to various factors, which include failures of the treatment process or human factors, breakdowns in the distribution network, secondary contaminations, hydraulic impulses in the pipeline. One of the most significant risk factors for microbial contamination of drinking water is the presence of pathogenic microorganisms, i.e. organisms that may cause diseases. If such an event occurs, timely identification of the pollution is necessary so that a suitable decontamination procedure can be selected. The aforementioned issues gain crucial importance with the existence and public knowledge of (owing to press and media coverage) epidemics from drinking water occurring within several years in the Czech Republic (Prague Dejvice—August 2014, Prague Újezd nad Lesy—September 2014, Prague Dejvice—May 2015, Trnová—2014 to 2015, Okrouhlá—August 2015, Nový Bor—August 2015, Prachatice—October 2015, etc.) [7].

According to the WSP, an essentially simple and undemanding form of inspecting produced water, with the goal of detecting threats and risks, is the performance of biological (hydrobiological) audits. The biological analysis makes a big difference when solving various problems, especially in water treatment facilities treating raw water from tanks and streams. In the water-supply practice in the Czech Republic, we have had an opportunity to verify the fact that biological analyses are important for the operators of water-supply facilities. Their significance increases if the analysis results are suitably interpreted so that efficient interventions and technological measures are performed based on them [3, 6, 8]. In the drinking water-supply system, it is appropriate to use methods of biological analyses of samples focused on the entire supplied area up to the end points in the distribution network at the consumer, including the raw water source, the water accumulation and treatment structures (technology arrangement suitability), the distribution network and water-supply tanks (accumulation). According to the experience when performing comprehensive audits of water-supply systems, the inclusion of microscopic image monitoring is also important, detecting the presence of organisms that are not detected during microbiological testing, and furthermore, it should be added that the microscopic findings can be accomplished in a substantially shorter time interval [9]. Of course, an integral and highly appropriate aspect of biological monitoring is to obtain an overview of which water quality microbiological indicators are present that would show organic pollution and faecal contamination of the water [6].

As far as the verification of the quality of drinking water is concerned, it is possible, for the needs of these Water Safety Plans to utilise methods, procedures, and tests complementing the regular tools of operative monitoring. Those are mostly methods for determining whether the drinking water supply is in accordance with the established parameters of Decree No. 252/2004 Coll. or whether a reassessment of the entire system is necessary. From the biological standpoint, the microbial quality of water is monitored. This is based on microbiological testing including mostly the determination of the presence of faecal indicator microorganisms, or a direct determination of a specific pathogenic, conditionally pathogenic or hygienically significant microorganism. And furthermore, swab samples from various wetted surfaces areas are taken with the aim of microscopic identification of the bioseston and abioseston present. From the water samples and the swab samples from the wetted surfaces of water-supply facilities, it is possible to evaluate the processes taking place in the transported water. Bioseston (microorganisms) and abioseston (particles) are important indicators not only of the quality of drinking water, but they also provide a picture of biological stability of whole distribution system in the context of the drinking water quality [2, 6, 8, 10, 11].

In this concept, the biologist (hydrobiologist) should focus on an essentially systemic, and thus also comprehensive in the sense of WSP, evaluation of the entire water-supply system. This includes the hydrobiological inspection of the accumulation structures (collecting reservoirs) and raw water supplies; the inspection of the first separation stage efficiency; the inspection of the second separation stage efficiency and also further technological stages (ozonation, disinfection, membrane separation, etc.). In the first stage efficiency are monitored the catching organisms in flocs, the information can be used for the selecting the optimum coagulant dose or changing the coagulant or flocculant. In the second stage efficiency can be evaluated the filtering efficiency, sand bed ingrowth, intensity, and duration of washing. The hydrobiological analysis of biofilms and attached organisms in all water-supply stages and in the distribution network gives information about biological stability. Also is recommended the evaluation of water quality changes due to regular desludging. Nevertheless, the assistance when solving organoleptic failures and last but not least suggestions for the prevention of microbiological contamination, and any recommendations or corrective actions, are valuable. Only then will the results from biological analyses (biological audit) be beneficial for the operator of the water-supply system. It is also important that the biological audit is performed over the course of one year so that it catches the seasonal changes and eliminates possibly incorrectly deduced conclusions and recommendations just based on a single series of sampling. The comprehensive biological audit offers an opportunity for discovering defects that cannot be detected using chemical analyses. With the help of basic and essentially also quick methods, the biological audit facilitates the identification and analysis of causes, or risks, of contamination in the distribution system [2, 3].

A simple procedure diagram and a biological audit plan are stated in the text below. The prerequisite for its execution is basic knowledge concerning the manner of sampling and the treatment and preservation of samples, and the principles of the hydrobiological analyses (determination of the microscopic image with the goal of finding the presence and number of bioseston and abioseston) and microbiological analyses (cultivation determination, screening methods).

7.2 Biological Methods Used During Monitoring

Microscopic determinations and microscopic analyses form an integral part of biological (hydrobiological) analysis and their procedure is also specified by many CSN standards accepted by operating laboratories. The analysis is both qualitative (in determining what taxon, particles, etc. are present) and quantitative (in determining the quantity). In the Czech Republic, procedures for determining the microscopic image according to standards CSN 75 7712 and CSN 75 7713 are used for the needs of a biological audit of a water supply. The results of microscopic analyses are suitable for evaluating the quality of surface water for water-supply treatment; determining the quantity of bioseston; determining the abioseston that are present; evaluating the relationships between individual components of the water ecosystem; evaluating the separation efficiency of the water-supply technology; evaluating the quality of the produced drinking water; and solving general problems caused by the presence of organisms [9].

The method according to CSN 75 7712 *Water Quality—Biological Analysis—Determination of Bioseston* is based on the thickening of organisms by centrifuging a certain volume of water with subsequent microscopic analysis [12]. The determination is a qualitative (taxonomic, determination, assessment of an organism) and a quantitative (numerical representation, semi-quantitative evaluation, abundance, cover-abundance) analysis of bioseston of various types of water (surface, drinking, ground, waste). 10 mL of a thoroughly mixed sample are measured into a centrifugal test tube. It is centrifuged at 2000 rpm for 5 min when using a swinging rotor with a radius of 0.08 m. After centrifuging, the water is poured from the test tube into a clean vessel by quickly turning it upside down, without whirling the sediment. The centrifuged residue is adjusted with a Pasteur pipette to a suitable volume (e.g. 0.1–1 mL). By repeated suction using a pipette, the contents are transferred after mixing onto a screen of the counting chamber Cyrus I and covered with a cover glass and subsequently fastened with clips (the defined volume is thus precise). A microscope is then used to determine which organisms (bioseston) are present. The prepared sample is inspected at low magnification. The organisms (unicellular or multicellular) are counted either in the entire chamber area or a part of its area according to the density of the organisms present. Of course, there is a difference in the numerical representation of organisms in drinking water and surface water. With drinking water analyses, the entire chamber is always evaluated. In case of standard CSN 75 7712, it is necessary to specify what a specimen means. A specimen means a separate cell, where its size does not matter, furthermore, a coenobium or colony up to the size of 100 μm and a fibre up to the length of 100 μm. Coenobiums, colonies, and fibres exceeding the stated dimensions are expressed as their multiples. With colonial ciliates, rotifers, and minor multicellular metazoans, each specimen is counted separately. If the analysis purpose requires it, e.g. when evaluating technological processes or determining the volumetric biomass, individual cells (colonial and filamentous cyanobacteria and algae) are counted. The identification of the biological condition (i.e. distinction of live organisms) is performed with autotrophic organisms based on autofluorescence of chlorophyll-a, with other organisms according to the morphological condition of the cells, the structure of the protoplast, and the signs of physiological activity, which are for instance movement or a reaction to stimuli. The microscopic analysis of the water sample provides information on the species structure and quantity of associations of the organisms present. If it is focused on the autotrophic component of biocoenosis (phytoplankton), then it is possible to state an association of organisms either by the expression of the number of cells or specimens (abundance) or as volumetric biomass. This method can be used for surface water, attached organisms, and sediments.

Identification of the particles of abioseston (abiotic particles) helps to determine the origin of the pollution and load sources of various types. The procedure is governed by the requirements of standard CSN 75 7713 *Water Quality—Biological Analysis—Determination of Abioseston* [13]. The determination serves for assessing the self-cleaning processes in water streams, ascertaining the effect of valley reservoirs, fish ponds, and other reservoirs and basins on the water stream or for evaluating the efficiency of individual stages of water treatment. The method is based on the determination of the cover-abundance of the field of view of a microscope with particles thickened by centrifuging a certain volume of water. In some cases, the sample is not centrifuged (a scum, sediment, or sludge sample). The adjustment of the sample by centrifuging and the preparation of the sample on the screen of the counting chamber Cyrus I takes place in the same manner as in determining the bioseston (see CSN 75 7712). The estimate of the cover-abundance may be implemented in several ways, among which are image analysis, cover-abundance estimate of individual particles, and comparison with the estimate scale. The determination is disturbed by a higher content of live organisms of the order of 10^4 to 10^5 specimens in a volume of 1 ml of the tested water. Among the particles of abioseston are, for example, inorganic particles, sand, iron and manganese coagula, corrosion products, oil beads, plant residues (alga envelopes, cell walls, plant trichomes, plant tissue residues), animal residues (envelopes, body residues, insect chitin, filtering combs), starch grains, pollen grains, bird feather residues, butterfly scales, oligochaetes bristles, detritus (unidentifiable organic residues), etc.

Microbiological water examination (analyses) is based on monitoring the possible occurrence of bacteria that indicate general (organotrophic) and faecal water pollution. The presence of bacteria is used for regular monitoring of drinking water quality in accumulations and distribution networks. Coliform bacteria signal by their presence non-conformity of water treatment technology, secondary, and additional water contamination by its distribution, increased quantities of nutrients in water or inefficiency of disinfection during water treatment and transport (distribution). The information on faecal pollution is specified by *Escherichia coli*. As a serious indicator of hygienic defects, one may see intestinal enterococci, which are more sensitive to external influences and indicate fresh faecal pollution in water (they do not reproduce in water) and are chlorine-resistant. With this resistance to chlorine, they indicate imperfect and ineffective chlorine disinfection (even in the case of simultaneous negative determination of coliform bacteria). A no less important indicator showing long-term faecal pollution is anaerobic clostridia (*Clostridium perfringens*) with high resistance to chemical and physical factors thanks to their spores.

7.3 Hydrobiological Monitoring of Raw Water Sources

Hydrobiological analyses are, within the scope of monitoring the quality of accumulated water in the sense of the possible prognosis of the condition and development of water quality, very important and have a long-term history of practice in the Czech Republic [10].

During the monitoring of water-supply reservoirs, at least three critical points are monitored, and these are the influx(es) into the reservoir, the actual reservoir itself (mixed sample, zoning), and the outflow from the reservoir. At the influx into the reservoir, the occurrence and creation of attached organisms, and the possible penetration of organisms causing organoleptic defects are evaluated. At the sampling horizon point, zoning sampling is performed on the reservoir, while the most suitable point for subsequent treatment (of less commonly occurring and difficult to separate organisms) is determined mainly through hydrobiological analyses. A quick handover back of analysis results is essential, with a comment on possible problems during organism separation. In this manner, it is possible to elaborate a certain design of the optimum water treatment technology with regard to the succession of the organisms present. It is important to state the share of individual taxons of cyanobacteria and algae (groups) in the overall volumetric biomass. Bioindication using the index of saprobity contributes to the assessment of pollution, the progression of eutrophication, and possibly also the self-cleaning process. During zoning sampling, it is, of course, appropriate to record the water transparency and colour in situ (Secchi depth), measure the value of pH, conductivity, oxygen concentration, and temperature.

At the site, also important is an evaluation of the orientation of the terrain and the surroundings, in the sense of assessing the possible negative effects and risks in the basin, e.g. the management system, agricultural activity, animal breeding, anthropogenic activity, fertilisation (with pesticides, animal faeces), and forest preserves (wildlife).

During hydrobiological monitoring of sources, the qualitative and quantitative analysis is applied with the objective of ascertaining the organisms present (bioseston, phytoplankton, zooplankton, etc.) and the abundance of taxons. Furthermore, the changes in the composition of biocoenosis are recorded, as well as the composition of the biomass, the activity of physiological functions, and the trophic preference being evaluated. When assessing the eutrophication, it is possible to utilise the indirect dependency between the transparency of the water column and the determined density of phytoplankton (the number of specimens, the number of cells, the size of the volumetric biomass, the concentration of chlorophyll-a).

A significant component of water biocoenosis is phytoplankton, which is very sensitive to the conditions of the environment, i.e. it reacts quite quickly to the changes in the ecosystem. As a biological component of quality (according to the Framework Directive 2000/60/EC), it reflects all the important characteristics of the biotope, including the weather effects, and shows a marked succession throughout the year. Therefore, it is possible to use the composition of the functional groups for an evaluation of the water's character and a prediction of the condition and development in the biotope. The biotope evaluation system based on phytoplankton is based on the presumption that every environment is inhabited by species having similar environmental properties [14, 15]. Phytoplankton has been divided into 31 associations (codons, groups), designated with an alphanumeric code, according to seasonal succession, water trophic levels, physical conditions, tolerance, or sensitivity to environmental conditions, resistance, or sensitivity to zooplankton predation. Within this context, when monitoring water biotopes, it is important to monitor the composition of the phytoplankton, its associations, to record the quantity, concentration of chlorophyll-a, the size of the volumetric biomass, which indicates the trophic conditions of water. Due to the vertical and horizontal stratification of phytoplankton, it is suitable to focus on sampling from the littoral zone with a sample container and depth sampling taken with a sampler of the Van Dorn type from preselected profiles with the goal of catching individual layers (epilimnion, metalimnion, and hypolimnion). So, phytoplankton with the dominance of diatoms, chrysophytes, green algae, dinoflagellates, and cyanobacteria is monitored, and each of these associations has various interpretations of the biotope condition and possible predictions of the biotope condition in the course of the vegetation season/year. Phytoplankton with the dominance of diatoms indicates cold mixed water. Diatoms react to the lengthening of the day at the end of winter. Sometimes, they dominate in the summer and replace the cyanobacteria water bloom (associations A, B, C, and D, mixed associations N and P). Phytoplankton with the dominance of chrysophytes indicates clean, oligotrophic, and cold waters and/or fish ponds with nutrients and in adverse conditions, dormancy can be encountered (associations E and U, mixed associations X2, X3, and W). Phytoplankton with the dominance of green algae is typical for clean and hypertrophic waters (associations F, J, X1, and G, mixed associations W and X3). Phytoplankton with the dominance of dinoflagellates can be found in oligotrophic and mesotrophic waters (associations L_0 and L_M). For phytoplankton, with the dominance of cyanobacteria, the indication is ambiguous because cyanobacteria assimilating nitrogen are typical for oligotrophic to hypertrophic waters. Moreover, they have the capability to utilise a minimal supply of light (chromatic adaptation) and to adapt to water layer mixing, cyanobacteria fall under multiple associations according to this concept (associations H1, H2, K, M, R, and S). Phytoplankton with the dominance of cryptomonads indicates mesotrophic and hypertrophic waters and falls under association Y. Picoplankton and nanoplankton are represented by minor species of algae that grow and reproduce rapidly, resist the grazing pressure, hydraulic conditions, and temperature (associations Z, X1, X2, and X3) [14, 15].

The treatment capacity of raw water is determined by the quantity and type of organisms present. It is related to their morphology, size, surface nature (envelope, slime layer, sheath), charge potential, active movement, development stage, and trophic preference. It is also important to assess organisms as a part of certain biocoenosis that is bound to the specific conditions of the environment and have specific demands for nutrition and adaptation to the environment (planktonic and attached forms). Increased occurrence of phytoplankton in raw water in the process of being treated is subsequently shown in the water treatment facility by mechanical defects in technologies (filtering problems, poor coagulation, penetration through the technological line), accumulation of silt content and contaminants (sensory defects), worsening of the physical and chemical properties of water, etc. It can be said that in case of a number of 3×10^3 organisms in 1 mL, treatment takes place without troubles, if the number of organisms detected in water being treated is above this value and reaches a maximum of 10^4 organisms in 1 mL, it is probable to expect shortening of the filtering cycles and clogging of sand filters. Higher numbers of organisms (5×10^4 organisms in 1 mL) lead to emergency conditions during treatment and are accompanied by organoleptic defects. In case of single-stage water treatment, emergency conditions occur already at the number of 5×10^3 organisms in 1 mL (the filtering cycle length is shortened).

Besides the evaluation of the quantity of cells/multicellular organisms in the treated water, the phytoplankton which is present is currently more and more often assessed also from the standpoint of organoleptic defects of water. This might be caused by the presence of higher concentrations of volatile organic compounds (VOC) expressed by earthy and spicy odours and flavours of water. Phytoplankton (cyanobacteria and algae) produces various organic substances that are responsible for the problems frequently being solved connected to achieving an appropriate quality of drinking water. E.g. chrysophytes of species *Synura petersenii* produce *trans,cis*-2,6-nonadienal with a cucumber odour, and a fish odour is caused by *Uroglena americana* by producing (*E,E*)-2,4-heptadienal and (*E,Z*)-2,4-heptadienal. Attention is focused on odour-forming metabolites, which occur in waters mostly in such concentrations that have no cytotoxic or genotoxic effect on humans. Nevertheless, they are negatively perceived and affect the final water consumption by the consumer (ng L^{-1}). Often discussed are geosmin and 2-MIB (pyrazines and terpenoids), whose strongly earthy (musty/muddy) odour is also solved by the shutdown of the supplied area for a longer period. Production of geosmin and 2-MIB has been found especially with bacteria (actinomycetes, streptomycetes, *Myxococcus xanthus*), nevertheless, phytoplankton representatives, cyanobacteria in this case, also participate in the production. The genera of cyanobacteria *Oscillatoria, Anabaena,* and *Microcystis* have been identified as producers of geosmin (*trans*-1,10-dimethyl-*trans*-9-decalol), 2-methylisoborneol (2-MIB), β-cyclocitral, sesquiterpenes, 3-methyl-1-butanol, and 6-methyl-5-hepten-2-one. Algae are indeed also significant originators of organoleptic defects in waters, but the production of geosmin and 2-MIB has not been confirmed with them yet(!). Mostly, this involves the production of substances having a similar fishy, earthy, mouldy, or cucumber odour. In general, the production of geosmin and 2-MIB is affected by many environmental factors, the composition of the associations which are present, weather conditions, the season of the year, water composition, nutrients, basin morphology; moreover, it is sometimes problematic to predict it [16–18].

Defects that can be perceived by the senses do not have to be connected with risks and health harmfulness of drinking water. Nonetheless, in case of an odour and a strange taste of water, the distrust of consumers in the quality of water unfortunately arises, the consumers are unsatisfied and complain. It is up to every operator of a water-supply network to start solving why that is the case, what causes the defect, and how can it be operatively and possibly also preventively remedied. The question is whether it is a temporary condition that will be solved by the currently taken sample or a thoroughly performed biological audit (monitoring) of the entire distribution system including the source and the technology or it is a long-term issue that will require larger investment costs in the technology and distribution of drinking water [17]. The data provided by the biological monitoring should generally be understood suitably and reacted to properly. If the data from the results of regularly performed microscopic analyses is available and it is apparent from historical records that at a certain number of specimens in 1 mL a manifestation of an organoleptic defect has occurred, then it is possible to control the prediction of what will happen in the technology from the microscope. It is best to determine a warning limit of the number of specimens (organisms) in 1 mL that will be crucial for the technologist at the water treatment facility and indicative in that the frequency of biological monitoring of the source is subsequently increased, allowing for an adequate reaction at the water treatment facility by complementing another separation process. The technology problems can be solved for example the complementing a layer of activated carbon, a washing mode change or shutting the technology down [19].

Phytoplankton present in raw treated water can also be viewed from the perspective of its separation from the treated water by coagulation and filtering. In this connection, categorisation of organisms difficult to remove by treatment is applied in the Czech Republic pursuant to standard TNV 75 5940 *Microscopic Assessment of the Separation Efficiency of Water-Supply Technology* [20]. Microorganisms have been divided into five categories according to their morphology (specimen, colony, coenobium, filaments) and physiology of cells (movement/occurrence preference), designated as A, B, C, D, and E. Categories A to C include mostly planktonic species but also species living among attached organisms that are often broken into short segments transported along the route. Category A includes organisms easy to defragment and smaller than 30 μm, an example are colonial cyanobacteria of the genus *Microcystis* and/or centric diatoms of genera *Stephanodiscus, Cyclotella*, and *Aulacoseira*, green coccal algae *Tetraedron* and *Oocystis*. Category B includes organisms of an elongated (acicular) shape, e.g. diatoms *Synedra acus*, *Fragilaria crotonensis*, *Nitzschia acicularis*, *Nitzschia holsatica*, and *Asterionella formosa*. Category C includes minor filamentous and elongated thin cells, e.g. of cyanobacteria *Limnothrix redekei*, *Synechococcus capitatus* and green algae *Klebsormidium*, *Monoraphidium*, *Koliella*, etc. Category D is the most problematic for water treatment. It is represented by phototrophic and phototactic species of flagellates, e.g. green algae of genera *Chlamydomonas, Chlorogonium, Phacotus, Pteromonas, Pyramimonas, Coccomonas, Carteria, Haematococcus, Trachelomonas, Euglena*, chrysophytes of genus *Chrysococcus*, cryptomonads of genera *Cryptomonas*, *Chroomonas*, and *Rhodomonas*. These flagellates with their active movement complicate the efficiency of coagulation and filtering. They are temporarily caught into flocs, but due to the supply of light through the windows in the water treatment facility halls, they are released from the flocs. Category E is represented by moving pennate benthic diatoms smaller than 30 μm, released from the attached organisms of water-supply facilities and causing organoleptic defects, e.g. genera *Navicula, Nitzschia,* and *Cymbella* [9].

With water-supply streams, in case of water off-take from weir basins, a quantitative evaluation of bioseston of free water is performed (CSN 75 7712 *Water Quality—Biological Analysis—Determination of Bioseston*) and with off-takes from a flowing section of a stream, attached organisms (procedure pursuant to CSN 75 7715 *Water Quality—Biological Analysis—Determination of Attached Organisms*) and benthic organisms (phytobenthos, zoobenthos, procedure pursuant to CSN 75 7714 *Water Quality – Biological Analysis—Determination of Benthos*) are evaluated [12, 21, 22]. Also, when monitoring water-supply streams, it is recommended to record organisms participating possibly in organoleptic defects of water (bacteria, cyanobacteria, algae), forming biofilms (attached organisms) on wetted surfaces of water-supply structures and releasing spores (cysts) that penetrate through the technological line (unit) during water treatment.

During the microscopic assessment of a sample of attached organisms (sediment, sludge, biofilm, water, etc.), a situation may occur where quick identification of bacteria (and/or micromycetes) cell vitality is needed without the necessity of performing a cultivation determination. An example is a sample of attached organisms with metabolic products of iron-oxidising bacteria of genera *Gallionella, Leptothrix,* etc. where precisely the products present (iron coagula) prevent the observation of vital cells (type/morphology of taxon). In this case, it is suitable to use fluorescent labelling of the target microorganism via Live/Dead kits (Molecular Probes™ LIVE/DEAD™, www.thermofisher.com) [23].

7.4 Hydrobiological Monitoring of the Water Treatment Technology

Hydrobiological monitoring of the water treatment technology includes the evaluation of all critical points that are a potential source of organisms, their reproduction, and cause technological problems. The biological audit includes the evaluation of raw water supplies and the individual sections of the technological process sequence in the water treatment facility.

Raw water inlets leading water into the water treatment plant (supply pipeline, collecting wells) are a suitable environment for the creation of biofilms (the absence of light excludes the occurrence of attached algae). Absolutely inappropriate is the building of sedimentation tanks without roofing because they are an ideal environment for the reproduction of phototrophic organisms. Monitoring of the raw water inlets into the water treatment structure is desirable. The solution is regular maintenance and mechanical removal of attached organisms or alluvia. In the littoral of standing and flowing water, phytobenthos and zoobenthos are present, such that together with the drift of inorganic and organic particles, they get up to the water treatment facility through the inlet mains. Very often, the off-take equipment before the entry into the water treatment facility includes screens, suction strainers, siphons, sedimentation tanks, and micro-sieves. The present organisms are cumulated on particles or in sediments. They survive in dead corners where they can even lead to severe technological problems which must be solved downstream. From experience of water-supply practice, it is known that emergency conditions can occur, connected with the occurrence of bryozoans, animal fungi, and/or bivalves, which are known to live among the attached organisms under sand filters and/or even in accumulations of water reservoirs. Filamentous green algae behave similarly, which gradually spread through water using fragments of thalli, propagated colonies of iron-oxidising and manganese-oxidising bacteria (e.g. *Crenothrix polyspora*, *Gallionella*, and *Leptothrix*).

In connection to the spring and autumn circulation, when the water treatment facility has no possibility of choosing another off-take horizon, the issues of occurrence of undesirable phytoplankton is solved, showing by the vegetation colouring of water (in a worse case by the formation of algal bloom). An increase of the number of organisms in the treated water leads to failures during coagulation (different results according to the prevailing kind of phytoplankton), filtering (filter clogging, penetration through the filtering material, shortening of the filter's working cycle), and an increased content of organic substances.

The course of coagulation/flocculation and catching of phytoplankton into flocs is influenced by the properties of the actual organisms. One of the aspects of the efficiency of the separation process is the construction and chemical composition of the outer cell membrane giving the cell the resulting electric charge. The cell casing of algae is represented by a membrane, a cell wall, or solid envelopes. Significant is also the substances secreted by cyanobacteria and algae (algogenic substances) that they have on their surface and that affect the coagulation process to a certain extent. These are siliceous envelopes of diatoms and scales of coloured flagellates (golden algae), mucopolysaccharides in the cell wall of cyanobacteria, cellulose, and hemicellulose with green algae (chlamydomonas, chlorophytes, euglenids). Organisms with active movement, representatives of coloured flagellates of genera *Rhodomonas* (cryptomonads), *Gymnodinium*, *Peridinium* (dinoflagellates) and chlorococcal algae are the most difficult to separate organisms. The vegetation colouring of chlorococcal algae and diatoms prevents the creation of a floc cloud. No less important aspects during the separation of microorganisms are their size, action radius, and shape. Articulate organisms with a large area are separated by coagulation very well, and it is also possible to remove by mechanical means of filtering with fine sand filters. The worsening of the function of the treatment process is significantly contributed to by the diatoms of genera *Asterionella, Aulacoseira, Fragilaria, Tabellaria, Synedra, Navicula, Nitzschia, Cyclotella,* and *Diatoma* (see the difficult to separate organisms, categories A, B, and E). Diatoms have a highly resistant envelope formed by silicon dioxide, and these envelopes clog the pores of the sand bed of the filter even after the necrosis of the protoplast. Complete clogging of the filters takes place, e.g. with numbers of 2×10^4 of cells of genus *Asterionella* in 1 mL of water. The biomass of diatoms is also a live substrate for secondary propagation of microorganisms. They are a good cultivating medium for the growth of bacteria and actinomycetes, which are a source of further unpleasant organoleptic and hygienic defects.

The efficiency of separation of organisms during water treatment can be evaluated using a microscopic image (bioseston, abioseston). Pursuant to TNV 75 5940 *Microscopic Assessment of the Separation Efficiency of the Water-Supply Technology* [20], the separation efficiency is assessed based on the size and nature of created flocs of coagulant on the raster of the counting chamber Cyrus I. and the presence of organisms caught in or outside the flocs (a water sample taken, e.g. after the reaction tank). This standard mentions four size categories of flocs (the size has been derived from the dimensions of the raster of the counting chamber Cyrus I). Flocs of category I. have a diameter of up to 60 μm, flocs of category II. have a size of 60–125 μm, flocs of category III. have a size of 125–250 μm, and flocs of category IV. reach dimensions over 250 μm. An optimum floc is solid, lumpy and compact. With round flocs, their diameter is to be considered, and with elongated flocs, their length. During the inspection of the organism separation efficiency at sand filters, the quantity of the flocs passing through, and the nature of the penetrating larger or allochthonous particles of abioseston is microscopically evaluated in the filtrate samples, and the organisms present are recorded (e.g. larger than 60 μm). With sand filters, an insignificant outflow of flocs of categories I. and II. is admissible. With a through-flow of flocs of categories III. and IV., separation is significantly impaired, and the organisms (biomass) subsequently penetrate further along the route and place a burden on further technologies [9]. The caught reproductive particles of some inconvenient organisms causing organoleptic defects can be found using samples created by leaching a fraction from the sand taken from the monitored filter. It is also possible to monitor the load of the sand bed during the filtration cycle, where the filtered water and the taken fraction of the sand bed (points situated along the length and width of the filter) are evaluated. Such adapted sampling and subsequent biological evaluation help with finding out whether the filtering surface is loaded evenly in the course of the filtering cycle and to what extent the filtering function affects the resulting effect in the quality of the treated water (a correct function of the declining rate system) [24].

This methodology of microscopic evaluation of the particles (size, shape) is also usable for assessing the separation efficiency using membrane technologies. Particle counters that discern the passing particles based on size are more sophisticated but do not determine the particle type (microscopic image).

During a biological audit, it is also appropriate to focus on the sampling of attached organisms (biofilms) that are formed on the walls of technological equipment. According to the level of securing of the technological line (light/dark, heat/cold) and structure maintenance (mechanical removal of attached organisms?), it is realistic to record an increased occurrence of filamentous green algae, cyanobacteria, diatoms, bacteria, and micromycetes. Although indirect light is not sufficient for the development of green algae, it is enough for the reproduction of diatoms, which are undesirable originators of organoleptic defects (volutin). Cases of secondary propagation of organisms on the walls of technological equipment (sand filters) situated in a hall with direct entry of solar radiation through the windows are known from the water-supply practice. It has been recommended by repeated (hydro)biological audits to secure the existing windows, e.g. by fitting suitable foils eliminating photosynthetically active radiation.

7.5 Hydrobiological Monitoring of the Distribution Network

Materials coming into contact with drinking water must show conformity to Decree No. 409/2005 Coll., on hygienic requirements for products coming into direct contact with water and for water treatment. In case of water supply, the nature of the surfaces including their negative effect on drinking water is specified in the Decree in §3 (production requirement) and §8 (surface finish by painting, tinning, and plastic coating), §12 is then especially dedicated to the description of the materials of the pipelines and water reservoirs (cementation). Pursuant to the Decree, the participation of the material in the pollution is ascertained by the leaching test, which determines the presence and concentration of components that are characterised as a natural part or a possible pollutant for the tested product and are risky from the standpoint of health protection of the population. The test sample of the product designed for contact with drinking water is tested gradually in three consecutive 72-h intervals at the testing water temperature of 23 ± 2 °C. From the leaches (extracts) of each time interval separately, chemical analyses are performed, and the values of concentrations of monitored indicators (TOC, $CHSK_{Mn}$, Mn, Cd, Pb, phenols, pH, colour, turbidity, odour, flavour, volatile organic compounds, etc.) are determined. The aforementioned implies that based on the requirements of Decree No. 409/2005 Coll., the selection of the materials used for making products designed for contact with drinking water is controlled based on the technical requirements and criterions affecting the water quality in the sense of substance release and water odour and colour influence. Microbiology is not being solved in this respect. Microorganisms in an aqueous environment have a tendency to attach to the surface of solid substances. The structural materials are suitable base. With regularly used materials, possible support of microbial growth and formation of biofilm are not evaluated, or the biological instability of drinking water in contact with the material is not considered. A hygienic (microbiological) problem may occur even in the case when the material supports microorganism propagation. Those are materials containing organic substances that may be utilised by the microorganisms and may lead to a deterioration of the organoleptic, physical, and microbiological quality of water [25].

From the water-supply practice, cases are known where an element was used as a structural material that conformed to Decree No. 409/2005 Coll., nevertheless, secondary hygienic defects of drinking water took place due to the nature and conditions of the environment. For instance, recycled granulate used for the production of a polyethylene pipe had the requirement for contact with drinking water declared, but the water showed organoleptic and microbiological defects. Or the already mentioned case of water tank reconstruction by applying a sealant that was not suitable for contact with drinking water (rich attached moulds in the gaps), etc. [4, 5].

During biological audits of the distribution networks, it has repeatedly been found that delaying the water in the pipeline negatively affects the microbial quality of water. Corrosion of the wetted surface of the pipeline appears, and particles and microorganisms are found that have got into the drinking water secondarily [3]. In the taken samples of drinking water, live microorganisms, conidia of micromycetes, cysts of protozoa, dormancies, and findings indicating air contamination are very often present. The formation of biofilms in drinking water distribution systems and the secondary propagation of microorganisms in water are considered to be manifestations of insufficient biological stability in the entire water-supply treatment process [5]. Biofilms have a significant effect not only on the hygienic safety of supplied water but also on its organoleptic properties and corrosion aggressiveness. A highly varied extracellular matrix with linked aerobic and anaerobic niches is created there. By limiting the transport of substances into the bottom layers of the biofilm, the speed of its growth decreases. The severity of microbial biofilms in water treatment facilities consists especially in the possible occurrence of pathogenic organisms, e.g. *Legionella, Mycobacterium,* and *Aeromonas*, or propagation of coliform bacteria. The disinfection techniques used are ineffective against bacteria in a biofilm. In the case of the biological instability of drinking water, air contamination is also significant [8].

In the distribution network, the distribution system is inspected, and water samples are taken using fittings located along the route. Microscopic assessment of the quality of the water transported by the pipeline is governed by the guidelines stated in standards TNV 75 5941 *Microscopic Assessment of the Quality of Water Transported by a Pipeline* [26], CSN 75 7712 *Water Quality—Biological Analysis—Determination of Bioseston* [12], and CSN 75 7713 *Water Quality—Biological Analysis—Determination of Abioseston* [13]. When desludging the network, it is appropriate to perform microscopic analysis of the concentrate from a larger volume of water and furthermore a sludge analysis. In the samples taken in the distribution network, inorganic particles, various residues, and envelopes of organisms evaluated as abioseston are mostly found. Organisms that passed through the water-supply treatment without damage may also occur. Live organisms are inadmissible; only dead organisms are acceptable (the limit is governed by the requirements of the valid legislation). Frequent reasons of occurrence of live microorganisms in distribution networks are mostly inappropriate and insufficient separation water treatment technology, imperfect function of chlorination, or unsuitable placement of the chlorinator, absence, or malfunction of additional chlorinating equipment along the route and exhaustion of chlorine. Also occurred live microorganisms indicate leaks in the structure of water-supply equipment, defects in the pipeline, incidental pollution and assemblies and air contamination of water reservoirs [5, 27, 28].

Within the framework of a hydrobiological audit of several important water distribution pipelines, samples collected within a relatively short period of desludging, in three series, three weeks apart were evaluated. During the analyses, the presence of iron-oxidising bacteria was confirmed, and their biological activity (metabolism) was detected based on biological activity tests. When considering a longer residence time and a decreased consumption of drinking water, there is a high probability of fatigue and impairment of the quality of the pipeline material (biocorrosion). Vital iron-oxidising bacteria participate with the metabolic activity in the consumption of the chlorination agent, thus allowing for secondary propagation of microorganisms, which are in some cases in a vital condition. Furthermore, they are aggressive with their metabolic activity against the wetted surface of the pipeline material, which is grey cast iron, cast iron, and steel. The desludging of the affected sections of the pipeline is important for maintaining the quality of water transported by the pipeline to the consumer. During desludging, corrosion products together with bacteria are flushed further away. In case of turbidity and an increased percentage of corrosion products and iron and manganese coagula, more frequent desludging or preventing of hydraulic impulses in the pipeline is usually recommended. Nonetheless, more frequent desludging is not a solution to the situation in question. Due to higher speeds during desludging, the formed build-up and sediments are loosened. Originally protected places are exposed in this manner, and corrosion processes are allowed to continue. The solution is the use of more suitable materials for distribution systems and/or their surface finishes [28].

During the reconstruction of the distribution networks and water-supply structures, it is needed to consider the selection of suitable materials including their surface finish so that they not only comply with the requirements of Decree No. 409/2005 Coll. but also respect the environment, in which they are subsequently applied. It is known that there is a mutual relationship between the potential growth of microorganisms during biofilm formation, the loosening of cells into water or surviving in the biofilm, and the consumption of disinfecting means. Due to some environmental factors, e.g. water temperature, the presence of disinfecting agents, the possible residual quantity of microorganisms, etc. the situation may be even amplified. The rate of growth of microorganisms between individual water-supply structures and hence also the possible negative effect on the quality and harmlessness of drinking water varies and depends on the character of the entire distribution network including the layout, operation, and the residence time of water within the system. From the operational point of view, it is a crucial question what the maximum residence time of drinking water in the pipeline and water reservoirs is still safe and harmless to health. This is related to the ratio of biofilm formation on the materials coming into contact with water and the material preference by the attached microorganisms.

Significant and several times verified in practice is also the monitoring of the effect of the materials on the possible support of microbial growth. The condition of the surface of the material affects the creation of the biofilm, while more suitable conditions are provided by a rougher surface of the material [29]. The hydrodynamic conditions affect the transport of nutrients, their transfer into the biofilm, the reaction rates in the biofilm, and its adhesion to the surface. The result of the interaction of the microorganisms with the construction materials is the creation of a more or less continuous layer of biofilm on the surface of the construction material. The consequence of the activity of microorganisms in the biofilm on the surface of metals is metallic corrosion. Either an initiation of a corrosion attack occurs in conditions where the attack would not occur if the biofilm did not exist or a change of the corrosion mode takes place (e.g. from even corrosion to a localised attack). Suitable places for microorganisms to be attached are pores in the weld metal, with rust-resistant steel; these areas are the most probable points of a corrosion attack. Measurement of the corrosion potential provides information on the condition of the activity of microorganisms in the biofilm. Materials with a developed biofilm have higher values of the corrosion potential than they would have in the absence of the biofilm. The study from 2012 [30] used exposure tests for monitoring the formation of biofilm and the progress of corrosion, during which the differences of microbial pollution on the inner surfaces between individual exposed materials were compared, and the influence of the enrichment of the test water with peptone water and sulphates on the development of the biofilm was evaluated simultaneously. There were used methods described in TNV 75 7121 *Water Quality—Requirements for the Quality of Water Transported by the Pipeline* (*With a Calculation Program*) [31]. (Exposure tests pursuant to TNV 75 7121 consist in preparing loops fitted with testing coupons of materials such as steel, cast iron, concrete, polyethylene, etc. the exposure time is 63 and 126 days.) The biological evaluation was limited to the determination of the microscopic image, the microbial activity based on ATP (adenosine triphosphate), and the microbiological analysis (cultivated microorganisms at 22 and 36 °C, iron-oxidising bacteria, micromycetes). The highest values of the corrosion potential were achieved with oxidised and blasted samples. As the most suitable surface for the attachment of microorganisms emerged to be the surface of rust-resistant steel with a surface finish of oxidation at 600 °C, blasting, and grinding, the least suitable was the surface of brass, titanium, and rust-resistant steel with a stained surface finish. During the visual evaluation, a spot attack was observed on the inner surface of rust-resistant steel with a surface finish of blasting and staining, and in two cases also with a surface finish of oxidation at 600 °C. The observed spot attacks confirm the presumption of the tendency of rust-resistant steel for spot corrosion in an aqueous environment containing chlorides. ATP luminescence method showed changes (drops, rises) in the quantity of microorganisms in free water and the biofilm similarly as cultivating methods [30]. The use of the ATP determination method is not usual in water-supply operations, but an inspiration for the use of this simple and applicable method were the results of the published study from 2003 [29].

The study from 2011 [32] is very interesting, based on the monitoring of the model simulating the conditions in the drinking water distribution networks. During the monitoring, the temperature, pressure, flow speed, and water composition were considered, moreover, iron, calcium, and bicarbonates were complemented. For 2 years, materials of stainless steel were tested, e.g. ferritic, austenitic, and duplex types, which were compared in practice with regularly used materials, e.g. copper, galvanised steel, polyethylene, and cement. After 15 months, a state of equilibrium was achieved with all materials in terms of surface contamination. The results proved that the evaluated materials significantly affect the composition of the microbial biofilms (abundance, species diversity, vitality) and that the maximum degree of pollution and the kinetics of the formation of the biofilm (*biofouling*) considerably depend on the material type. For instance, the heterotrophic bacteria (cultivated) reached a stable level relatively quickly with rust-resistant steel. Lower contamination with heterotrophic bacteria was recorded with copper, apparently due to the bacteriostatic nature. At the end of the study, the lowest degree of contamination was found with copper and galvanised steel. Enterobacteria were found in all biofilms, apart from copper and two types of stainless steel (austenitic and duplex), significant quantities of enterobacteria were detected on cement and polyethylene. Nitrification bacteria, which may support the population of heterotrophic bacteria by creating enough organic carbon for further microbial populations from the inorganic carbon present, were found on all materials. After two years of exposure, it was ascertained by SEM (scanning electron microscopy) that the biofilms cover the entire surface of all materials with the exception of galvanised steel.

The methods that significantly focus on assessing materials from the standpoint of support of microbial growth are specified in standard ISO 16421 *Influence of Materials on Water for Human Consumption—Enhancement of Microbial Growth* (EMG). Attention is considerably focused on non-metallic materials, which are evaluated using three methods in the exposure tests, namely the method utilising the measurement of the ATP concentration, the method measuring the formation of the biofilm, and the method evaluating the microbial activity based on the decrease of concentration of dissolved oxygen. To estimate the microbial growth on exposed materials, a natural mixture of water organisms is used. The methods have their limitations they do not provide information and recommendations concerning the physical and chemical nature of the sample, toxicological action, or resistance to detergents and disinfectants. Also, they do not solve possible pathogenicity of the microorganisms, the numbers of which may increase during the test by releasing nutrients from the tested material.

The aforementioned biological indicators affect the hygienic harmlessness of the supplied drinking water, its organoleptic properties, and corrosion aggressiveness. The water transported by the pipeline to the consumer may no longer have the quality as produced in the water treatment facility. One of the criteria of the WSP which could be applied is the inspection of the produced water throughout the production chain with the goal of finding the threat or risk and prevent it if possible. Therefore, it is necessary to focus on the monitoring of the possible growth of microorganisms between individual water-supply structures of the spinal mains and also on the possible negative effect on the quality and safety of drinking water and its safety from the WSP standpoint. Also, it is necessary to focus on the aggressive processes of drinking water, chemical and biological instability, condition of the material of the distribution network, and secondary contamination in structures. Based on suitably adapted exposure tests, it is also possible to recommend a suitable (better) material including its surface finish.

7.6 Hydrobiological Monitoring of Water Reservoirs

Water reservoirs are strategically important structures that may affect the quality of drinking water by not meeting the construction-structural requirements for operation and attendance, which often allows for the supply of allochthonous particles and organisms participating in the biological instability of water. Biological audits of the distribution network have proven in the samples of the taken water and the swab samples from the coupon the presence of particles of abioseston and bioseston that are of allochthonous origin. These particles are indicators of secondary air contamination and the only possible path where they have got into the distributed water are points along the route. These points are mainly unsecured water reservoirs. Water reservoirs were monitored for several years in a very detailed manner from the perspective of secondary contamination, structural arrangement, size, type of used materials, operation, handling, maintenance, etc. So that the result of the biological audit has a practical impact on the solution of the ascertained defects of the biological origin, the direct participation of an erudite (hydro)biologist at the site of the sampling implemented is desirable. The (hydro)biologist, as a completely independent entity, often assesses the possible risks, which may result from a structure of unsuitable construction and design, more critically and appropriately. It is desirable to evaluate the structure concerning the construction-technical aspect (outside/inside), to detect defects potentially participating in the biological problem and microbial contamination, and to sample the water and take swab sample from the wetted walls of the water reservoir tanks [2, 3, 5, 6, 27].

Biological audits of water reservoirs, where water samples were taken directly from the tanks during operation (near-surface layer), and swab samples were taken from tank walls, confirmed the negative effect of air contamination on the repeated supply of micromycetes, allochthonous particles, and organisms. The existing legislation of the Czech Republic does not consider the hygienically significant indicator of the swab sample type (biofilm, attached organisms, etc.), nevertheless, this indicator has an indispensable significance for the inspection of the biological stability of drinking water. In the swab samples from the wetted surfaces of unsecured water reservoir structures, conidia and hyphae of micromycetes, starch grains, pollen grains, butterfly scales, bird feathers, plant and animal residues and envelopes, plant trichomes, insect chitin residues, and detritus are mostly present. Allochthonously present organisms and particles subsequently significantly participate in the degradation of the quality of the accumulated drinking water and are the substrate for further trophically dependent microorganisms. Biofilm formation is supported on the wetted walls of tanks (changes in water level, substrate, and nutrients supply) via air contamination. The wetted walls are often highly underestimated sources of an organic substrate [6].

By systematic biologic monitoring of various types of water reservoirs (size, type, shape, arrangement, location in the consumption area) during operation, the presumption of the negative impact of unsecured ventilation vents on the formation of biofilms in the near-level layer was confirmed [8]. The presence of particles of abioseston, conidia, and hyphae of micromycetes (often also pathogenic micromycetes) indicating air contamination was confirmed. For the confirmation of the presumption of the negative effect of secondary contamination by air on the quality of accumulated water, a water reservoir was selected that was monitored during operation in the course of one year. Already after a quarter of a year after regular cleaning of the accumulation tanks, increased activity of microorganisms was recorded in the near-level layer of water and furthermore on the wetted surface of the tank walls at the point of the culmination of the water level. By evaluating the swab (scraping) samples, it was ascertained that there is a direct proportion between the degree of contamination and the character of securing/non-securing and maintenance of the internal space of the accumulation tanks [6, 33]. If a pipe in the structure leads through the ceiling directly outside into vegetation and is not covered or secured in any way, then the degree of water pollution in the near-level layer and the thickness of the biofilm at the point of level fluctuation is significantly larger and biologically more harmful than in case of protected openings in the walls. The movement of the water level is related to the supply and discharge of air that is allowed by the ventilation system design (natural or forced). It is needed to secure the quality of the supplied (extracted) air into (from) the water reservoir and secure its inspection, e.g. by fitting filtering fabrics into the ventilation openings and their timely replacement [34].

From the monitoring of the water reservoirs during their operation and their regular cleaning, the severity of the partial causes of the changes in the quality of the accumulated drinking water in the water reservoirs, caused by secondary air contamination, became apparent [3, 6]. Pursuant to the WSP and the risk analysis, it is needed to focus on efficient manners of minimising the occurrence of secondary biological revival and the subsequent formation of biofilms in water reservoirs (accumulations with drinking water). By testing the catchment of dust particles, abioseston (pollen, starch, plant residues, butterfly scales, etc.) and micromycetes in individual filtration layers, the filtration unit consisting of six filtration layers was designed, and adapted so that it can be easily applied into any diameter in the wall (air vent) [2]. The requirements for the filtration unit (air filter) are specified in CSN 75 5355 *Water Reservoirs* and in the Technical Recommendation I-D-48 *Design Arrangement, Operation, and Maintenance of Water Reservoirs* [35].

The recommendations for water reservoir operators, which are mostly based on biological audits, involve the necessity of focusing on construction modifications and the design of the premises so that the secondary occurrence of organisms is minimised or completely eliminated. Besides air contamination, the presence of window openings and access of solar radiation are also crucial, which is negative concerning the secondary propagation of microorganisms (cyanobacteria, algae, micromycetes) on the damp walls of the structures. In the Technical Recommendation I-D-48 *Design Arrangement, Operation, and Maintenance of Water Reservoirs* [36], all specifications and recommendations are stated in detail, which should form a part of the technical and biological audits of the structures (informative appendix A, appendix B of the technical recommendation). During regular inspections of the structure condition by the operator or random inspections by the reviewer (auditor), it is important to record the condition and nature of the construction parts of the structure.

The lower structure, the bearing structure, roofing, entrances, viewing points, windows, vents, stairs, ladders, doors, gates, downpipes, inlet and outlet design in the structure, use of suitable construction materials, satisfactory surface finish of floors, and the separation of the handling and accumulation chamber are evaluated. Part of the evaluation is the assessment of the structure's surroundings, securing (fencing), labelling (water-supply structure, 'No unauthorized access' signs), and cleaning (mowing). Concerning the biological aspect, the presence of sediments and incrustation from/on the pipeline, the quality of the supplied water (analyses, swab samples, biofilms, and level of microbial contamination with paddle testers with a contact medium) are evaluated. Thanks to checking biological analyses of water and swab samples, it is possible to determine the schedule and procedure for efficient cleaning and disinfection of water reservoirs or to optimise the existing schedule [6].

7.7 Model Examples of Biological Audits

At the site, in connection to the issues of applying (hydro)biological audits in practice, it is certainly appropriate to state model examples of performed biological audits in the water supply. For illustration, audits of distribution networks were selected, where the supply was groundwater or surface water of various trophic levels, respecting the zoning of the tank and the possibility of the selection of a suitable profile, various arrangements of the distribution network and technological separation. The findings from the biological audits and the recommendations stated below have been accepted and gradually applied by the operators of water-supply networks. Upon the subsequently performed checking biological audits, after applied recommendations, a very good condition of the operated distribution network was stated, including the structures situated on it. The results of the biological audits make the critical and thus risk points (places) in the distribution network within the entire supply system visible, and the fact of the continuous system is confirmed (in the sense of contamination spreading along the route).

A model example of a biological audit of the water-supply network 1: The source of groundwater goes into the water treatment facility, from where the water after hygienic securing by chlorination is drawn through the pumping station into the water reservoir.

For the biological audit, the methods of microscopic and microbiological analysis were used, and water and swab samples (from the wetted walls of the tanks) were taken. At the water treatment facility, the quality of raw water (inlet or level, according to the situation in the treatment facility), the water from filters (sampling below the level or a swab sample from the wetted surface of the walls), and the water below the filters (sampling below the level) was monitored. Based on the biological analyses of water, a very good separation efficiency of the water-supply technology (TNV 75 5940 [20]) and quality of treated water was found. In the swab samples (scrapings) from the filters, indicators of air contamination (conidia, bird feathers, mycelia of micromycetes) and attached organisms (diatoms) were found. The organisms, whose occurrence was expected in the water samples (underground source), were iron-oxidising (manganese-oxidising) bacteria of genera *Gallionella* and *Leptothrix*. Furthermore, indicators of secondary contamination (micromycetes, diatoms) were found, which were not present in the underground source and had entered the water due to the manner of operating the technology at the water treatment facility (unsecured vents, unsecured windows, etc.). In the water samples taken further along the route in the distribution network, in this case at a pumping station, an increased occurrence of heterotrophic flagellates, ciliates, and diatoms was found. According to the legislative requirements of the Czech Republic for the quality of drinking water, the presence of live microorganisms (microscopic image) is inadmissible. The presence of live microorganisms indicates, among other things (technological problems, accidents, biological instability of drinking water), insufficient hygienic securing in the entire supply system (chlorination, insufficient dose, biological instability). In the water sample, the presence of abioseston particles (starch, pollen grains) and blooming mycelia of micromycetes (unquantified bioseston) were detected, which indicates air contamination. In the water reservoir situated further downstream, water and swab samples from accumulation tanks were taken. In the water samples, heterotrophic protozoa (ciliates and flagellates) were recorded, showing the insufficient hygienic securing and presence of organically degradable substrate and iron-oxidising bacteria (metabolically active and participating in the consumption of chlorine by oxidation processes). The abioseston found (pollen grains, starch, and plant residues) indicated air contamination. In the swab (scraping) samples from the walls of the water reservoir accumulations, the presence of a relatively high quantity of filamentous green algae of genus *Klebsormidium* was recorded. This indicates long-term contamination of the structure (possibly due to air contamination, attachment to the wall and propagation, or light entering). In the swab samples, hyphae of micromycetes were also presented. The results of the biological analyses were linked to the results of the in situ investigation during the individual assessment of the structures. The crucial recommendations resulting from the biological audit were directed to the necessity of solving a more suitable place, manner, and frequency of chlorination and securing the structure against air contamination and light supply.

A model example of a biological audit of the water-supply network 2: The source of surface water from a valley reservoir into the water treatment facility with sedimentation tanks and sand filters. The treated water after hygienic securing by chlorination is drawn through the pumping stations into the water reservoirs.

For the biological audit, the methods of microscopic and microbiological analysis were used, and water and swab samples (from the wetted walls of the tanks) were taken. At the valley reservoir, water samples were evaluated, taken from four available off-take horizons (vertical stratification, zoning) with the objective of finding a suitable off-take horizon for water treatment. In the samples taken, representatives of cyanobacteria (*Plantkothrix, Woronichinia*), dinoflagellates (*Peridinium, Katodinium*), chrysophytes (*Synura, Dinobryon, Chrysococcus*), cryptomonads, diatoms (*Cyclotella, Nitzschia, Navicula, Diatoma, Asterionella*), chlorococcal algae (*Chlamydomonas, Scenedesmus, Eudorina*), furthermore ciliates and small rotifers (*Lecane, Rotaria, Cephalodella*) were found. If the water treatment facility treats raw water from a horizon where higher biological contamination (count of cells) is found, and abundant occurrence of planktonic cyanobacteria is possibly recorded, the recommendation is to move to an off-take from a horizon with lower biological contamination (count of cells). The aforementioned organisms belong to a group of those difficult to remove by water-supply treatment pursuant to TNV 75 5940 [20]. Upon their occurrence, possible separation problems should be expected at the water treatment facility (it was subsequently confirmed by water analyses at the water treatment facility). At the water treatment facility, presence of organisms belonging to a group difficult to remove by water-supply treatment was recorded in raw water, especially phototactic coloured flagellates (*Dinobryon, Chrysococcus, Synura, Cryptomonas, Peridinium, Katodinium*), acicular types of pennate diatoms (e.g. *Nitzschia, Navicula*), small species of diatoms (e.g. *Cyclotella*), and/or easily disintegrating fragments of diatoms (e.g. *Asterionella, Diatoma, Aulacoseira, Melosira*). It is known about these organisms that they are caught into coagulant flocs with difficulty, or that they are caught only partially. They are loosened very often and penetrate through the sand filters into the treated water. The biomass present shortens the working phase of sand filters (overloading of the sand bed, clogging, and frequent washing). In the samples taken from the sedimentation tank, the character of the coagulant flocs was evaluated (size, shape, organism catchment), the flocs formed were large (several mm in diameter). However, unfortunately, a low percentage of the organisms present was caught into the created flocs and their agglomerates. The acicular and disintegrating types of diatoms were mostly present in the inter-floc spaces. Furthermore, the filter unit was evaluated, where medium and small coagulant flocs penetrated to the filters. As for the abioseston, plant and animal residues were recorded and furthermore particles of an allochthonous origin indicating air contamination (conidia of *Alternaria* sp.). Some organisms containing polyphosphate granules of volutin may also cause organoleptic defects upon their propagation (diatoms, chrysophytes, green algae), the recommendation in such case is to focus on frequent mechanical maintenance of the tank walls (filtering, sedimentation). Another part of the audit was a biological inspection of samples taken from the pumping stations and water reservoirs. At the inlet into the pumping station, representatives of acicular types of diatoms were presented in the free water sample, which penetrated already during the water treatment (penetration through filters), the present conidia found indicate air contamination. In the following two evaluated water reservoirs along the route, live colourless flagellates were detected in water samples, which indicate biological instability of drinking water. An alarming finding was a larger catchment of conidia of micromycetes in the swab samples, which indicate air contamination and biological instability of drinking water. A water reservoir where cleaning was underway was included in the biological audit. The accumulation chambers of the water reservoir were assessed also concerning the technical and construction aspect (materials, the condition of the walls, ceiling, and corrosion). From the water reservoir, swab samples from the bottom of the structure and its columns were taken. Cyanobacteria and their reproductive particles (hormogonia) and abioseston particles (starch grains, plant residues, bird feathers) were present in these samples in an increased quantity. On the concrete structures of the water reservoir column, iron, and manganese coagula were present in swab samples of black-brown colour and gooey structure which were permeated by microorganisms. Ciliates and nematodes were present, which indicates a decreased biological stability of drinking water, consumption of chlorine during accumulation, insufficient securing of the structure during operation. Alarming were the higher numbers of coliform bacteria detected (paddle tester imprint) in the swab samples (scrapings) from the ceiling and the column of the accumulation tank. Because the structure was secured against ingress of natural light, the secondary supply of organisms (cyanobacteria) can be assumed to be by air (confirmed by the detected particles of abioseston). The abioseston and bioseston found in the water sample from the tap at the inlet into the water reservoir which has been cleaned proved the supply of organisms and particles along the route. Plant residues and insect chitin residues were found, which indicates not only secondary air contamination but also penetration of particles and organisms along the route and/or whirling of the sediments settled in the distribution network or present in biofilms. Further examination of the water reservoir ascertained that the entrance into one of the accumulation tanks was not sealed enough to prevent secondary contamination. In the vicinity of the manhole, visible traces of snails and actual specimens were present (the explanation of the positive finding of coliform bacteria). The crucial recommendations resulting from the biological audit were directed to the necessity of solving a more suitable place, manner, and frequency of chlorination and securing the structure against air contamination and light ingress. The presence of windows and access to natural light is absolutely undesirable at the accumulation point. More efficient securing of the seals at the doors (entrances) into individual accumulation chambers and equipping the vents with efficient filtering (air filters) can effectively eliminate air contamination. Also maintenance and optimisation of cleaning, i.e. timely desludging, mechanical cleaning, disinfection, including regular inspection of the condition of the structures and their securing can help eliminate biological contamination.

A model example of a biological audit of the water-supply network 3: The surface water source goes from a valley reservoir with the possibility of selecting three off-take horizons into the water treatment facility. The technological line is a system of sedimentation tanks and sand filters. The treated water after hygienic securing by chlorination is then drawn through the pumping stations into water reservoirs.

For the biological audit, the methods of microscopic and microbiological analysis were used, and water and swab (scraping) samples (from the wetted walls of the tanks) were taken. Low biological rejuvenation of the order of 10^2 to 10^3 specimens per 1 mL (an expectation of minimum technological problems and defects during water treatment) was recorded during a biological evaluation at the water reservoir over the course of a year. With regard to the technological line placed downstream (coagulation, filtration), cyanobacteria and other difficult to remove organisms (TNV 75 5940 [20]) were not detected. Raw water was supplied via cascades into the water treatment structure. At the point of entry of the raw water into the water treatment structure, biological inspection of a sample of the supplied raw water and a swab (scraping) sample from the cascade walls was performed. In the swab sample, microorganisms were found in an increased quantity, namely attached organisms (green algae, cyanobacteria, rhodophytes). The recommendation that resulted from this finding was the necessity of mechanical cleaning of the cascade walls and securing the windows of the structure with suitable foils eliminating photosynthetically active radiation. During an inspection of the technological line sections downstream, this recommendation was significantly confirmed by biological findings in water samples and in swab samples from the tank walls. In the coagulation tank, organisms mostly difficult to catch into flocs were recorded. In the swab sample from the coagulation tank wall, filamentous organisms were present (green algae, cyanobacteria, rhodophytes) that were torn off downstream and passed to other technological stages. The recommendations resulting from the biological findings at this separation stage were aimed at minimising the algae growth (cyanobacteria) on the tank walls not only by frequent mechanical maintenance of the tank walls (spillway edges) but also by securing the windows of the structure, e.g. by fitting foils eliminating the photosynthetically active radiation. In the water samples from the filtration downstream, an increased occurrence of heterotrophic elements and multicellular organisms was found (rotifers, crustaceans, nematodes), indicating a longer period of water delay at the filters and a longer working cycle of filtration. In order to find the efficiency of filtration, biological inspection of the water below the filters was performed. The penetration of pennate diatoms and rotifers through the sand filter into the treated water was recorded. In the water treatment structure, inspection of the treated and hygienically secured water by chlorination was carried out. From the abioseston, particles indicating air contamination were present, especially plant residues, conidia, and pollen grains (fragments of hyphae of micromycetes). Furthermore, bristles of oligochaetes were found, which also indicates a possible presence of actual specimens. From the bioseston, live protozoa and rotifers were recorded, which indicates insufficient chlorination and low biological stability of water. In the distribution network, water reservoirs were monitored within the audit, where free water and swab samples from the wetted walls of chambers were taken. Diatoms, chrysophytes, and chlorococcal algae were found to be present, which may have been caused by the penetration into the technological line. In the swab samples, abioseston indicating air contamination was abundantly represented, as well as a significant quantity of starch and pollen grains together with attached bacteria and hyphae of micromycetes, plant residues, and bird feathers. Penetration of filtering combs of crustaceans was also recorded, but it is not necessarily caused by air contamination. The actual specimens or fragments of their bodies may penetrate during treatment, or secondary propagation along the route may also take place, etc. Among the attached organisms, filamentous cyanobacteria of the genus *Phormidium* were present, whose presence was later explained by a breakdown at the distribution mains (pipeline repair and suction of surface water into the water mains during the repair). At the route, the shaft was monitored, where the occurrence of rotifers and diatoms was recorded in the water, indicating insufficient biological stability of drinking water and their passing through the water-supply line. The ascertained number of organisms did not comply with the limit stated in Decree No. 252/2004 Coll. regarding the requirements for the quality of drinking water. The crucial recommendations resulting from the biological audit were aimed at a more suitable solution and optimisation of the operation of the existing technological line at the water treatment facility, for example, complementing the separation stage, shortening the filtration cycle time and more frequent washing. Also securing the structures against air contamination and light ingress can help in better drinking water quality of accumulated water. More efficient securing of the seals at the doors into individual accumulation chambers and fitting efficient filtering into the vents (air filters) can help with the air contamination. Also is desirable the maintenance and optimisation of cleaning of water plans including regular inspection of the structures condition and how secure it is.

7.8 Interpretation of the Results of Biological Analysis

Directive 98/83/EC, which forms the basis of the legislation of the Czech Republic in the area of drinking water hygiene, sets the minimum requirements. However, it is needed to pay special attention to them (in 2015, the EU Directive 2015/1784 changes Appendices II and III). Due to the importance of biological indicators for drinking water with regard to the assessment of the risks and the monitoring programme, the interpretation of the results of microbiological and hydrobiological (microscopic) analyses is important. It is crucial to know what is indicated by each of the water quality indicators and also to react in a correct and timely way to the arisen situation upon finding the presence of an indicator organism or pathogen in the drinking water-supply system [4, 37].

It is always important to consider the matrix and origin of a sample during biological evaluation. A water sample shows the current condition, while a swab sample informs about possible long-term problems that may occur in the water-supply system. Swab (scraping) samples from the surface of technological equipment throughout the system have an irreplaceable role when interpreting the results found by analyses.

Reliable indicators of the microbiological safety of drinking water are coliform bacteria and *E. coli*, intestinal enterococci, and clostridia, while *E. coli* can be designated as indicator organisms (in the sense of faecal contamination and danger). The presence of coliform bacteria does not have to mean faecal contamination. In many cases, these findings are caused by problems in the distribution system. Their detection may often mean that a failure in the drinking water-supply system's integrity took place. They indicate the presence of biofilms (hydraulic impulses), the influence of secondary contamination, or technological defects. They are also used as an indicator for monitoring the system operation (HACCP, WSP). Such situations in the water-supply network must be solved immediately, e.g. the presence of nutrients that will support the growth of microorganisms, or when an insufficient concentration of the disinfectant or a long residence period of the water in the distribution network is found.

Based on the significance of the indication of faecal contamination of drinking water, and frequent presence of *E. coli* and coliform bacteria during accidents and technological interventions in the distribution network, we have unequivocally chosen the bacteria *E. coli* as the indicator organism to be monitored. The thermotolerant coliform bacteria *E. coli* clearly indicate severe faecal pollution in drinking water. Upon their detection, it is necessary to find the contamination source immediately. It is known that *E. coli* also indicates a possible occurrence of further intestinal pathogens, such as *Salmonella*, *Shigella*, and/or *Campylobacter*.

For the detection of the indicator organisms, specifically for determining *E. coli* and coliform bacteria in drinking water, multiple methods are used. The traditional methods of detection include membrane filtration with subsequent cultivation on various specific media using various cultivation conditions, followed by confirmation tests. These methods are limited by the time of incubation, lack of specificity, and moreover, slowly growing or unculturable microorganisms are difficult to detect. The time for the results to be obtained using these methods is within 1–2 days. The cultivation determination of *E. coli* and coliform bacteria can be alternatively replaced with IDEEX determination, the so-called method of a defined substrate, where it is possible to obtain the result already after 18 h [38]. From the standpoint of quick identification of microbial contamination and hygienic harmfulness of drinking water, the aforementioned time of cultivation of the indicator organism is crucial. Instead of classic cultivation methods, alternative methods can also be applied for the detection of indicator organisms. Unfortunately, they do not always, despite good advertising by the manufacturer and the distributor, truly provide data that is related to the real conditions at the distribution network [39]. Besides classic cultivation methods, microbial contamination of drinking water can also be monitored in the current microbiological practice using alternative methods. An example of some alternative methods used in practice at various levels are methods focused on biochemical activity (the ATP method—adenosine triphosphate, H_2S strip test), microscopic methods utilising fluorescent marking (DAPI, FISH), and/or fluorochromes attacking the structures of vital and dead cells of bacteria (LIVE/DEAD™), or molecular biology methods based on PCR [40].

Intestinal enterococci are still significant indicators of drinking water quality from the viewpoint of faecal contamination indication. Their significance is due to their sensitivity to environmental conditions, i.e. they propagate in water very rarely, which means that their finding indicates fresh contamination. Additionally, they are quite resistant to disinfectants (chlorine), i.e. they indicate insufficiently performed disinfection, e.g. with a lower dose of the disinfectant (coliform bacteria and *E. coli* do not have to be detected by an analysis in the simultaneous presence of enterococci).

In water supply and subsequently in water hygiene, often discussed issues are the occurrence and detection of epidemiologically significant parasitic protozoa of genera *Cryptosporium* (oocysts with a size of 4–6 μm) and *Giardia* (cysts with a size of 8–15 μm). In water microbiology, the water quality evaluation system for detection of parasites is limited to the indicator system using alternative organisms indicating faecal pollution. The bacteria are mostly very sensitive to the disinfecting procedures used in water supply. Unfortunately, in the case of parasites and viruses, we have to take into account some certain resistance. The legislative regulations address the risks related to the occurrence of these organisms by monitoring the type indicator organism (it is not perfect!), which is the bacterium *C. perfringens* [41, 42].

When interpreting the occurrence of *C. perfringens* species in drinking water, it is necessary to consider whether the organism occurs as vegetative cells or as spores. This indicator is considered to be a very useful indicator of faecal pollution. It is present in the intestinal tract both in the form of vegetative cells and spores. Vegetative cells do not survive for a long time in the water, whereas spores do (for several months). They survive much longer than other indicator bacteria of faecal pollution. The spores of clostridia are resistant to hygienic securing of water (both physical and chemical means and procedures). *C. perfringens* is thus a useful indicator of the effectiveness of hygienic securing of drinking water. Clostridia indicate the efficiency of treatment to remove viruses, and they show the potential presence of pathogenic microorganisms (parasitic protozoa) in addition to being used for the evaluation of water treatment efficiency/effectiveness.

The cultivated microorganisms with growth specification at 22 and 36 °C are ubiquitous organotrophic microorganisms with low hygienic importance. The higher occurrence of these organotrophic organisms in water participates in the organoleptic defects of water. They indicate the formation of biofilms and the presence of organically degradable substrate. The environmental conditions can be responsible for their presence in the distribution system quite significantly, e.g. hydraulic conditions, temperature, biofilms, pipeline material, the biological stability of water, or disinfectant used. This means that they are important for the operators of the technological line and distribution networks.

It is always desirable to find the source of ingress of particles of abioseston and bioseston (by water, by air) and go through all the related sections in the network. Then it is possible to take a right corrective and preventive measures, to consider the possible reconstruction of the structure, fitting, hydrant, network, etc. and/or to choose adequate barriers of their secondary entry (elimination of air fallout by insulating the windows and vents). Abioseston, i.e. inanimate particles present in water, if determined correctly, indicate the nature and manner of operation of the water-supply system. It is possible to utilise this indicator during regular monitoring or as required. In many cases, they indicate secondary contamination, breakdowns or accidents, or technological failures. Corrosion products found during microscopic analysis of water (taken at the distribution network) indicate an ongoing corrosion process, breaches of the surface of the pipeline, and whirling of the sediments. They cause water turbidity, the possibility of organoleptic defects of the water, and in the case of their increased occurrence, it is also possible to consider the presence of metabolically active iron-oxidising bacteria. Vital and metabolically active iron-oxidising bacteria participate in the decrease of the concentration of the disinfectant (chlorination) and thus allow for the manifestations of the biological instability of drinking water, which are biofilms and microbiological activity (cultivated microorganisms). Pollen grains, starch grains, and plant residues get into the network by air contamination through any accessible places in the structures, where they are the substrate for organotrophic bacteria and cause decreased biological stability of drinking water. Plant residues (fibres, trichomes) occur significantly in the summer season when there is open access to the locally affected structure (water reservoir) at mowing periods of grassy areas. In general, animal residues (envelopes of rotifers, filtering combs of crustaceans, bristles of oligochaetes, insect chitin residues) indicate the penetration through the water-supply technological line. Depending on the nature of the particle and possibly the vitality of the organism, it is also possible to consider a technological defect, accident, or secondary contamination (by air, by water).

Bioseston, i.e. microorganisms provide crucial information on water quality in the case of qualitative and quantitative determination. The type of bioseston found indicates the nature of the raw water, the secondary contamination, and the ingress of allochthonous particles, and it determines the suitability of the technological line arrangement. Live organisms are not desirable in drinking water of course (the hygienic aspect). Conidia and hyphae of micromycetes are an indication of air contamination on the distribution system route. Sometimes they also indicate the long-term presence in the biofilm (insufficient hygienic securing), and with their occurrence, they warn about the possible hygienic harmfulness of drinking water, possible decreases in the biological stability of drinking water, and availability of the substrate for bacteria and other organisms. Dead diatoms (pennate, centric), tiny chlorococcal algae, etc. phototrophic organisms in general that are difficult to remove by water-supply treatment, indicate penetration through the water-supply technological line. Live organisms indicate an accident, penetration of surface water, or insufficient hygienic securing depending on the distance from the water treatment facility. Heterotrophic protozoa (flagellates, ciliates) indicate the ingress of the biologically degradable substrate and increase the expectation of a possible occurrence of organotrophic bacteria. They increase the quantity of disinfectant required and participate in the possible hygienic degradation of the drinking water quality. The presence of rotifers, crustaceans, nematodes, and other representatives of Metazoa is already a warning signal that should not be underestimated and requires a more rigorous solution in the distribution network [38].

7.9 Conclusions and Recommendations

Microscopic analysis can provide information, which is inexpensively and rapidly obtained, on many drinking water quality failures, to supplement other commonly used analytical checks. This kind of information is difficult to obtain by chemical or microbiological analyses (cultivation techniques). For this reason, it is suitable for inclusion in the HACCP system. However, an essential precondition is to have people well trained in the determination of microorganisms.

Biological analysis is very significant in solving various problems during the treatment of raw water from reservoirs and streams. Biological analysis offers in very short time indirect verification of potential hygienic defects. Hydrobiologists should focus on the hydrobiological monitoring of water-supply structures and raw water influent, checking the effectiveness of the first separation step processes (capture of organisms in flocs, choice of optimal coagulant dose, alternation of coagulant or flocculant), checking the effectiveness of the second separation step processes (filtration efficiency, growth of microorganisms in sand bed, intensity, and duration of washing), the hydrobiological analysis of biofilms and swab samples in all of the water treatment stages and in the distribution pipelines, the evaluation of water quality changes due to routine sludge removal, and suggestions or recommendations regarding the prevention of pollution, and the checking of the effectiveness of equipment and processes. Besides the quality control of treated water, monthly evaluations are necessary of the water quality in storage tanks at water treatment plants, in reservoirs, in the distribution network, and in its endpoints.

The water quality of the raw, untreated water source and of both point and diffuse pollution sources in the catchment area can be easily assessed with the aid of benthic, periphytic, or planktonic bioindicators. For example, the presence of cyanobacteria, algae or other surface water organisms in groundwater is evidence of the direct influence of surface water on ground water. Mass development of phytoplankton in eutrophic reservoirs causes problems with its removal during treatment processes, resulting in phytoplankton passing through into the treated water. Some of them can adversely affect drinking water quality due to taste and odour problems. Microscopic examination can give information on the phytoplankton species present and on their quantity. The penetration of microorganisms and particles into drinking water is evidence of ineffective water treatment. Autotrophic microorganisms penetrating into treated water may be used as indicators of ineffective disinfection (autofluorescence of chlorophyll). Likewise the presence of motile colourless flagellates, ciliates, or other organisms indicates disinfection ineffectiveness. Abioseston of various origins, microorganisms and their resting stages, micromycetes and macroinvertebrates occurring in the swab samples in water storage tanks, can be taken as an indicator of airborne contamination. The presence of rusty precipitation and iron bacteria (*Gallionella*, *Leptothrix*), in some cases, can indicate corrosion processes. The presence of living cyanobacteria, algae, and fragments of planktonic species skeletons and loricae in the drinking water network can indicate contamination by surface water.

References

1. Davison A, Howard G, Stevens M, Allan P, Fewtrell L, Deere D, Bartram J (2006) Plány pro zajištění bezpečnosti vody. Řízení kvality pitné vody od povodí k odběrateli. Česky vyšlo zásluhou VAS, a.s. na CD jako příloha ke sborníku konference Pitná voda Tábor
2. Říhová Ambrožová J, Hubáčková J, Čiháková I (2009) Drinking water quality in Czech Republic. Czech J Food Sci 27(2):80–87
3. Říhová Ambrožová J (2009) Zajištění zdravotně nezávadné a bezpečné pitné vody v distribuční síti. Chem Listy 103(12):1041–1046
4. Kožíšek F, Kos J, Pumann P (2007) Hygienické minimum pro pracovníky ve vodárenství, 1st edn. SZÚ, Praha
5. Hubáčková J, Slavíčková K, Říhová Ambrožová J (2006) Změny jakosti pitné vody při dopravě, Výzkum pro praxi, SEŠIT 53. Výzkumný ústav vodohospodářský T. G. Masaryka, Praha, pp 1–96
6. Říhová Ambrožová J, Říha J, Hubáčková J, Čiháková I (2010) Risk analysis for accumulation of drinking water. Czech J Food Sci 28(6):557–563

7. Kožíšek F (2016) Mediálně známé havárie a epidemie na vodovodech v ČR v období 2014–2015: Jaké jsou jejich příčiny a co nám mají říci. In: Říhová Ambrožová J, Pecinová A (ed). Sborník konference, Vodárenská biologie 2016. Praha, pp 34–41
8. Říhová Ambrožová J, Říha J, Hubáčková J, Čiháková I (2008) Minimalizace vzdušného spadu v objektech s akumulací pitné vody. In: Sborník přednášek XII. mezinárodní vodohospodářské konference, VODA ZLÍN 2008, Zlín, pp 167–172
9. Ambrožová J, Pumann P, Sládečková A, Kolář K (2003) Microscopic indicators of drinking water quality. In: Korhonen LK, Miettinen IT, Slobodnik J, van den Hoven T (eds) Proceedings of the 1st Weknow annual drinking water conference in Europe, first we know annual conference on drinking water in Europe, review of problems, current knowledge and knowledge needs, Kuopio, Finland, pp 42–45
10. Ambrožová J (2001) Trend vývoje biologických rozborů. SOVAK, 10(6):25/173–28/176
11. Ambrožová J (2001) Biologie—prvotní indikace stavu vodárenských procesů. In: Sborník semináře, Aktuální otázky vodárenské biologie. Praha, pp 29–31
12. ČSN 75 7712 (2013) Kvalita vod—Biologický rozbor—Stanovení biosestonu. ÚNMZ, Praha, 20 p
13. ČSN 75 7713 (2015) Kvalita vod—Biologický rozbor—Stanovení abiosestonu. ÚNMZ, Praha, 36 p
14. Reynolds C, Huszar V, Kruk K, Naselli-Flores L, Melo S (2002) Review. Towards a functional classification of the freshwater phytoplankton. J Plankton Res 24(5):417–428
15. Padisák J, Crossetti L, Naselli-Flores (2009) Use and misuse in the application of the phytoplankton functional classification: a critical review with updates. Hydrobiologia 621:1–19
16. Wnorowski AU (1992) Tastes and odours in the aquatic environment. A review. Water SA 18:203–214
17. Říhová Ambrožová J (2016) Organoleptické závady pitné vyvolané přítomností geosminu a 2-MIB v surových vodách. SOVAK 25(10):6/258–9/261
18. Munzar T, Říhová Ambrožová J, Kosina J (2017) Geosmin a 2-MIB: Častí původci organoleptických závad pitné vody. In: Říhová Ambrožová J, Pecinová A, (ed) Sborník konference, Vodárenská biologie 2017, Praha. Praha, pp 34–39
19. Říhová Ambrožová J, Munzar T, Vejmelková D, Kosina J, Hrušková P, Brabenec T, Říha J (2018) Případová studie za účelem monitoringu původců organoleptických závad. Vodní hospodářství 68(5):16–19
20. TNV 75 5940 (1999) Mikroskopické posuzování separační účinnosti vodárenské technologie. ČNI, Praha, 13 p
21. ČSN 75 7715 (2015) Kvalita vod—Biologický rozbor—Stanovení nárostů. ÚNMZ, Praha, 12 p
22. ČSN 75 7714 (2015) Kvalita vod—Biologický rozbor—Stanovení bentosu. ÚNMZ, Praha, 16 p
23. Říhová Ambrožová J (2012) Využití fluorescenčního značení pro detekci bakterií a kvasinek v technické mikrobiologii. Vodní hospodářství 62(1):13–15
24. Ambrožová J (2002) Průběh zatížení pískové filtrace v extrémních podmínkách. Sborník semináře. Aktuální otázky vodárenské biologie, Praha, pp 58–60
25. Říhová Ambrožová J, Hubáčková J, Čiháková I (2007) Složení nárostů a charakter sedimentů ve vodojemech. In: Sborník přednášek XI. Mezinárodní vodohospodářské konference, VODA ZLÍN 2007, Zlín, pp 155–160
26. TNV 75 5941 (2004) Mikroskopické posuzování jakosti vody dopravované potrubím. ČNI, Praha, 28 p
27. Ambrožová J, Bigar J, Říha J (2002) Optimalizace provozu vodovodního řadu. In: Sborník mezinárodní konference, Pitná voda, Trenčianské Teplice, pp 13–18
28. Říhová Ambrožová J, Hedbávný J (2008) Využití výsledků hydrobiologického auditu distribučních sítí k jejich rekonstrukci. Vodohospodársky spravodajca 51(11–12):8–11
29. van der Kooij D et al (2003) Assessment of the microbial growth support potential of products in contact with drinking water. Development of a harmonised test to be used in the European acceptance scheme concerning CPDW, EC, CPDW project, EVK1-CT2000-00052, pp 1–115

30. Bystrianský J, Říhová Ambrožová J (2012) Kritéria volby materiálů do výměníků tepla, pracujících v chladicích vodách. Sborník mezinárodní konference CHEO 9:200–231
31. TNV 75 5121 (2010) Jakost vod. Požadavky na jakost vody dopravované potrubím (s výpočetním programem). ČNI, Praha, 18 p
32. Allion A, Lassiaz S, Peguet L, Boillot P, Jacques S, Peultier J, Bonnet MC (2011) A long term study on biofilm development in drinking water distribution system: comparison of stainless steel grades with commonly used materials. Revue de Métallurgie 108:259–268
33. Říhová Ambrožová J, Říha J (2009) Možné negativní dopady sekundární kontaminace vzduchem na jakost akumulované pitné vody. Vodní hospodářství 59(2):66–72
34. Říhová Ambrožová J, Říha J, Martínek R (2011) Kapacita filtračních náplní vzduchových filtrů. In Sborník XV. mezinárodní konference, Pitná voda 2011, Trenčianské Teplice, pp 83–90
35. ČSN 75 5355 (2011) Vodojemy. ÚNMZ, Praha, 20 p
36. Technické doporučení I-D-48 (2008) Konstrukční uspořádání, provoz a údržba vodojemů. Hydroprojekt CZ, a.s., Praha, 60 p
37. Baudišová D (2007) Současné metody mikrobiologického rozboru vody. Příručka pro hydroanalytické laboratoře, 1st ed. Výzkum pro praxi, sešit 54; VÚV T.G.M., Praha
38. Říhová Ambrožová J, Vejmelková D, Čiháková P (2017) Technická mikrobiologie a hydrobiologie, 1st edn. Praha, VŠCHT
39. Vavrušková L, Kolář K, Kabátová J, Vejmelková D, Říhová Ambrožová J (2016) Možnosti řešení alternativních metod při řešení výjimečných stavů mikrobiální kontaminace. In: Sborník konference, Pitná voda 2016, Tábor, pp 241–246
40. Vejmelková D, Říhová Ambrožová J (2016) Alternativní přístupy k rychlé detekci mikrobiální kontaminace pitných vod. In: Říhová Ambrožová J, Pecinová A (ed) Sborník konference, Vodárenská biologie 2016, Praha, 2016. Praha, pp 42–50
41. Kožíšek F, Pumann P, Šašek J, Dolejš P (2005) Metodické doporučení—parazitičtí prvoci v pitné vodě. (Online). http://www.szu.cz/tema/zivotni-prostredi/metodicke-doporuceni-paraziticti-prvoci-v-pitne-vode?highlightWords=cryptosporidium. Accessed May 3 2018
42. Pumann P, Pouzarová T (2014) Stanovení parazitických prvoků Giardia a Cryptosporidium. In: Baudišová D, (ed) Sborník příspěvků a dalších konferenčních materiálů, Mikrobiologie vody a prostředí. Praha, pp 12–18

Chapter 8
Gray Water Reuse in Urban Areas

Jakub Raček

Abstract The reduced quality and yield of surface and groundwater due to droughts and changing climatic conditions lead to an increase to reusing wastewater (WW) in the Czech Republic. Reusing WW can be implemented at the wastewater treatment plant or in the building. The concept of reused WW in buildings is based on the treatment of low polluted gray water (GW) by gray water treatment system (GWTS) which produces whitewater. GW represents WW from showers, basins, washing machines, kitchen sinks, and dishwashers. Whitewater does not meet such strict parameters as drinking water and can be used for toilet flushing, irrigation or for other use. The WW reuse with GW in urban areas is designed and implemented in the guidelines of a number of countries: UK, Germany, Australia, Canada, USA, and internationally by WHO. Use of treated GW and RW in the buildings in the Czech Republic has not yet been legislatively approved. The characteristic domestic WW as rainwater, gray water, and black water were described. The dark GW (from kitchen and laundry), light GW (from the bathroom), and non-separated GW were reported on their chemical, physical, and microbiological properties. The production of GW represents 70% of total domestic WW, and the toilet flushing represents 30% of the demand for water in the household building. The selected chemical–physical and microbiological indicators of whitewater were described. The groups of GW treatment processes were reported: simple, extensive, chemical, physical, biological and MBRs treatment. The MBRs represent a modern treatment of WW and also are associated with biological treatment and separation of solid and liquid substances. The MBR is directly installed in the biological activated process and replaces settlement tank. The buildings with a high level of GW production and with the high level of demand for whitewater represent the suitable solution for ecological and economical management of WW. GWTS has been provided data on MBRs technologies (microfiltration or ultrafiltration) that seem to be the acceptable solution in the building. In the CR, the system reuses GW only sporadically, but present droughts and changing climatic conditions represent a challenge in the disposal of WW reuse management.

J. Raček (✉)
Faculty of Civil Engineering, AdMaS Research Centre, Brno University of Technology, Purkyňova 651/139, 612 00 Brno, Czech Republic
e-mail: racek.j@fce.vutbr.cz

M. Zelenakova et al. (eds.), *Management of Water Quality and Quantity*, Springer Water, https://doi.org/10.1007/978-3-030-18359-2_8

Keywords Wastewater reuse · Treatment · Membrane filtration · Gray water · Whitewater

8.1 Introduction

The reduced quality and yield of surface and groundwater caused by droughts and changing climatic conditions lead to an increase of reusing wastewater (WW) [1]. The increased water availability [2] is connected to the steady rise in the world's population which also increases WW production [3] or with climatic changes [4].

Reusing of WW can be implemented at the municipal wastewater treatment plant (WWTP) [5, 6] or locally with a connection to the sewer as sewage mining [7] or in the buildings [8] as part of decentralized water system [9]. The potential reused water sources can be municipal WW (domestic or industrial) or rainwater (RW) from the roofs and the paved area. Based on the literature [10–12], reused municipal WW can be associated with energy recovery from WW for heating and cooling in the buildings. The concept of reused WW in the buildings is based on the treatment of low polluted gray water (GW) by gray water treatment system (GWTS) which produces whitewater which represents recycling of GW. Generally, GW represents WW from showers, basins, washing machines, kitchen sinks, and dishwashers. Whitewater does not meet strict parameters as drinking water. Several authors [13–16] described GWTS installed in the buildings where whitewater was used for toilet flushing, irrigation, or for other use.

The WW reuse with GW in urban areas is designed and implemented in the guidelines of a number of countries reported in Table 8.1: UK, Germany, Australia, Canada, USA, and World Health Organization (WHO).

In the Czech Republic (CR), project No. TA01020311, titled "The Use of Grey and Rainwater in Buildings" was solved by Brno University of Technology and by ASIO spol. s.r.o. under the ALFA program of the Technology Agency over the

Table 8.1 Guidelines for GW reuse of a number of countries

Country	UK	Germany	Australia	Canada	USA	WHO
Guidelines	British Standard BS 525-2:2011 [17, 18]	Service water reused for toilet flushing [19]	Australian domestic greywater treatment systems accreditation guidelines [20]	Canadian guidelines for domestic reclaimed water for use in toilet and urinal flushing [21]	EPA guidelines for water reuse [22]	Guidelines for greywater reuse for different purposes [23]

period 2011–2013. The draft of the Czech standard "ČSN 75 6780 Use of treated GW and RW in the buildings and the adjacent land" [24] was result of the project No. TA01020311 which has not been legislatively approved yet. Therefore, in the CR the GW reuse system is only sporadical but present droughts and changing climatic conditions represent a challenge in the disposal of WW reuse management.

8.2 Water Consumption

Generally, the low level of quality drinking water is becoming a global problem and the reduced quality resources of water increase the investment and operating cost. Thus, this leads directly to an increase in the price of drinking water and limitations: water saving battery, water saving toilet flushing system, water saving household appliances such as washing machines and dishwashers.

8.2.1 Current Consumption of Water

At present, in the CR, there is the specific water demand of 88.7 $L{\cdot}person{\cdot}day^{-1}$ [25] in 2017 which represents a significantly lower than 171 $L{\cdot}person{\cdot}day^{-1}$ in 1989 [26]. The specific water demand in the CR is lower than Western countries of Europe [27–29]. Reducing water consumption is due to the limitation of industrial production and the implementation of water safe measures in the industry. Total consumption of water has increased markedly with the increase in the world's population, from 1940 to 2006, water consumption on Earth has increased four times [31].

In the CR, water (1.4 EUR in 2017) and sewage (1.3 EUR in 2017) rates are cheaper than the Western European countries, despite the increasing tendency [30]. The CR, unlike other countries, still complies with the principle recommended by the World Health Organization and the World Bank, pointing out that water and sewage prices should remain socially bearable for the population. Household water expenditure should not exceed 2% of its gross average income [31].

Another indicator of water management, the comparison of water quantities and prices in selected countries of the world has shown that states with lower water prices have a higher demand for water than those with higher prices [32]. However, this does not apply to Australia where there is a lower share of large raw water resources compared to increased water demand for watering gardens. Based on Global Water Intelligence (GWI) in 2009 [32], water and sewage prices in the world are still rising. The average cost of water and sewage in the 266 major cities of the world, as measured by the 2009 GWI survey, is about 1.4 EUR. The largest increase can be seen in Eastern Europe and Central Asia caused by the rapid devaluation of local currencies.

8.2.2 *Alternative Sources of Water*

Generally, the quality and supply of raw water are limited and not available in many countries, and this raw water is not available for treatment and to produce drinking water thus, WW reuse with alternative sources has been the solution in the water supply.

In the past, RW was used as a reservoir for many communities throughout the world for many thousands of years. In the past centuries, raw water from rivers and often underground sources has often been contaminated by fecal pollution, which in many cases has led to the outbreak of cholera or dysentery. Thus, the collection and RW reuse was taken a less hazardous source from the point of view of hygienic safety. Probably for these reasons, at present, RW reuse is perceived for many people like a clean, fresh source of water. In the CR, the RW reuse systems are used in family houses or gardens colonies where RW is collected and used for watering the garden.

In coastal states, alternative sources of water are associated with desalination of salt water, and this reused water is used for household and industry water consumption. These systems are economically demanding and cause a number of operational problems associated with additional dosing of calcium and magnesium.

Another alternative source of water is the reuse of domestic or industrial WW. Significant alternative water sources include low polluted GW from the operation of selected objects. Treated GR called whitewater is used for toilet flushing, irrigation, or for other use.

This paper describes the state-of-art GW reuse in urban areas for conditions in CR and focuses on the design of the GW treatment system.

8.3 Domestic Wastewater

Generally, WW is water that changes physical or chemical properties when used and this WW represents hazardous for surface or groundwater. Domestic WW from the place of origin of the waste (sink, sink, toilet, etc.) is collected by the internal piping system of the building. Frequently, this domestic WW is connected to a sewer, non-effluent WW tank, or domestic WWTP.

8.3.1 *Characteristic of Domestic Wastewater*

The domestic WW can be categorized into three groups by origin: RW, black water (BW) and GW. The RW represents rainfall collected from building roofs and paved areas. The BW from toilets and urinal consists of fecal pollution. The GW represents

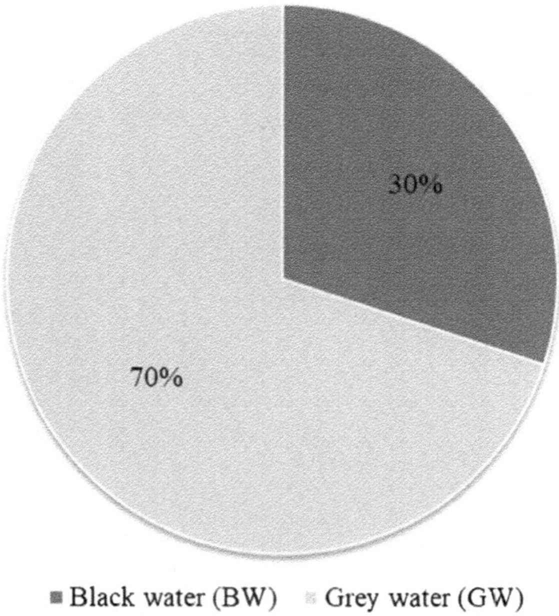

Fig. 8.1 Production of BW and GW in the household building [34]

WW from bathing, hand wash, kitchen sink, laundry, clean industrial water, WW usually does not consist of fecal pollution.

The main content of pollutants and up to 80% of organic matter in WW is coming from urine and feces [33]. Thus, GW can represent less polluted sources of WW to easily treated and reuse in the building.

WW can also be characterized by the production of BW and GW. The percentage of production WW [34] is shown in Fig. 8.1. The production of GW represents 70% of total domestic WW.

On the other hand, the demand for water by use in the household building [35] is illustrated in Fig. 8.2. Toilet flushing, bathing, and showering represent the highest level of water consumption. The graph reveals the production of the GW corresponding to about 50% of the total water demand.

RW can be collected from impervious surfaces, especially roofs of the buildings. Generally, RW contains free substances from the air and other substances of the roof. These substances represent pollution of RW and dependent on the material of the roof. RW from the roof and impervious surfaces can be contaminated by microbiological pollution. RW from impervious surfaces can be polluted by petroleum products. The treatment of RW is carried out by basic purification processes such as sedimentation, filtration, and disinfection. Therefore, the RW system is strongly dependent on rainfall, climatic conditions, and storage tank size. For the above reasons, RW and other WW should be used in a combined reuse system in the buildings.

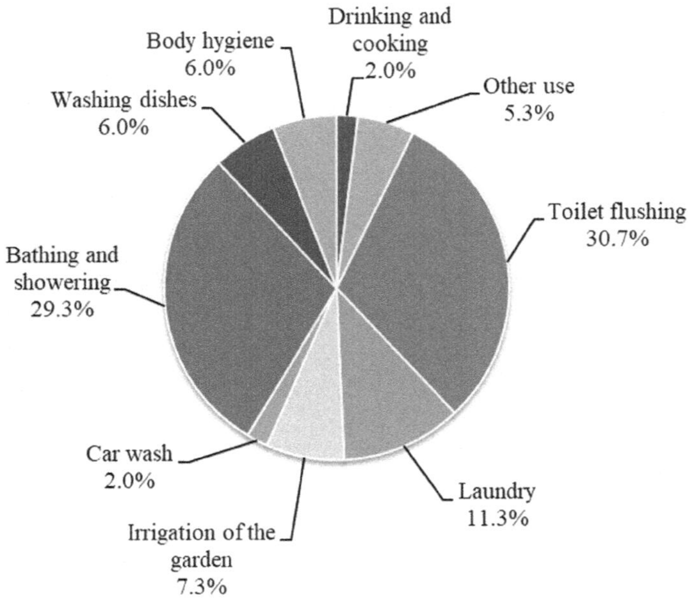

Fig. 8.2 Demand for water in the household building [35]

The Czech standard ČSN EN 12056-1 to 5 [36] defines BW as WW with fecal and urine. Based on literature review, BW has characterized to brown and yellow WW [37, 38]. Brown water is from toilets and represents fecal pollution. Yellow water is from the urinal and consists of urine pollution.

The human body produces daily between 0.6 and 2.0 L of urine, an average of 1.5 L of dry matter of 60 g which is little less water than a person consumes per day [33]. This yellow water represents pollution from the metabolism of the human body and the pollution associated with use in a common toilet bowl. This yellow water was mixed with fecal, and thus yellow water is no longer sterile. Compared with brown water, the yellow water does not consist of bacteria, fungi, and viruses. The color of urine depends on several factors, particularly on the composition of urea CH_4N_2O, dissolved salts and organic matter [39]. Generally, yellow water contains phosphorus and nitrogen, which could be used in agriculture as a fertilizer.

Brown water is from the toilet, and this contains fecal pollution which is hazardous for organisms. The human body produces 120–330 g per day with 30–75 g dry solids (DS). This DS of brown water consists of 90% organic matter [33].

WW contains nutrients: inorganic nitrogen compounds, phosphorus, and potassium. Differences are illustrated in Fig. 8.3 with a percentage distribution of yellow water, brown water, and GW. The phosphorus content is predominantly in the form of sulfate ions. Nitrogen from yellow water is easily degraded to ammonium and nitrates [39]. Nutrients from yellow water are biologically available for plants, and therefore, urine is generally considered a natural fertilizer.

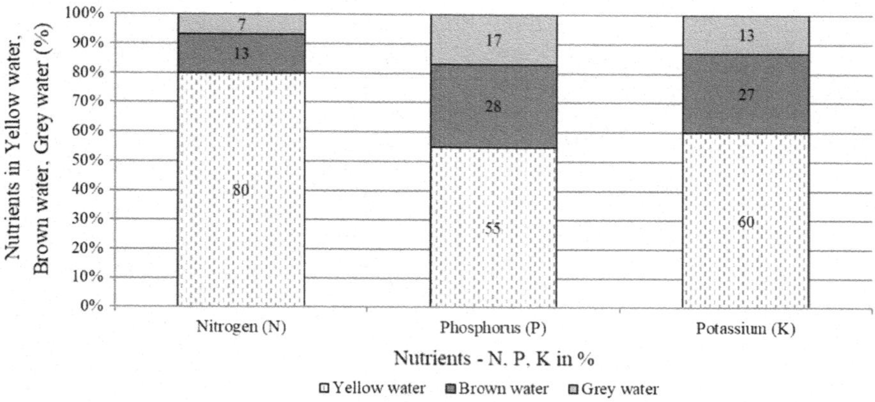

Fig. 8.3 Nutrients N, P, K in yellow water, brown water, and gray water, in % [39]

Generally, the reuse of brown and yellow water is represented as an interesting and useful topic. Unfortunately, in connection with high pollution organic level, this topic is associated with the current economy and high technology treatment which is not recommended.

8.3.2 Gray water Reuse

The Czech standard ČSN EN 12056-1 to 5 [36] and German standard DIN 4045 Abwassertechnik—Grundbegriffe [40] define GW as WW without fecal and without urine. GW is low polluted WW from showers, basins, washing machines, kitchen sinks, and dishwashers. In the building, it is necessary to install a separate second GW piping system. GW is collected to the GWTS. The GWTS can consist of input pipe, piping station, treatment tank, a reservoir with automatic pump station, output pipes. Treated GW, thus whitewater will be used for toilet flushing, irrigation, or other use. This whitewater needs separately pressure piping system.

Generally, GW based on the origin can be characterized as follows categories:

- GW from the kitchen sink, dish washer;
- GW from the washing machine, laundry;
- GW from basin, bath, shower;
- Other GW;
- Non-separated GW.

Several authors [41–43] characterized only three categories:

- dark GW (DGW) from kitchen and laundry;
- light GW (LGW) from the bathroom;
- non-separated GW (NSGW).

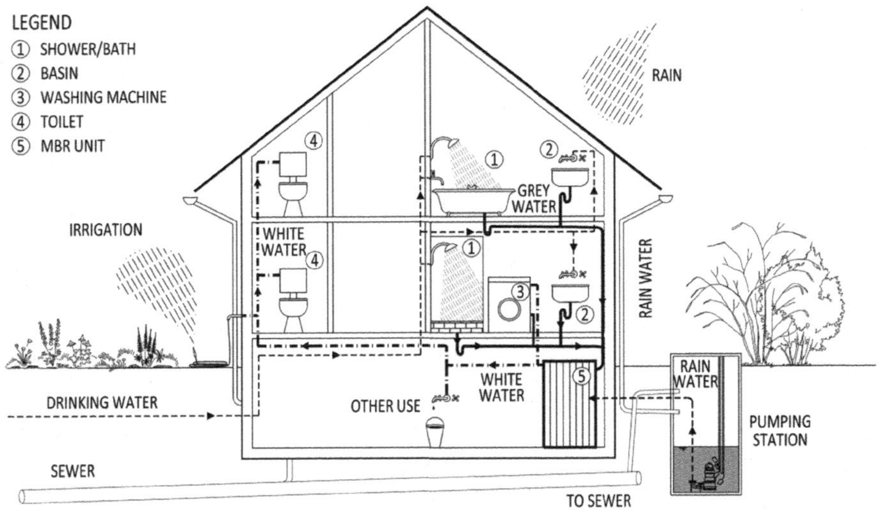

Fig. 8.4 Gray water reuse in the building

GW reuse in the building is described in Fig. 8.4. GW from the place of origin of the waste (basin, bath, shower, washing machine, kitchen sink, dish washer, etc.) is collected by an internal GW piping system in the building. GW is collected in the GW treatment, also located in the basement of the building. The treatment of GW is associated with physical, chemical, biological, or combination of these treatment processes. An automatic pressure station with disinfection unit is usually installed in the output of the treatment system. Cleaned water, thus whitewater is used for toilet flushing, irrigation, or for other use in the building. The important component of the GW reuse system is a safety overflow from the GW treatment that it is connected into the sewer.

GW reuse system in the buildings can be divided according to the demand of whitewater as follows categories:

- non-contact use: for irrigation system;
- low-contact use: for toilet flushing, laundry, floor washing where is minimal contact with human skin;
- contact use: After quality treatment and disinfection, this whitewater is used for bathing, floor washing where it is in contact with human skin;
- other use: Whitewater is used in industries such as large laundries [44].

Relatively large amounts of GW are produced in the buildings with the high level of water demand: hotels, schools, restaurants, hostels, college buildings, administrative buildings, and other similar objects. Laundries seem to be an interesting source of GW which also there is the high level of whitewater demand.

High GW production is not only one benchmark for design GW reuse system. Therefore, it is necessary to consider the need for whitewater for further use. In the case of larger amounts of whitewater, this water would not be consumed, and unnecessary treatment would mean an increase in operating costs. On the other hand, the low level of whitewater has been solved by the supply of drinking water from the public water supply system or by the addition of rainwater from the roof or the paved surfaces.

Other suitable installations for the use of gray water can include sites where there is a shortage of raw or drinking water sources or where the price for water from the public water supply is considerably high.

8.4 Quality Gray and Whitewater

The current water strategies in the CR do not strictly address the WW reuse because there are no needs to find new possibilities of water sources without conventional raw water sources. The problem has not only the technical and economic aspects but also the negative public opinion about the reuse of treated WW. On the other hand, the current dry weather accompanied by extreme weather fluctuation in the years 2015–2017 provides the necessity to deal with the use of WW. In the CR, there is no direct regulation which would limit the WW reuse.

According to Decree no. 252/2004 Sb., with amendment no. 293/2006 Sb. [45], laying down hygienic requirements for drinking and hot water and the frequency and scope of drinking water control. An assessment of WW reuse would be undertaken individually taking into account the water quality requirements in the area of use [46], i.e., the specific needs in the selected buildings.

At present, GW reuse system in the selected building is technologically and economically acceptable. In the Czech Republic, GW reuse systems are already demonstrated [47–49], and it is necessary to define all technical and legal aspects and to ensure the acceptability of such GW reuse.

8.4.1 Chemical–Physical Properties of GW

Based on the literature [41, 50–63], the selected chemical–physical properties of GW (DGW, LGW, NSGW) are summarized in Table 8.2.

The higher pH values of 9.3–10.0 are presented in DGW from the washing machine, laundry. In contrast, DGW from the kitchen sink, dish washer has a low level of acidic conditions with pH 6.3–7.4. The temperature is variable and generally higher for DGW from the washing machine and basin and shower. The color and the turbidity level are higher for LGW from basin, bath, and shower. The floating substances are reported for DGW from the kitchen and represent organic pollution which needs more demanding GW treatment. BOD and COD ration is described 4:1, thus higher than 2:1 for standard municipal WW [65]. This higher ratio represents a higher level of biodegradable organic matter in GW.

Table 8.2 Selected chemical and physical properties of GW [41, 50–63]

Chemical–physical parameter	Unit	Dark GW (DGW)		Light GW (LGW)	Non-separated GW (NSGW)
		Kitchen sink, dishwasher	Washing machine, laundry	Basin, bath, and shower	
pH	–	6.3–7.4	9.3–10.0	5.0–8.6	6.1–8.4
Temperature	°C	–	28–32	18–38	–
Color	Pt C^{-1}	–	50–70	60–100	–
Turbidity	NTU	–	14–296	20–370	–
Floating substances	mg L^{-1}	134–1300	79–280	7–120	–
Suspended solids (SS)	mg L^{-1}	–	–	–	45–330
BOD_5	mg L^{-1}	669–756	48–682	19–200	41–194
COD	mg L^{-1}	26–1600	375	64–8000	495–623
N	mg L^{-1}	13.1–16.0	6.2	0.6–46.4	8.1
P	mg L^{-1}	3.1–10.0	0.06–57.0	0.11–2.20	3.3–11.0

The total nitrogen content in GW is characterized as organically bound (in protein). The high content of total phosphorus is in dishwasher detergents and DGW from washing machines [51]. In many countries including the Czech Republic, use of phosphate products is prohibited or restricted.

Generally, the low polluted GW represents LGW from basin, bath, and shower. This LGW is primarily suitable to GW treatment system in the building.

8.4.2 Microbiological Pollution of GW

Several authors [41, 50–64] focused on microbiological pollution of GW (DGW, LGW, NSGW) which are summarized in Table 8.3.

Based on the reported data in Table 8.3, microbiological pollution can be caused by hand washing, urine when showering (especially by children) and food processing. Washing hands and showers may result in higher concentrations of microbiological contamination than from washing in the short term.

Table 8.3 Selected microbiological pollution of GW [41, 50–64]

Microbiological parameter	Unit	Dark GW (DGW)		Light GW (LGW)	Non-separated GW (NSGW)
		Kitchen sink, dishwasher	Washing machine, laundry	Basin, bath, and shower	
Fecal coliforms	CFU 100 mL^{-1}	–	10^1–10^4	10^1–10^6	10^2–10^6
Total coliforms	CFU 100 mL^{-1}	–	10^1-10^8	10^1-10^9	10^5–10^8
Escherichia coli	CFU 100 mL^{-1}	10^5–10^8	10^1-10^6	10^1-10^7	10^1-10^2
Streptococci	CFU 100 mL^{-1}	10^3–10^8	10^1-10^7	10^1-10^6	102
Total number of colonies	CFU 100 mL^{-1}	–	–	10^2–10^8	–
Pseudomonas aeruginosa	CFU 100 mL^{-1}	–	–	N-10^3	10^2–10^5
Salmonella	CFU 100 mL^{-1}	–	N	N	–
Cryptosporidium	CFU 100 mL^{-1}	–	N	N	N
Giardia intestinalis	CFU 100 mL^{-1}	–	N	N	–
Enterococci	CFU 100 mL^{-1}	–	N	N	10^3–10^5

N non-identified

8.4.3 Chemical–Physical Indicators of Whitewater

The WW reuse with GW in urban areas is designed and implemented in the guidelines of a number of countries [17–23] and is focused on chemical–physical indicators of whitewater which are summarized in Table 8.4.

Based on the reported data in Table 8.4, the selected chemical–physical indicators of whitewater are pH, turbidity, SS, BOD, and chlorine residual. The strict indicators are for BOD_5 which needs to be less than 10.0 $mg{\cdot}L^{-1}$ (5.0 $mg{\cdot}L^{-1}$ for BOD_7) and for SS less than 10.0 $mg{\cdot}L^{-1}$.

8.4.4 Microbiological Indicators of Whitewater

The guidelines of a number of countries [17–23] describe microbiological indicators of whitewater which are summarized in Table 8.5.

Based on the reported data in Table 8.5, the selected microbiological indicators of whitewater are Fecal coliforms, Thermoresistant coliforms, Total coliforms, *Escherichia coli*, *Pseudomonas aeruginosa*, and Enterococci. These microbiological indicators represent the need to design high efficient GW treatment system or highly effective disinfection of whitewater.

Table 8.4 Selected chemical–physical indicator of whitewater [17–23]

Chemical–physical indicator	Unit	Region guidelines/reuse					
		UK[a]	Germany[b]	Australia[c]	Canada[d]	USA[e]	WHO[f]
		Toilet flushing	Toilet flushing	Toilet flushing and laundry	Toilet flushing	Unlimited urban reuse	Toilet flushing
pH	–	5.0–9.5	–	–	–	6.0–9.0	–
Turbidity	NTU	<10.0	–	–	≤2.0**	≤2.0	–
Suspended solids (SS)	mg L^{-1}	–	–	<10.0*	≤10.0***	–	≤10.0
BOD_5	mg L^{-1}	–	<5.0 (BOD_7)	<10.0*	≤10.0***	≤10.0	≤10.0
Chlorine residual	mg L^{-1}	–	–	–	≥0.5	≥1.0	–

*<10.0 mg L^{-1} (90% of samples) and <20.0 mg·L^{-1} (maximum)
**≤2.0 mg L^{-1} (median) and <5.0 mg·L^{-1} (maximum)
***≤10.0 mg L^{-1} (median) and <20.0 mg·L^{-1} (maximum)
[a]UK: British Standard BS 525-2:2011 [17, 18]
[b]Germany: Service Water Reused for Toilet Flushing [19]
[c]Australia: Australian Domestic Greywater Treatment Systems Accreditation Guidelines [20]
[d]Canada: Canadian Guidelines for Domestic Reclaimed Water for Use in Toilet and Urinal Flushing [21]
[e]USA: EPA Guidelines for Water Reuse [22]
[f]WHO: Guidelines for Greywater Reuse for Different Purposes [23]

Table 8.5 Selected microbiological indicator of whitewater [17–23]

Microbiological indicator	Unit	Region guidelines/reuse					
		UK[a]	Germany[b]	Australia[c]	Canada[d]	USA[e]	WHO[f]
		Toilet flushing	Toilet flushing	Toilet flushing and laundry	Toilet flushing	Unlimited urban reuse	Toilet flushing
Fecal coliforms	CFU·100 mL^{-1}	–	<10	–	–	Non-detectable	–
Thermoresistant coliforms	CFU·100 mL^{-1}	–	–	–	≤200	–	≤10
Total coliforms	CFU·100 mL^{-1}	1000	<100	<10*	–	–	–
Escherichia coli	CFU·100 mL^{-1}	250	–	–	≤200	–	–
Pseudomonas aeruginosa	CFU·100 mL^{-1}	–	<1	–	–	–	–
Enterococci	CFU·100 mL^{-1}	100	–	–	–	–	–

*<10 (90% of samples) and <20.0 (maximum)
[a]UK: British Standard BS 525-2:2011 [17, 18]
[b]Germany: Service Water Reused for Toilet Flushing [19]
[c]Australia: Australian Domestic Greywater Treatment Systems Accreditation Guidelines [20]
[d]Canada: Canadian Guidelines for Domestic Reclaimed Water for Use in Toilet and Urinal Flushing [21]
[e]USA: EPA Guidelines for Water Reuse [22]
[f]WHO: Guidelines for Greywater Reuse for Different Purposes [23]

8.5 Gray water Treatment

The treatment of GW in the 1970s was achieved on the basic treatment systems: mechanical treatment such as coarse filtration or membrane filtration, often with disinfection [3]. Gradually, the biological principles of treatment began to be used for GW. Followed by modern treatment technology represents the membrane bioreactors which represents biological and physical membrane filtration. Despite technological advances in sewage treatment technologies, at present, we can see cheaper and simple GW treatment systems such as reed bed and root WWTP. However, these systems are limited by a large area and are unable to achieve stringent treatment limits, as is the case with MBRs technologies [66].

The groups of GW treatment process are summarized in Table 8.6: simple, extensive, chemical, physical, biological, and MBRs treatment.

8.5.1 GW Simple Treatment

GW simple treatment usually needs a simple treatment process: sedimentation, coarse filtration, or combination of these systems. Based on the simple treatment, it is necessary to install disinfection of whitewater. The effectiveness of these simple GW treatment systems is considerably limited in terms of organic and inorganic pollution. Generally, the GW simple treatment is used usually in small scale, especially for houses. The LGW is treated, and whitewater is used for irrigation and for toilet flushing [14, 67–69].

8.5.2 GW Extensive Treatment

GW natural extensive treatment is usually characterized by the use of wetland, root WWTP or reed bed. The first treatment step represents the removal of large particles by the pretreatment process, followed by sand filtration. The most common plant is common reed (Phragmites australis). Generally, the GW extensive treatment is usually used in small scale due to the need for large area requirement. The LGW is treated, and whitewater is used for irrigation and toilet flushing [15, 70–78].

8.5.3 GW Chemical Treatment

GW chemical treatment is carried out by the processed: coagulation, electrocoagulation, or chemical oxidation by hydroxyl radicals. These processes can be used in specific industrial cases such as laundry GW [16, 79–85].

Table 8.6 Summary of GW treatment process [14–16, 66–102]

Group of GW treatment	Description of treatment technology	Advantage	Disadvantage
GW simple treatment	Course filtration	Simplicity treatment process, low operating cost	Low treatment efficiency, limited disinfection efficiency
GW extensive treatment	Wetland, root WWTP, reed bed	Simplicity treatment process, low operating cost	Large area requirement, not control the treatment process; complicated treatment in the winter
GW chemical treatment	Coagulation, electrocoagulation, chemical oxidation by hydroxyl radicals	For specific cases with high treatment efficiency, industrial GW such as laundry	High investment cost and high operating cost
GW physical treatment	Sand filtration	Simplicity treatment process, low operating cost	Not complete a reduction of pollution, it can also clog and congestion of the treatment system
	Granular activated carbon filtration (GAC)	Simplicity treatment process, effective treatment	High investment cost, the GAC filtration removes some pollution
GW biological treatment	Aerobic biological treatment	The system responds flexibility to the amount and volume of pollution, high treatment efficiency of organic pollution	High investment cost and high operating cost, not remove all pathogens
GW MBRs treatment	Membrane bioreactor	Effective treatment, low space requirements, possible to operate at high sludge loads, biological removal of difficult to decompose substances	The requirement for high-quality pretreatment, regular membrane cleaning, high investment cost and high operating cost
	Physical membrane filtration	Effective treatment, low space requirements	The requirement for high-quality pretreatment, regular membrane cleaning, high investment cost and high operating cost, only for low organic polluted WW

8.5.4 GW Physical Treatment

The sand filtration, granular activated carbon (GAC), or combination of both is represented GW physical treatment. Physical treatment involves a basic purification process based on the adsorption of suspended solids on the sand filter. This process represents volumetric filtration, thus filtration through a layer of granular material. The sand filtration can be used separately or in combination with GAC and with disinfection [86–89].

8.5.5 GW Biological Treatment

GW biological treatment is associated with a wide range of treatment processes that can be successfully used in the treatment process. GW biological treatment includes a pretreatment process, mostly mechanical treatment process. There are installed screens, sedimentation tank, filtration with biological treatment tank, and mostly activation tank. The principle of biological pollution activation is to aerate the activated sludge in the activation tank of WWTP. The activated sludge is composed of a mixed culture of microorganisms supporting biological purification. GW biological treatment system is generally used for high pollution of GW with the high degree of organic pollution removal. These are large office buildings, hotels, multi-story buildings, and colleges [90–95].

8.5.6 GW MBRs Treatment

The MBRs represent a current WW treatment and also is associated with biological treatment and separation of solid and liquid substances. The MBR is directly installed in the biological activated process and replaces settlement tank. The effluent quality is not affected by insufficient sedimentation properties of secondary sludge or unfavorably formed foam on the surface this is one of the main advantages of MBRs technology. MBRs technologies can also be operated at high biomass concentrations up to 15 $kg{\cdot}m^{-3}$ [96]. This has the positive effect of reducing tank volumes and reducing sludge production.

Generally, the membrane is situated after the mechanical and biological treatment process. After this process, pretreatment water is treated by MBRs where the solid particles are separated from the liquid parts. The treated WW does not contain all the particles greater than the pore size of the membrane. The treated WW by MBRs process does not change thermal, biological, or chemical properties. Depending on the size of the pores, the membranes can be divided into different categories: microfiltration, ultrafiltration, nanofiltration, and reverse osmosis [97]. The MBRs can replace

Table 8.7 Advantages and disadvantages of the MBRs treatment process [96]

The advantage of the MBRs	The disadvantage of the MBRs
Small area requirements, small build area	Higher investment costs, increased operating, and maintenance costs
Possibility of the reuse of permeate	Complicated machinery and associated higher operating costs compared to separation in relatively simple settlement tanks
Eliminating the influence of sludge quality on separation efficiency, solving problems with fibrous microorganisms	At higher concentrations of activating sludge than 15 kg m^{-3} there may be problems with aeration or foam formation
High biomass concentration (up to 15 kg m^{-3}), possibility to significantly reduce the volume of tanks	Water flowing to the membrane must be properly pretreatment otherwise there is a risk of clogging the MBRs
Higher sludge age and higher sludge concentrations allow the removal of even biologically degradable substances	Need for a relatively balanced flow on the membrane
Simplicity of operation with unstable properties and composition of activated sludge	Need for regular cleaning and regeneration of membranes by chemical agents

the settlement tank at the WWTP and can also be used for tertiary treatment, and the appropriate selection of MBRs can remove microbiological contamination.

Based on the literature [98–102], the most frequency used type of MBRs GW treatment represents microfiltration and ultrafiltration with the working pressure of 0.5 MPa. Pore size does not release microparticles, colloids, bacteria, and some viruses. The membrane fibers can be polymeric (flat or hollow) or inorganic (ceramic, carbon).

The microfiltration and ultrafiltration membranes can be divided according to the direction of flow of the filtered suspension to be distributed as follows: MBRs with dead-end filtration and MBRs with cross-flow filtration [96].

The main advantages and disadvantages of the MBRs treatment process are reported in Table 8.7.

8.6 Gray water Treatment System in the Buildings

Based on the previous parts, it is possible to define a design gray water treatment system GWTS in the building. The buildings with the high level of production GW and with the high level of demand whitewater (hotels, schools, restaurants, hostels, college buildings, administrative buildings, laundries, and other similar objects) represent the suitable solution for ecological and economical management of WW. The literature review of GW treatment system has been provided data of MBRs technologies that seem to be the acceptable solution for the design of GWTS in the building.

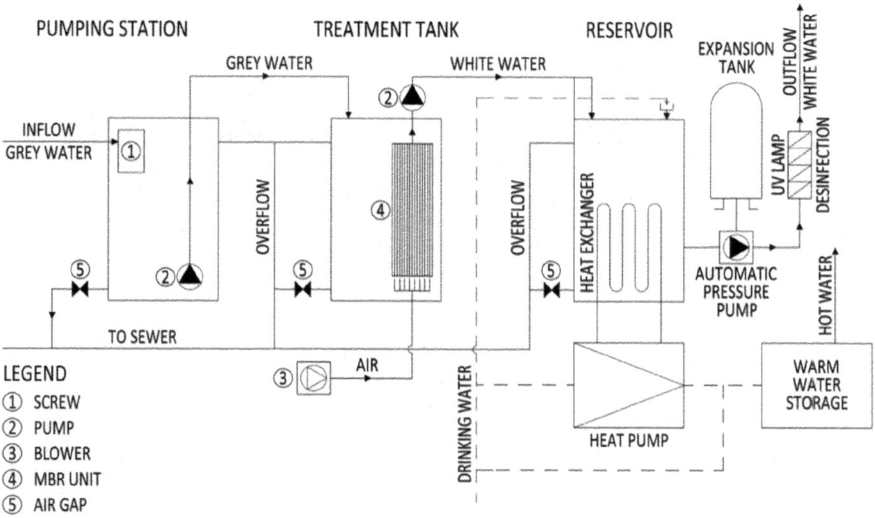

Fig. 8.5 Gray water MBRs treatment system

The schematic diagram of GWTS with MBRs technology is illustrated in Fig. 8.5. GW is collected at the pumping station with the screw and the overflow with a connection to sewer. GW without suspended solids is pumping by sewage submersible pump to the treatment tank.

There is GW treated biological or physical treatment process. In the case of biological inflow pollution, representing in particular GW from the kitchen sink and dishwasher, the activated sludge is located in the treatment tank. GW treatment is carried out by MBRs: microfiltration or ultrafiltration. Pollution GW with low biological load pollution is performed MBRs physical process. The air for the activated sludge and the membrane flushing is filled with a blower. The treatment tank with MBRs includes overflow and bottom drainage pipe with the connection to sewer. The permeate pump is pumping whitewater to the reservoir.

The reservoir is designed for balancing the demand for whitewater and can be supplement with the heat exchanger. The heat from whitewater can be used by the heat pump for heating water in the building. In the case of missing whitewater, the reservoir is suitable to connect to the rainwater or drinking water through the air gap. The whitewater is pumped by automatic pressure pump with expansion tank through disinfection by UV lamp to toilet flushing, irrigation, or for other use.

8.7 Conclusion and Recommendations

The current water strategies in the CR do not strictly address the WW reuse because there are not needs to find new possibilities of water sources without conventional raw water sources. The problem is not only the technical and economic aspects but also the negative public opinion about the reuse of treated WW. On the other hand, the last dry whether accompanied by extreme weather fluctuation in the years 2015–2017 provides the necessity to deal with the use of WW. In the CR, there is no direct regulation which would limit the WW reuse.

It is therefore expected that they will further increase water and sewerage rates to allow a smooth recovery of predominantly obsolete utilities and the rehabilitation or construction of new water resources. This increase creates conditions for return on investment in the GWTS in the selected buildings.

The drought, poor quality water resources and the increasing cost of water and sewage lead to reuse of WW. Given the current continuing trend of reducing the cost of treating technology, the GWTS will be designed in practice. And these will improve to increase conditions for further research and development in this area of WW reuse management.

Acknowledgements This chapter has been worked out under project No. LO1408 "AdMaS UP—Advanced Materials, Structures and Technologies", supported by Ministry of Education, Youth and Sports under the "National Sustainability Programme I." And this contribution has been prepared within project No. TA01020311, titled "The use of grey and rainwater in buildings", which is being implemented with the financial aid of the TA ČR.

References

1. Raclavský J, Biela R, Vrána J, Hluštík P, Raček J, Bartoník A (2013) Greywater and rainwater management in buildings. Adv Mater Res 649:195–198
2. United States Environmental Protection Agency, EPA (2004) Guidelines for Water Reuse. USEPA, Washington, DC, USA, Report EPA/625/R-04/108, pp 1–480
3. Pidou M, Memon FA, Stephenson T, Jefferson B, Jeffrey P (2007) Greywater recycling: treatment options and applications. In: Proceedings of institution of civil engineers-engineering sustainability, vol 160, pp 119–131
4. Solomon S (2011) Global water issue. Bureau of International Information Programs, United States Department of State, pp 1–152
5. Djemil W, Hannouche M, Belksier MS (1968) Reuse of treated wastewater in agriculture: physicochemical quality and environmental risks. Case of wastewater treatment plant of Baraki and Beni Messous. Algeria. In: AIP Conference Proceedings, p 020011. https://doi.org/10.1063/1.5039170
6. Ilias A, Panoras A, Angelakis A (2014) Wastewater recycling in Greece: the case of Thessaloniki. Sustainability 6:2876–2892. https://doi.org/10.3390/su6052876
7. Kapuku Bwapwa J (2018) Review on main issues causing deterioration of water quality and water scarcity: case study of South Africa. Environ Manag Sustain Dev 7:14. https://doi.org/10.5296/emsd.v7i3.13156

8. Friedler E, Kovalio R, Galil NI (2005) On-site greywater treatment and reuse in multi-storey buildings. Water Sci Technol 51(10):187–194
9. Capodaglio AG, Callegari A (2016) Domestic wastewater treatment with a decentralized, simple technology biomass concentrator reactor. J Water, SanitIon Hyg Dev 6:507–510. https://doi.org/10.2166/washdev.2016.042
10. Ma H, Yin L, Shen X et al (2016) Experimental study on heat pipe assisted heat exchanger used for industrial waste heat recovery. Appl Energy 169:177–186
11. Lian HK, Li Y, Shu GYZ, Gu CW (2011) An overview of domestic technologies for waste heat utilization. Energy Conserv Technol, pp 123–128
12. Raček J, Úterský M, Ševčík J, Dufek Z, Hlavínek P (2017) Energy recovery from wastewater for heating and cooling of multifunctional building in Brno: Modeling the connection. In: 17th International multidisciplinary scientific geoconference SGEM 2017. Bulgaria: SGEM, pp 519–526. ISBN: 978-619-7408-29-4. ISSN: 1314-2704
13. March JG, Gual M, Orozco F (2004) Experiences on greywater re-use for toilet flushing in a hotel (Mallorca Island, Spain). Desalination 164:241–247
14. Lazarova V, Hills S, Birks R (2003) Using recycled water for non-potable, urban uses: a review with particular reference to toilet flushing. Water Sci Technol: Water Supply 3(4):69–77
15. Gross A, Shmueli O, Ronen Z, Raveh E (2006) Recycled vertical flow constructed wetland (RVFCW)—a novel method of recycling greywater for irrigation in small communities and households. Chemosphere
16. Sostar-Turk S, Petrinic I, Simonic M (2005) Laundry wastewater treatment using coagulation and membrane filtration. Resour Conserv Recycl 44(2):185–196
17. UK. British Standard BS 8525-1:2010 (2010) Greywater systems—Part 1: Code of practice. BSI British Standards, 46p
18. UK. British Standard BS 8525-2:2011 (2011) Greywater systems—Part 2: Domestic greywater treatment equipment. Requirements and test methods. BSI British Standards, 32p
19. Germany (2003) Service water reused for toilet flushing, Berlin Senate Department for Urban Development
20. Australia (2005) Australian domestic greywater treatment systems accreditation guidelines, Part 4, Clause 43(1), Local Government (Approvals) Regulation
21. Canada (2010) Canadian guidelines for domestic reclaimed water for use in toilet and urinal flushing
22. United States Environmental Protection Agency USEPA (2012) Guidelines for Water Reuse 26:642. EPA16251R-921004
23. World Health Organization WHO (2006) Guidelines for greywater reuse for different purposes, WHO/CEHA
24. Proposal of Czech standard ČSN 76 6780 (2014) Use of treated gray and rainwater in the buildings and the adjacent land
25. Czech Statistical Office (2018) We consume less than 89 liters of water a day, an estimate for 2017. https://www.czso.cz/csu/czso/denne-spotrebujeme-necelych-89-litru-vody
26. Melounová M (2010) Drinking water supply in the Czech Republic in 2008. In: Conference proceedings drinking water 2010, pp 25–30. W & ET Team, České Budějovice. ISBN: 978-80-254-6854-8
27. Webster Systems (2010) LLC dba Data360: Average water use per person per day. Data360.org. http://www.data360.org/dsg.aspx?Data_Set_Group_Id=757
28. Water Footprint Network (2008) National water footprints—United Kingdom. waterfootprint.org. Water Footprint Network. http://www.waterfootprint.org/?page=files/UnitedKingdom
29. European Environment Agency (2011) Water pricing and household water use in Spain Water pricing and household water use in Estonia. eea.europa.eu. http://www.eea.europa.eu/data-and-maps/figures/water-pricing-and-household-water
30. Czech Statistical Office (2017) Water and sewage system in 2017—water and sewerage again on the rise. https://www.czso.cz/csu/xj/vodovody-a-kanalizace-v-roce-2017-vodne-a-stocne-opet-na-vzestupu

31. The most used water in the US to pay the most in Denmark (2018) Vodarenstvi.cz. http://www.ovodarenstvi.cz/clanky/nejvice-vody-spotrebuji-v-usa-nejvice-zaplati-v-dansku
32. Global Water Intelligence (2011) Global water tariffs continue upward trend 12(9). globalwaterintel.com http://www.globalwaterintel.com/archive/12/9/market-profile/global-water-tariffs-continue-upward-trend.html
33. Pitter P (1999) Book hydrochemistry, VŠCHT, UNITISK, s.r.o., 568 p, ISBN: 80-7080-340-1
34. Khalaphallah R (2012) Greywater treatment for reuse by slow sand filtration: study of pathogenic microorganisms and vage survival, These de Doctorat. Nantes Ecole des Mines, dans le cadre de Sciences pour l'ingénieur, géosciences, architecture. Yves Andres
35. Asio, spol. sr.o. (2018) Treatment gray water and use energy from it. Asio.cz http://www.asio.cz/cz/153.cisteni-sedych-vod-a-moznost-vyuziti-energie-z-nich
36. Czech standard ČSN EN 12056-1 to 5 (2001) Internal sewage—Gravity systems—par 1 to part 5. Prague: Czech Standardization Institute, pp 12, 14, 22, 35, 36
37. Topinfo (2008) Wastewater as source. tzb-info.cz. Plotěný K. http://www.tzb-info.cz/4955-odpadni-vody-jako-zdroj-surovin
38. Impea (2010) Decentralization and reuse wastewater in practise. e-voda.cz. Mifková T, Hlavínek P. http://www.e-voda.cz/clanek/69/decentralizovane-odvadeni-a-opetovne-vyuziti-odpadnich-vod-v-praxi
39. ASB-Portal (2013) New methods to disposal of wastewater. asb-portal.cz. Mifková T. http://www.asb-portal.cz/tzb/zdravotni-technika/nove-metody-nakladani-s-odpadnimi-vodami-2687.html
40. German Standard DIN 4045 (2003) Abwassertechnik – Grundbegriffe (Wastewater engineering), 44p
41. Eriksson E, Auffarth K, Henze M, Ledin A (2002) Characteristics of grey wastewater. Urban Water 4(1):85–104
42. Li F, Wichmann K, Otterpohl R (2009) Review of the technological approaches for greywater treatment and reuses. Sci Total Environ 407(11):3439–3449
43. Matos C, Sampaio A, Bentes I (2012) Greywater use in irrigation: characteristics, advantages and concerns
44. Biela R (2012) Greywater: quality and reuse. Sovak, pp 11–13. ISSN 1210-3039
45. Act No. 252/2004 Sb., Laying down the sanitary requirements for drinking and hot water and the frequency and scope of drinking water control (2006). Amendment No. 293/2006 Sb., Amending Decree No. 252/2004 Coll. Ministry of Health, pp 5402–5422. Collection of Laws of the Czech Republic
46. Šrámková M, Wanner J (2010) Reuse of cleaned wastewater. In: Proceedings of the conference drinking water 2010, pp 259—264. ISBN 978-80-254-6854-8
47. Materials for buildings (2010) Mosaic house—the first installation of reuse and recuperation of greywater in the Czech Republic. https://imaterialy.dumabyt.cz/rubriky/aktuality/projekty/mosaic-house-prvni-instalace-recyklace-a-rekuperace-sede-vody-v-cr_102143.html
48. The Rozmarýnek open new eko-building in Jundro (2014) iBrno.cz. https://www.ibrno.cz/zajimavosti/60536-rozmarynek-otevira-novy-ekodum-v-jundrove-.html
49. Greywater reuse and energy (2016) Asio.cz. https://www.asio.cz/cz/560.recyklace-sedych-vod-a-jejich-energie
50. Raček J (2016) Methodology of use greywater in the selected objects. Dissertation thesis. 198 p. Brno University of Technology. Faculty of Civil Engineering
51. Bartoník A, Holba M, Vrána J, Ošlejšková M, Plotěný K (2012) Greywater: possibilities of reusing and energy efficiency, treatment and reuse. Water management, pp 60–65. ISSN 1211-0760
52. Burrows WD, Schmidt MO, Carnevale RM, Schaub SA (1991) Non-potable reuse: development of health criteria and technologies for shower water recycle. Wat Sci Tech 29(9):2169–2184
53. Butler D, Friedler E, Gatt K (1995) Characterising the quantity and quality of domestic wastewater inflows. Wat Sci Tech 31(7):13–20

54. Casanova LM, Gerba ChP, Karpiscak M (2001) Chemical and microbial characterization of household greywater. J Env Sci Health A 36(4):395–401
55. Christova-Boal D, Eden RE, McFarlane S (1996) An investigation into greywater reuse for urban residential properties. Desalination 106:391–397
56. Elmitwalli T, Mahmoud N, Soons J, Zeeman G (2003) Characteristics of greywater: Polderdrift, the Netherlands. In: 2nd international symposium on ecological sanitation, Melbourne
57. Jefferson B, Judd S, Diaper C (2001) Treatment methods for greywater. In: Lens P, Zeemann G, Lettinga G (eds) Decentralised sanitation and reuse. IWA Verlag, London
58. Oldenburg M, Otterpohl R (2005) Untersuchungen zur anaeroben Behandlung von Fäkalabwasser mit Bioabfällen und weiteren Cosubstraten sowie der Grauwasserreinigung am Pilotprojekt Lübeck-Flintenbreite. Abschlussbericht Deutsche Bundesstiftung Umwelt
59. Palmquist H, Hanaeus J (2005) Hazardous substances in separately collected grey-and blackwater from ordinary Swedish households. Sci Tot Env 348:151–163
60. Rose JB, Sun G, Gerba CP, Sinclair NA (1991) Microbial quality and persistence of enteric pathogens in greywater from various household sources. Wat Res 25(1):37–42
61. Santala E, Uotila J, Zaitsev G, Alasiurua R, Tikka R, Tengvall J (1998) Microbiological greywater treatment and recycling in an apartment building. In: Advanced wastewater treatment, recycling and reuse: selected proceedings of the 6th international conference on advanced wastewater treatment, recycling and reuse, Held in Milan, Italy
62. Shin HS, Lee SM, Seo IS, Kim GO, Lim KH, Song JS (1998) Pilot scale SBR and MF operation for the removal of organic and nitrogen compounds from greywater. Wat Sci Tech 38(6):79–88
63. Siegrist R, Witt M, Boyle WC (1976) Characteristics of rural household wastewater. J Env Eng Div 102:533–548
64. Ottosson J (2003) Hygiene aspects of greywater and greywater reuse. Royal Institute of Technology (KTH), ISBN 91-7283-436-6
65. Biela R (2011) Topinfo. Greywater quality and reuse. tzb-info.cz. http://voda.tzb-info.cz/8097-kvalita-sedych-vod-a-moznost-jejich-vyuziti
66. Pidou M (2006) Hybrid membrane processes for water reuse, Cranfield university, School of Applied Science, Department of Sustainable Systems, Centre for Water Science, PhD Thesis, 2006, supervisor: Dr. Bruce Jefferson
67. Brewer D, Brown R, Stranfield G (2000) Rainwater and greywater in buildings: project report and case studies. Report 13285/1, BSRIA Ltd., Bracknell, UK
68. Gerba CP, Straub TM, Rose JB, Karpiscak MM, Foster KE, Brittain RG (1995) Water quality study of greywater treatment systems. Water Resour J 18:78–84
69. Mars R (2004) Case studies of greywater recycling in Western Australia. In: Proceedings of the 1st international conference on onsite wastewater treatment & recycling, Fremantle, Western Australia
70. Frazer-Williams R, Avery L, Jeffrey P, Shirley-Smirley C, Liu S, Jefferson B (2005) Constructed wetlands for urban greywater recycling. In: Proceedings of the first conference on sustainable urban wastewater treatment and reuse, Nicosia, Cyprus
71. Dallas S, Scheffe B, Ho G (2004) Reedbeds for greywater treatment—case study in Santa Elena-Monteverde, Costa Rica, Central America. Ecol Eng 23:55–61
72. Dalllas S, Ho G (2004) Subsurface flow reedbeds using alternative media for the treatment of domestic greywater in Monteverde, Costa Rica, Central America. Water Sci Technol 51(10):119–128
73. Shrestha RR, Haberl R, Laber J, Manandhar R, Mader J (2001) Application of constructed wetlands for wastewater treatment in Nepal. Water Sci Technol 44(11–12):381–386
74. Li Z, Guylas H, Jahn M, Gajurel DR, Otteroporhl R (2003) Greywater treatment by constructed wetlands in combination with TiO_2-based photocatalytic oxidation for suburban and rural areas without sewer system. Water Sci Technol 48(11–12):101–106
75. Schonborn A, Zust B (1997) Underwood E. Long term performance of the sand-plant-filter Schattweid (Switzerland). Water Sci Technol 35(5):307–314

76. Borin M, Cossu R, Lavagnolo MC, Gandini M (2004) Phytotreatment of greywater with yellow water addition from an aesthetic approach. In: Proceedings of the international conference on-site wastewater treatment and recycling, Murdoch University, Western Australia
77. Fittschen I, Niemczynowicz J (1997) Experiences with dry sanitation and greywater treatment in the ecovillage Toarp. Water Sci Technol 35(9):161–170
78. Grunther F (2000) Wastewater treatment by greywater separation: outline for a biologically based greywater purification plant in Sweden. Ecol Eng 15:139–146
79. Dušek L (2010) Wastewater chemical oxidation treatment b hydroxyl radicals. Chemistry papers, 846–854
80. Alaton IA, Balcioglu IA, Bahnemann DW (2002) Advanced oxidation of a reactive dyebath effluent: comparison of O_3, H_2O_2/UV-C and TiO_2/UV-A processes. Water Res 36(5):1143–1154
81. Muruganandham M, Swaminathan M (2006) Advanced oxidative decolourisation of Reactive Yellow 14 azo dye by UV/TiO_2, UV/H_2O_2, UV/H_2O_2/Fe^{2+} processes comparative study. Sep Purif Technol 48:297–303
82. Duran A, Monteagudo JM (2007) Solar photocatalytic degradation o reactive blue 4 using a Fresnel lens. Water Res 41:690–698
83. Peternel I, Koprivanac N, Kusic H (2006) UV based processes for reactive azo dye mineralization. Water Res 40:525–532
84. Patsons SA, Bedel C, Jefferson B (2000) Chemical versus biological treatment of domestic greywater. In: Proceedings of the 9th international Gothenburg symposium on chemical treatment, Istanbul, Turkey
85. Lin CJ, Lo SL, Kuo CY, Wu CH (2005) Pilot-scale electrocoagulation with bipolar aluminium electrodes for on-site domestic greywater reuse. J Environ Eng, pp 491–495
86. Itayama T, Kiji M, Suetsugu A, Tanaka N, Saito T, Iwami N, Michuochi M, Inamori Y (2004) On site experiments of the slanted soil treatment systems for domestic gray water. Water Sci Technol 53(9):193–201
87. Hypes W, Batten CE, Wilkins JR (1975) Processing of combined domestic bath and laundry wastewaters for reuse as commode flushing water. Report NASA TN D-7937, Langley research centre, Hampton, USA
88. Prathapar SA, Ahmed M, Al Adawi S, Al Sidiari S (2006) Design, construction and evaluation of an ablution water treatment unit in Oman: a case study. Int J Environ Stud 63(3):283–292
89. CMHC (Canada Mortgage and Housing Corporation) (2002) Final assessment of conservation Co-op's greywater system. Technocal series 02-100, CHMC, Ottawa, Canada
90. International Technology Service (2000) Water reuse and recycling in Japan. Report of Overseas Mission of Science and Technology Experts in Japan
91. Imura M, Sato Y, Inamori Y, Sudo R (1995) Development of a high efficiency household biofilm reactor. Water Sci Technol 31(9):163–171
92. Friedler E (2004) Quality of individual domestic greywater streams and its implication for on-site treatment and reuse possibilities. Faculty of Civil and Environmental Engineering, Technion, Haifa 32000 Israel
93. Bino MJ (2004) Greywater reuse for sustainable water demand management. In: Proceedings of the international water demand management conference, Amman, Jordan
94. Nolde E (1999) Greywater reuse systems for toilet flushing in multi-storey buildings—over ten years experience in Berlin. Urban Water 1(4):275–284
95. Jenssen PD, Moehlum T, Krogstad T, Vrale L (2005) High performance constructed wetlands for cold climates. J Environ Sci Health 40:1343–1353
96. Chorvátová M (2006) Biological wastewater treatment—study material. VSCHT
97. Judd S, Judd C (2006) The MBR Book: principles and applications of membrane bioreactors in water and wastewater treatment, ISBN 978-1-85617-481-7
98. Liu R, Huang H, Chen L, Wen X, Qian Y (2005) Operational performance of a submerged membrane bioreactor for reclamation of bath wastewater. Process Biochem 40(1):125–130

99. Andersen M, Krinstensen GH, Brynjolf M, Gruttner H (2001) Pilot-scale testing membrane bioreactor for wastewater in industrial laundry. Water Sci Technol 46(4–5):67–76
100. Lesjean B, Gnirss R (2006) Greywater treatment with a membrane bioreactor operated at low SRT and low HRT. Desalination 199:432–434
101. Laine AT (2001) Technologies for greywater recycling in buildings. PhD Thesis, Cranfield University, UK
102. Lodge B (2003) Membrane fouling during domestic water recycling. EngD thesis, Cranfield University, UK

Part IV
Water Supply and Sewerage Systems

Chapter 9
The Necessary Documents for the Design Documentation for Water Supply and Sewerage Systems in the Czech Republic

Vojtěch Václavík and Tomáš Dvorský

Abstract This chapter is devoted to the preparation of the data necessary for the individual stages of design documentation concerning the design of sewerage system, sewer connection, water supply, and water-service pipe. It does not deal with the specific design of the above-mentioned constructions, but it introduces the minimum requirements resulting from the existing legislation in force in the Czech Republic. It also outlines the possibilities of various permission modes for these constructions and deals with the obligations of the owners of these constructions and the possible sanctions for administrative offenses in the Water Act section.

Keywords Necessary documents · Design documentation · Water supply · Sewerage system

9.1 Basic Terms

A definition of the character of the construction of water supply, sewerage systems, water-service pipes and sewer connection is defined in the provisions of § 55 of Act No. 254/2000 Coll., on Water and on Amendments to Some Acts, which came into effect on January 1, 2002, (hereinafter the "Water Act") [1]. The constructions of water supply and sewerage systems that are permanently used by at least 50 natural persons or where the average daily production of an annual average of drinking or wastewater per day is 10 m^3 or more, and the constructions are defined as water supply and sewerage systems for public use. These systems follow Act No. 274/2001 Coll., on Water Supply and Sewerage Systems for Public Use and on Amendments to Certain Acts (Water Supply and Sewerage Systems Act) [2], as amended (hereafter only "Water Supply and Sewerage Systems Act").

V. Václavík (✉) · T. Dvorský
Faculty of Mining and Geology, Institute of Environmental Engineering, VŠB—Technical University of Ostrava, 17. listopadu 15/2172, 708 33 Ostrava-Poruba, Czech Republic
e-mail: vojtech.vaclavik@vsb.cz

M. Zelenakova et al. (eds.), *Management of Water Quality and Quantity*, Springer Water, https://doi.org/10.1007/978-3-030-18359-2_9

9.1.1 Sewerage System

A sewerage system is an operationally independent set of constructions and facilities including sewers used to drain waste and rainwater together or to drain wastewater and rainwater separately, sewerage facilities, wastewater treatment plants, as well as constructions used to purify wastewater before its discharge into the sewerage system. If wastewater and rainwater are drained together, it is a single-pipe system and the rainwater entering the sewerage system directly or via a connection becomes wastewater. If the wastewater is drained separately and so is the rainwater, it is a separate sewerage system. According to § 55 paragraph 1 letter c) of the Water Act and § 2 paragraph 2 of the Water Supply and Sewerage Systems Act [2], sewerage system is a waterwork.

Internal sewerage system is a pipeline designed to drain wastewater or rainwater from the building to its outer face. In cases where wastewater or rainwater is drained from the building, and the land outside the building, the end of the internal sewerage system is represented by the last connection of the external pipes. These places are also the beginning of the sewer pipeline.

9.1.2 Sewer Connection

Sewer connection is an independent structure consisting of a pipeline section from the outlet of the building internal sewerage system or land drainage to the sewage system. Sewer connection is not waterwork according to § 3 paragraph 2 of the Water Supply and Sewerage System Act [2] and § 55 paragraph 3 of the Water Act [1].

9.1.3 Water Supply System

Water supply system is an operationally independent set of constructions and facilities including the water mains and water supply facilities, which are, in particular, structures for the abstraction and diversion of surface water or groundwater, its treatment and collection. According to § 2 paragraph 1 of the Water Supply and Sewerage System Act [2] and § 55 paragraph 1 c) of the Water Act [1], the water supply system is waterwork.

Internal water supply system is a pipeline intended for water distribution across land or building that is connected to the end of the water-service pipe. According to § 2 paragraph 7 of the Water Supply and Sewerage System Act [2] and § 55 paragraph of the Water Act [1], the internal water supply system is not waterwork.

9.1.4 Water-Service Pipe

Water-service pipe is an independent construction consisting of a pipeline section from the water mains branching to the water meter, and, if there is no water meter, to the inner shut-off valve of the connected land or building. The branching with a shut-off valve is part of the water supply system. According to § 3 paragraph of the Water Supply and Sewerage System Act [2] and § 55 paragraph 3 of the Water Act [1], the water-service pipe is not waterwork.

9.2 Basic Requirements

Sewerage systems must be designed and built in such a way not to adversely affect the environment, to ensure sufficient capacity for drainage and treatment of wastewater from the drained area and to ensure the continuous drainage of wastewater from the customers of this service. At the same time, it must be ensured that the pollution of the recipients caused by rainfalls is limited. Sewerage systems must be constructed as waterproof constructions, protected from freezing and from damages caused by external effects. Further requirements for wastewater treatment, including the requirements for the design documentation, construction, and operation of sewerage systems and treatment plants are laid down in the implementing legal regulation, namely the Decree No. 428/2001 Coll. [3], which implements Act No. 274/2001 Coll., on Water Supply and Sewerage Systems for Public Use and Amendments to Certain Acts (the Water Supply and Sewerage System Act).

Sewers draining wastewater, except rainfall sewers, as well as sewer connections, must be installed deeper than the water pipes for drinking water distribution in case of their crossing or parallel position. An exemption according to the provisions § 12 of the Water Supply and Sewerage Systems for Public Use Act [2] may be granted by the Water Authority, provided that the technical measures will prevent the possibility of contamination of drinking water with wastewater during normal operation, even in case of sewerage system failure.

Water supply systems must be designed and constructed in such a way to ensure sufficient amount of harmless drinking water for public use and in the defined area and to ensure a continuous supply of drinking water to the customers. If the water supply system is the only source for the supply of firewater, it must meet the requirements of fire protection to provide water for firefighting if this is technically possible.

The **water supply pipeline** for public use, including its connections and the internal distribution systems connected to it, shall not be connected to water supply pipelines from a source other than the public water supply system.

Water supply systems must be protected against freezing, damage caused by external effects, external and internal corrosion, and against intrusion of harmful microorganisms, chemicals and other substances degrading the quality of drinking water. Other technical requirements for the construction of water supply systems are laid down in an implementing legal regulation, which is Decree No. 428/2001 Coll. [3], which implements Act No. 274/2001 Coll., on Water Supply and Sewerage Systems for Public Use and Amendments to Certain Acts (the Water Supply and Sewerage System Act).

9.3 Requirements for Sewerage (Sewage) System Design Documentation

Drainage of wastewater is designed according to the amount of wastewater, the calculated amount of drained rainwater and the single-pipe or separate sewerage system. The design and construction of sewage systems are based on standard values: ČSN 75 6101 Sewer Systems and house connections [4], ČSN EN 752 Drain and sewer systems outside buildings—Sewer system management [5], ČSN EN 1091 Outdoor vacuum systems of sewage systems, ČSN EN 1671 Pressure sewerage systems outside buildings [6]. Sewage system is designed as gravity, pressure, and vacuum type or a combination thereof. Sewers and objects on sewers are designed as watertight structures. The pipe connections must be watertight. This water-tightness is demonstrated according to the standard values in compliance with: ČSN EN 1610 Construction and testing of drains and sewers [7], ČSN 75 6909 Testing water-tightness of drains and sewers [8], ČSN 75 0905 Water supply and sewerage tanks—Testing of water-tightness [9].

With a pipeline slope of up to 10‰, the height deviation in the sewer position can be ±10 mm, with a slope above 10‰, it can be ±30 mm in comparison with the bottom spot height determined by the design documentation. Counter-slope of the pipeline is not allowed. Straight sewer segments between two shafts may have a directional deviation from a straight line of up to 50 mm in case of internal diameter of up to 500 mm, or up to 80 mm in case of larger diameters. In the event that a new part of the sewerage system draining rainwater from new buildings in the built-up area will be connected to a single-pipe or separate sewerage system draining rainwater, a new calculation will be carried out. A new calculation is included in the design documentation in order to verify the capacity of the sewerage system to drain the increased amount of these waters. This calculation is the basic document of the owner of the sewerage system or, where applicable, of the operator, if authorized by the owner, used to allow or reject such a connection. The distance of the inspection and access manholes in the straight line of impassable sewers does not exceed 50 m, in the case of the passable sewers, it is up to 200 m. The inspection and access manhole and the well hole cannot be placed outside the sewer route.

9.4 Technical Requirements for the Construction of Water Supply Systems

Water supply network and the supply pipelines are designed for maximum hourly water demand. The pipelines of other water mains are designed for maximum daily water demand. The water supply pipeline is designed according to standard values: ČSN 75 5401 Designing of water pipeline [10], TNV 755402 Water pipeline construction [11]. The maximum overpressure at the lowest points of the water supply network of each pressure zone shall not exceed 0.6 MPa. In justified cases, it may increase to 0.7 MPa. When installing on two above-ground floors, the hydrodynamic overpressure in the distribution network must be at least 0.15 MPa at the connection point of the water-service pipe connection. When installing in a building with two above-ground floors, it must be at least 0.25 MPa. The water supply pipeline must be protected against external and internal corrosion with regard to the properties of the pipe material, the quality of the water to be transported, and the environment in which it will be installed. The water pipeline with the inner diameter of up to 200 mm is designed with a longitudinal slope of at least 3‰. The water pipeline from an inner diameter of 250 mm to an internal diameter of 500 mm is designed with a slope of at least 1‰. A pipe with an internal diameter of 600 mm and more is designed with at a slope of at least 0.5‰. The water meter shaft must be secured against the ingress of impurities, underground and surface water and must be ventilated and accessible. The shafts on the water supply pipeline must be designed in such a way to make sure the fittings installed on the pipeline are adequately protected against frost. The requirements for materials used chemicals and products in direct contact with drinking water are stipulated in a special legal regulation 13).

9.5 Legislative Background

The basic principles of building permissions, their modifications as well as permissions for the use of buildings are defined in the following list of basic legal enactments:

- Act No. 500/2004 Coll., the Administrative Procedure Code, as amended (hereinafter referred to as the "Administrative Procedure Code") [12].
- Act No. 183/2006 Coll., On Land-use Planning and the Building Code, as amended (hereinafter referred to as the "Building Act") and its implementing regulations [13].
- Act No. 254/2001 Coll., On Water and Amendments to Certain Acts, as amended (hereinafter referred to as the "Water Act") and its implementing regulations [1].
- Act No. 274/2001 Coll., On Water Supply and Sewerage Systems for Public Use and on the Amendment of Certain Acts, as amended and its implementing regulations [2].

The Administrative Procedure Code is a general, basic legal regulation defining the actions of the bodies of executive power, bodies of territorial self-governing units and other bodies, legal entities and natural persons, if they exercise their force in the field of public administration (they are referred to as administrative authorities). Both the Building Act and the Water Act are special laws with respect to the Administrative Procedure Code. Similarly, the Water Act is a special law in relation to the Building Act. This means that, in relation to the different phases of the permission procedure for sewerage and water supply systems, it is necessary to proceed according to the Water Act. In the event that this act does not deal with this procedure, according to the Building Act. In the event that it is not dealt with in this act either, according to the Administrative Procedure Code.

9.6 Subject Matter and Local Jurisdiction

The subject matter and local jurisdiction are generally described in the Administrative Procedure Code in provisions § 10 and 11 [12]. The local jurisdiction means that the administrative authorities can only act and decide on matters that have been entrusted to them by the law or on the basis of the law. In relation to water supply and sewerage systems, the local jurisdiction is determined by the location of the construction of a particular water supply and sewerage system.

As already mentioned above, water supply and sewerage systems are constructions of waterworks (provision of § 55 of the Water Act [1]).

The General Building Authority is the competent administrative body as far as the subject matter jurisdiction for the location of the construction of water supply, and sewerage systems are concerned. The list of the General Building Authorities is presented in the provision of § 13 of the Building Act [13]. The competent administrative body as far as the subject matter jurisdiction of the other stages of the permission procedure for water supply and sewerage systems is concerned is the Water Authority. It performs the function of special building authority for waterworks (which include the constructions of water supply and sewerage systems)—provisions of § 15 paragraph 4 of the Water Act [1] and provisions of § 15 paragraph 1 of the Building Act [13]. The special building authorities have the competence of the Building Authority, except the authority for the sphere of territorial decision-making within the area in question—Water Authorities have the competence in the field of waterworks. A list of the Water Authorities is presented in the provision of § 104 of the Water Act [1].

As already mentioned above, the local jurisdiction is given by the location of the construction of particular water supply and sewerage system.

9.7 Design Documentation and Designing Activity in Construction

9.7.1 Design Documentation

The design documentation in relation to water supply and sewerage systems is the documentation required for:

- the issuance of a building permit,
- the conclusion of a public contract,
- an assessment by an authorized inspector,
- the construction change before its completion in relation to previously mentioned constructions,
- the previously mentioned constructions, if they are subjected to repeated building proceedings or an additional building permit,
- the execution of construction,
- the necessary modifications, and documentation for announced wastewater treatment plants.

These facts make it clear that the documentation for the issuance of a zoning and planning decision and the documentation for the conclusion of a public contract replacing the zoning and planning decision are not designed documentation within the meaning of the Building Act.

The scope of the individual documentation is set out in Decree No. 499/2006 Coll., On Construction Documents, as amended [14]. This Decree determines the structure of the documentation and the contents of the individual sections.

The preparation of the design documentation, as well as the preparation of the documentation for the issuance of the zoning and planning decision and for the conclusion of a public contract replacing the zoning and planning decision, represents a selected activity in construction. Such an activity may be performed only by natural persons who have been authorized to perform it under a special legal regulation, namely Act No. 360/1992 Coll., on Professional Practice of Authorized Architects and Authorized Engineers and Technicians Active in Construction, as amended [15]. An authorization for a designing activity as a selected activity in construction is not an authorization to perform entrepreneurial activity according to the Trade Act or to carry out a trade of designing activity in construction, or designing simple and small buildings, their changes and removals. That is why a designer cannot show the Building Authority an authorization to execute the designing activity on the basis of a trade certificate but on the basis of the relevant authorization certificate.

The documentation shall be submitted in two copies; if the Building Authority is not a municipal office at the location of the building, in three copies; if the builder is not the owner of the construction, one additional copy of the design documentation is attached. If the waterwork under the permission proceedings concerns boundary waters, the design documentation shall be submitted in a minimum number of two copies.

9.7.2 Construction Designing Activity

A designer is responsible for the accuracy, entirety, and completeness of the documentation prepared for the issuance of the zoning and planning decision, especially while respecting the requirements for the protection of public interests and their coordination. A designer is obliged to comply with the legislation and to act in cooperation with the relevant land-use planning authorities and the authorities concerned. A designer is also responsible for the integrity, completeness, correctness, and safety of the construction carried out according to the design documentation prepared by him and the feasibility of the construction according to this documentation, as well as the technical and economic design level of the technological equipment, including the environmental impact. A designer is obliged to comply with the legal regulations and general construction requirements related to the particular building plan and to act in cooperation with the relevant authorities concerned. Static or other calculations must be carried out in such a way to be controllable. If the designer is not competent to prepare any part of the design documentation himself/herself, he/she must invite a person authorized for the relevant branch or specialization, and this person is responsible for the elaborated design. The designer's responsibility for the design documentation for the construction as a whole is not affected by that.

9.8 Construction Location

The location of the water supply and sewerage system constructions and their changes can only be made on the basis of a zoning and planning decision or zoning and planning consent. The zoning and planning decisions may, under certain conditions, be replaced by a zoning and planning consent or a public contract. The zoning and planning decision is not issued if it is replaced by a regulatory plan. The Building Authority may also jointly conduct the area management and building proceedings. If it is possible to replace the zoning and planning decision with a public contract and, at the same time, it is possible to replace the building permit with a public contract, a public contract, which will replace both the zoning and planning decision and the building permit, can be concluded.

The competent administrative authority is represented by the municipal building authority.

9.8.1 Decision on Construction Location—Zoning and Planning Decision

The decision on construction location is determined by the building land. It locates the designed construction, determines its type and purpose, the conditions for its location, for the preparation of design documentation necessary for issuing the building permit, the announcement of the construction and for the connection to the public transport and technical infrastructure. The issuance of the zoning and planning decision is preceded by the territorial proceedings. The application for the zoning and planning decision includes, among other things, the documentation necessary for the issuance of the zoning and planning decision. Its preparation is a selected activity in construction. The scope of this documentation is defined in Decree No. 499/2006 Coll., On Construction Documents, as amended [14]. By issuing the zoning and planning decision, the Building Authority approves the designed plan and determines the conditions for the use and protection of the territory, the conditions for further preparation and implementation of the plan, especially for the designing preparation of the construction.

The zoning and planning decision on the location of the construction is valid for two years from the effective date unless the term is extended for a longer period by the Building Authority, but it will not exceed 5 years. The Building Authority can extend the validity of a zoning and planning decision upon a reasonable request. The zoning and planning decision shall not cease to be valid if a final building permit or another similar decision pursuant to this Act or special legal regulations have been issued on the basis of an application submitted during its effectiveness. If on the basis of a notification of a construction plan assessed by an authorized inspector submitted during its effectiveness, the right to execute the building plan came into being. And if a public contract has been concluded on the basis of a proposed public contract replacing the building permit and submitted at the time of its effectiveness and this public contract has come into force.

The zoning and planning decisions may be changed at the request of the beneficiary if the land-use planning documentation or other documents for the zoning and planning decision or the conditions in the area have changed, by replacing the former part with a new zoning and planning decision.

9.8.2 Simplified Territorial Proceedings

Systematic pressure to simplify the procedures according to the Building Act has resulted, among other things, in the development of simplified territorial proceedings, which involve the issuance of the zoning and planning decision without prior administrative proceedings in the entirety, but provided that certain legislative conditions are met.

9.8.3 Public Contract

The Building Authority will conclude a public contract concerning the location of the construction with the applicant, which will replace the zoning and planning decision. The applicant will, among other things, attach documentation to the public contract to the same extent as in case of an application for the zoning and planning decision and other details to the same extent as in case of the application for the issuance of the zoning and planning decision. The Building Authority shall review the draft of the public contract and either accepts or rejects the draft and notifies the applicant of the reasons for the refusal. If the Building Authority accepts the draft of the public contract, it will indicate the effectiveness of the contract and its graphic attachments.

The effects of the public contract expire two years after the day of its effectiveness unless a longer term is agreed, but not longer than five years. The effects of the public contract shall not cease to be valid if a final building permit or another similar decision pursuant to this Act or special legal regulations have been issued on the basis of an application submitted during its effectiveness. If on the basis of a notification of a construction plan assessed by an authorized inspector submitted during its effectiveness, the right to execute the construction plan came into being. And if a public contract was concluded on the basis of a proposed public contract replacing the building permit and submitted at the time of its effectiveness, and this public contract came into force. The effects of the public contract can be extended (a proposal for an extension must be submitted before the effects of the public contract expire).

An applicant may withdraw from the public contract on the basis of a notice to the Building Authority stating that the applicant is abandoning his plan—this does not apply if the execution of the plan has already been launched.

9.9 Permit to Commence Construction

9.9.1 Building Permit

A building permit is an administrative decision, the virtue of which entitles the builder to commence and execute the construction. According to the Water Act [1], the Water Authority (which is the relevant administrative authority) has to issue a permit to carry out waterworks (i.e., water supply and sewerage systems). A permit to execute or change waterworks to be used for water management (which also requires a permit of the Water Authority) may only be granted if the corresponding water management has been permitted or the water management is permitted along with the permit to execute the waterworks.

Building permits and announcements are not required in case of construction modifications (changes of the completed construction, which maintains the external ground and height limits of the construction) of water supply and sewerage systems unless their route is changed (either vertically or horizontally).

The Water Authority has to be notified about the renewal of waterworks destroyed by natural disasters or accidents and maintenance work that could negatively affect the environment or the stability of the waterworks. In the case of maintenance work on the construction, i.e., work that ensures its good construction condition to avoid degradation of the construction and to prolong its usability as long as possible. The maintenance work is, therefore, an activity that does not increase the value of the construction.

Any other construction of water supply and sewerage systems (waterworks) is subject to the permission of the Water Authority, as already mentioned above.

The building processes which preceded the issuance of a building permit are only commenced at the request of the builder. The application for a building permit shall be accompanied by the documents stipulated by the relevant implementing regulation—including, inter alia the following (see Sect. 9.7):

- the document on the basis of which the construction was located,
- the document through which the builder demonstrates the right to execute the construction on land not owned by him,
- the binding opinions or decisions of competent authorities,
- the views of the public transport and technical infrastructure owners on the possibility and method of connection or on the conditions of the affected protection and safety zones,
- the design documentation prepared by the designer.

The submission of design documentation not prepared by the designer is a reason for stopping the proceedings.

The Building Authority will review the application and the attached documents from the point of view of whether the construction can be executed. It will check whether the design documentation is prepared in accordance with the zoning and planning decision or a public contract replacing the zoning and planning decision or the zoning and planning approval, or with a regulatory plan to the extent in which it replaces the zoning and planning decision. The authority will also check whether the design documentation is complete, well arranged and whether the general construction requirements are addressed to the right extent (the scope of the design documentation is determined by Decree No. 499/2006 Coll. [14]). It will also check whether there is access to the construction site, whether the technical or other equipment necessary for the proper use of the construction required by a special legal regulation was built in time, and whether the submitted documents comply with the requirements imposed by the authorities concerned. The Building Authority will also verify the effects of the future use of the construction.

If the application contains the required parts, the Building Authority shall notify the known parties to the proceedings and the authorities concerned of the commencement of the building proceedings and shall set a term within which they may raise their objections, evidence, or binding opinions. The Building Authority disregards, among other things, the objections that were or could be raised during the territorial proceedings or the preparation of the regulatory plan. Then the Building Authority decides on the merits.

In the building permit, the Building Authority shall define the conditions for the execution of the construction and, if necessary, for its use as well. It also defines the conditions in order to ensure the protection of public interests and provides, in particular, the continuity with other constructions and facilities, observance of the general requirements for construction, including the requirements for barrier-free use of the construction, or the technical standards. It may also stipulate that the building may be used only on the basis of a certificate of occupancy. In addition, it may impose the execution of a trial operation. It is in the case of a construction containing technological devices for which the eligibility for safe use and the compliance with the building permit conditions. It is also if an integrated permit according to a special legal regulation must be verified (in this case, the construction use may be permitted only in the certificate of occupancy mode). In the building permit, the Water Authority may require the submission of the operational regulation for the waterworks.

The Water Authority shall require the submission of an operating permit together with the application for the issuance of the certificate of occupancy. It is in a case of the building permit or in the public contract relating to the construction of water mains, water facilities, water treatment plants, sewers, including sewerage facilities or wastewater treatment plants, which are part of public water supply or sewerage systems for public use.

The building permit shall expire if the construction had not started within two years from the date when the permit became final. The Building Authority may extend the validity of the building permit based on a reasoned request from the builder submitted before its expiration. The building permit will also expire on the day when the Building Authority receives a notice from the building contractor stating that the project was abandoned—this is not valid if the construction has already started.

9.9.2 Public Contract

In case of constructions requiring building permits, the Building Authority may conclude a public contract for the execution of the construction with the builder, which replaces the building permit. This Building Authority for water supply and sewerage systems (waterworks) is represented by a special building authority—the Water Authority. The builder will, among other things, attach the design documentation and other documents to the public contract draft with the same scope as required for an application for the building permit. The Building Authority shall review the draft of the public contract and either accepts or rejects the draft and informs the builder of the reasons for refusal. The draft of the public contract will always be refused if the design documentation is not elaborated by the designer. If the draft of the contract is accepted, the Building Authority it will indicate the effectiveness of the contract, verify the design documentation and make a label containing the building permit identification data ("building permit" label).

The effects of a public contract expire two years after its effective date if the construction has not started within this period. The effects of a public contract can be extended (an application for an extension must be submitted before the effects of the public contract expire).

The builder may withdraw from a public contract on the basis of a notice to the Building Authority stating that he abandons the plan—this does not apply if the construction has already begun.

9.9.3 Authorized Inspector

An authorized inspector is a person who, to a certain extent, plays the role of the Building Authority. According to the Water Act, the waterworks are not eligible for inspection by an authorized inspector—this does not apply to the constructions of water supply mains, sewers, and sewerage facilities that do not require a water management permit. It is therefore evident that, in the case of water supply systems, the assessment by an authorized inspector is possible, in the case of sewerage systems, it is so only if a water management permit is not required in relation to the sewerage system.

If the builder signs a contract with an authorized inspector to check the design documentation of the construction the builder intends to carry out, the authorized inspector may assess the design documentation instead of the Building Authority provided that: (1) it is not a construction that is designated by a special legal regulation as ineligible for an assessment by an authorized inspector, (2) it is not a construction that has significant environmental effects or effects on other lands, (3) it is not a construction with a common boundary with the building land. The authorized inspector must notify the Building Authority about the conclusion of the contract without undue delay together with the required documents—including the design documentation. The Building Authority will announce the building plan on the official notice board for at least 30 days. Within this time limit, the persons who would be the participants of the building proceedings may object to the notified construction plan, only provided that it does not comply with the documents on the basis of which they had granted their consent or because their consent had not been acquired. Within the same term, the Building Authority or the authority concerned may object against the notification of the building plan, if:

- they think that the construction is ineligible to be assessed by an authorized inspector,
- the announcement does not meet the conditions and does not contain the necessary particulars,
- the authorized inspector has violated the prohibition of activity when issuing the certificate,
- the construction assessment has not met the requirements or if the building proceedings have not been legally closed yet.

The submission of objections or reservations has a deferring effect, and the right to execute the construction will not be granted. The Building Authority will then notify the builder of the objections or reservations and will submit the matter to the administrative authority that would otherwise be competent to deal with an appeal against the building permit. The administrative authority shall review the notification of the building plan and make a decision—it shall refuse the objections or reservations or decide that the notice has no legal effect.

The builder will have the right to execute the notified construction plan upon the expiry of the term for the submission of objections or reservation or on the day following the day when the rejection of objections or reservations was announced to him. The right to execute the construction will cease to exist if the construction has not commenced within two years. The Building Authority may extend this term upon a reasoned request made by the builder before its expiration.

9.10 Construction Change Before Its Completion

The builder is obliged to execute the construction with a permit. A building permit, a public contract, a notified issuance of authorization—on the basis of a certificate of an authorized inspector, a repeated building permit and an additional building permit are considered as permits for construction in relation to water supply and sewerage systems.

A change of construction before its completion can be permitted before the commencement of the construction or during the execution of the construction—this procedure cannot be applied in cases where the change has already been made. A construction change before its completion may be permitted only in compliance with the zoning and planning decision or another act replacing the zoning and planning decision. If the construction change before its completion requires a change of the zoning and planning decision, it is possible to decide on the change in the joint proceedings, otherwise, after the change of the zoning and planning decision has been resolved. The Building Authority may, at the request of the builder or his/her legal successor, permit a construction change before its completion. The application shall be accompanied by the construction change documentation or, where appropriate, a copy of verified design documentation in which the designer shall indicate the proposed changes. The construction change before its completion is negotiated adequately according to the provisions concerning the construction proceedings and only within the extent of the proposed changes.

You can change the notified construction on the basis of an announcement. The procedure is similar in this case.

The competent authority for permitting a construction change of water supply and sewerage systems (waterworks) is the competent Water Authority (whether for the permit proceedings or the notification proceedings).

A construction change before its completion can be done on the basis of a public contract; it can also be carried out on the basis of notification to the Building Authority—a certificate of an authorized inspector.

In the event that the construction change does not affect the rights of the other participants in the construction proceedings, the Building Authority may approve the construction change before its completion by a decision issued on site during the inspection of the construction. It may do so only if the change does not affect the conditions of the zoning and planning decision, public interests protected by special legal enactments or in cases where the competent authority concerned agrees with the change in writing or through a statement in the protocol.

9.11 Construction Use

The competent administrative authority for the use of the construction of water supply and sewerage systems—waterworks is the Water Authority.

The use of constructions involves both permanent and temporary constructions (constructions where the Building Authority limit their duration in advance), completed constructions or their completed parts that can be used independently, as well as constructions before their completion, and constructions the functionality and properties of which are to be verified.

9.11.1 Final Inspection Certificate, Certificate of Occupancy, and Notification of the Use of Constructions

Completed constructions of water supply and sewerage systems or completed parts of water supply and sewerage systems capable of independent use (whether they are temporary or permanent constructions) may be used on the basis of a notice to the Building Authority or a certificate of occupancy. They can be used regardless of the fact whether or not they required a building permit or a notice, whether they were executed on the basis of a public contract or a certificate of an authorizes inspector, whether they were completed under a repeated building permit or an additional building permit. The notification form for the building used to the Building Authority can be used in those cases where it is not a construction which requires a certificate of occupancy. A certificate of occupancy is required for constructions whose properties cannot be influenced by the future users, as well as for constructions that have been subjected to trial operation, and for construction changes of cultural monuments. The certificate of occupancy will also be granted for construction where it was by the Building Authority in the building permit.

When putting the construction into use, the Building Authority shall examine whether the construction has been carried out in accordance with the decision of its location or another act replacing the zoning and planning decision (including its conditions) and the building permit (including its conditions) and documentation, or verified design documentation, in compliance with the opinions or decisions of the authorities concerned, if issued in accordance with special legal regulations, and if the general construction requirements are followed. It also examines whether the actual execution or use of the construction will not endanger the life and public health, animal life or health, and the safety or the environment. For the assessment of these facts, the builder submits to the Building Authority the documents listed in the relevant implementing regulation. For water supply and sewerage systems it will, among other things, include:

- measurement of the actual execution of the construction,
- documentation of the actual execution of the construction,
- evidence of the performed tests—pressure test, hydrant flow capacity, water analyzes, functionality of the water supply system signaling conductor, water-tightness tests of the sewerage pipeline and shafts (according to standard values), etc.,
- certificates of the used materials, followed by a binding opinion of the Public Health Protection Authority in the case of water supply systems used for drinking water, etc.

If the Building Authority concludes that the above-mentioned facts have not been met or that the construction was carried out in contravention of the building permit or notification, or that the construction is used without prior notice, or the construction has defects preventing its safe use the next will follow. The Building Authority will prohibit the use of the construction until the builder notifies the Building Authority in writing of the removal of the deficiencies, and the Building Authority then verifies that this notification reflects the real state. The Building Authority then issues the relevant permit for the use of the construction.

In both authorization proceedings dealing with the use of construction, the Building Authority may approve minor changes to the construction. Subject to the issuance of the certificate of occupancy, a written agreement in which the owners of the water supply systems or operationally connected sewerage systems or their operationally connected parts shall adjust their mutual rights and obligations. So as to ensure the quality and smooth operation of the water supply or sewerage system is a condition for the issuance of the certificate of occupancy.

Completed constructions of water supply and sewerage systems or completed parts of water supply and sewerage systems capable of independent use (regardless of whether the constructions are temporary or permanent) may also be used on the basis of the final inspection certificate known from the previous Building Act No. 50/1976 Coll., on Land-use Planning and Building Code, as amended [16]. Although the Building Act effective from January 1, 2007 [13] does not use this term, you can come across it. In its transitional provisions, the Building Act states that buildings legally authorized before the effective date of this Act (i.e., until December 31, 2006 inclusive) shall be subject to a building permit proceedings in accordance with the existing legislation (i.e., the former Building Act No. 50/1976 Coll., on Land-use Planning and Building Code, as amended [16]). The assessment of the construction in order to permit its use is essentially the same as for the certificate of occupancy described above and the notices of the use of the construction.

9.11.2 Premature Use

At the request of the builder, the Building Authority may issue a time-limited permit for premature use of the construction before its completion, if it does not have a significant effect on the usability of the construction, it does not endanger the safety and health of persons or animals or the environment. These facts shall be assessed by the Building Authority from the submitted documents, which are defined by the relevant implementing regulation. Upon completion of the construction, a standard procedure leading to the permit of the use of the construction will be carried out—certificate of occupancy, notices of the use of the construction, final inspection certificate.

9.11.3 Trial Operation

A trial operation of the construction verifies the functionality and properties of the construction according to the documentation or design documentation. The Building Authority may authorize a trial operation at the request of the builder or order it at the request of the authority concerned, or in another justified case. The evaluation of the trial operation shall be attached to the application for the issuance of the trial operation or the final inspection certificate. A notification of the use of the construction is excluded here.

9.12 Construction Removal Permit

The Building Act allows, under certain conditions, removal of construction on the basis of notification of the intention to remove the construction submitted to the Building Authority or on the basis of a permit of the Building Authority. As mentioned above, the Water Act is the special law for the Building Act, and it states that the Water Authority is required to issue a permit to remove the construction of waterworks. The Water Authority is the competent authority for the removal of water supply and sewerage systems, i.e., waterworks. It is also clear from the Water Act that the notification method cannot be applied in relation to the removal of sewerage and water supply system.

Only the owner of the construction can be allowed to remove it. One of the documents to be submitted with the application under the applicable implementing regulation is the document in which the applicant proves the ownership of the construction. Also, the demolition work documentation is also submitted with the request for the removal of the construction. The demolition work documentation is not, according to the Building Act, the design documentation and its preparation is not a selected activity in construction. The scope of the submitted documentation of demolition works is defined in Decree No. 499/2006 Coll., On Construction Documents, as amended [14].

9.13 Construction Passport and Documentation of Actual Execution

For the entire duration of the construction, the owner of the construction is obliged to keep verified documentation corresponding to its actual execution according to the issued permits. In cases where the documentation of the construction has not been made, has not been preserved or is not in proper condition, the owner of the construction is obliged to prepare the documentation of the actual execution of the construction. Upon a change of the ownership of construction, the owner will submit the documentation to the new owner of the construction. If the owner of the construction fails to meet this obligation, the Building Authority will order him to obtain the documentation of the actual execution of the construction. If the complete documentation of the actual execution of the construction is not necessary, the Building Authority will only order the elaboration of simplified documentation of the construction (construction passport), if the builder has not done so himself.

The scope of documentation of the actual execution of the construction and the construction passport is stipulated in the Building mentioned above Documentation Decree. From the above-stated facts, we can see that the documentation of the actual execution and the construction passport is not the design documentation and, therefore, a selected activity in construction, i.e., such documentation does not have to be prepared by an authorized designer.

The documentation of the actual execution of the construction shall also be submitted by the builder to the Building Authority if there were minor deviations from the building permit, the public contract, the certificate of an authorized inspector or the verified project documentation during the execution of the construction.

9.14 Obligations

The obligations of the owners of waterworks, i.e., in this case, the owners of water supply and sewerage systems, are defined in the provisions of § 59 of Act No. 254/2001 Coll., On Water and on the Amendments to Certain Acts, as amended [1]. The list of these obligations of the owners of waterworks in this provision is not final, because the owner of the waterworks also has general obligations as the owner of a construction stipulated by Act No. 183/2006 Coll., On Land-use Planning and the Building Code, as amended [13], especially to inform the Building Authority without delay about the defects of the construction that endanger the lives or health of persons or animals. These are obligations toward the owners of waterworks, i.e., the owner of the waterworks is responsible for their observance, although, for example, a breach of the obligations occurred due to the fault of third parties. The owner is then entitled to claim compensation for the incurred damage from this third party.

The owners of water and sewerage connections as owners of general constructions are bound by the provisions of Act No. 183/2006 Coll., On Land-use Planning and Building Code, as amended [13], which contain lists of obligations of owners of "general constructions."

9.14.1 Obligations of the Owners of Water Supply and Sewerage Systems

The rights and obligations of the owner of water supply or sewerage systems are specified with respect to the fact that this service is "for public use," and the owners of the water supply or sewerage systems are not only municipalities but also business entities that could completely disable water supply or sewerage system without any alternative operation. The provisions of § 8 of the Water Supply and Sewerage Systems Act in relation to the fact that they are constructions in case of which it is not allowed to inactivate the water supply or sewerage systems outside the proceedings since the parties to the proceedings are all the concerned persons, represented by the customers. In conjunction with the provisions of § 22 of the Water Supply and Sewerage Systems Act [2], the "public service obligation" ensures smooth and safe operation of the water supply and sewerage systems. The quality and continuous operation mean ensuring all activities listed in the definition of "operation" according to the provisions of § 22 paragraph 3 of the Water Supply and Sewerage Systems Act [2]. Only a fraction of the statutory duties is presented below just to give you an idea of the scope.

The owners of water supply or sewerage systems, for example, have an obligation to ensure their smooth and safe operation, to create a reserve of funds for their renewal and to document their use for these purposes. In the case of constructions of operationally connected water supply and sewerage system, the owners of such constructions are obliged to adjust their mutual rights and obligations by a written agreement so as to ensure the quality and smooth operation of the water supply or sewerage system. This agreement is a condition of the certificate of occupancy according to the Building Act. The agreement must at least include—the subject matter of the agreement, including the indication of the ownership of the water supply system, the definition of the supply conditions, such as the quantity, quality, pressure conditions, measurement, functionality, cost specifications, their controls, payment methods, contractual sanctions, and others. They are also obliged to allow the connection of water supply or sewerage systems of another owner, if the capacity and technical capabilities allow it. At the same time, the owners of these sewerage and water supply systems will sign a written agreement. The possibility of connection to the sewerage and water supply systems must not be conditioned by requiring any financial or other performances. The costs of the connection are covered by the owner who is allowed to connect the water supply or sewerage system (at the same time, an agreement according to § 8 paragraph 3 of the Water Supply and Sewerage System

Act [2] must be concluded). The owner of the water supply or sewerage system or the operator, if authorized, is also obliged to conclude a written contract for the supply of water or the discharge of wastewater with the customer. The obligations under this contract are transferred to the legal successor of the owner of the water supply and sewerage system and to the legal successor of the operator. One of the other obligations of the owners of water supply or sewerage systems is to ensure the continuous maintenance of the property and to keep operational records at their own expense. Selected data from the property and operational records, which are stipulated by the implementing regulation, must be submitted free of charge by the owners of the water supply and sewerage systems in electronic form in the prescribed format to the locally competent Water Authority, each year within 28 February for the previous calendar year.

Public water supply and sewerage systems can only be operated by a person who applied for the issuance of a permit to operate a water supply or sewerage system at the regional authority. The owner of such a water supply or sewerage system or the operator, if authorized to do so, is also obliged to publish a comparison of all price calculation items. Price calculation is according to the water and sewage tariff price regulations and the actual figures achieved in the previous calendar year by the 30 April of the calendar year, in a manner specified in the implementing regulation.

The obligations of the owners of water supply or sewerage systems, as the owners of constructions, are also defined in the Building Act (general obligations of building owners) and in the Water Act, as already mentioned above.

The ownership relations to water supply or sewerage systems are not registered in the Real Estate Cadastre.

9.14.2 Obligations of the Owners of Water-Service and Sewerage Connections

These are obligations resulting from the Building Act. They include, for example, obligations of the owner of the sewerage connection to ensure the sewerage connection is made as watertight and also to make sure the flow profile of the sewer into which it is leading is not reduced. The owner of a water-service connection is obliged to ensure that the water-service connection is made and used in such a way to make sure the water in the water main could not be polluted. The municipal authority may, in the delegated jurisdiction, decide that the owners of the building site or constructions on which the wastewater is or may appear are obliged to set up a connection to the sewerage system, if it is technically possible. It is in a case if there is a sewerage system used to drain wastewater at the real estate site, and the owner of the land or buildings in which wastewater originates or may be generated is not connected to this sewerage system.

The ownership relations to water and sewerage connections are not recorded in the Real Estate Cadastre.

9.15 Unauthorized Water Withdrawal, Wastewater Disposal

The specification of unauthorized water withdrawal and unauthorized wastewater discharge is necessary due to the sanctions for such behavior.

Unauthorized water withdrawal from the water supply system is a withdrawal in front of the water meter, without a signed written contract for water supply or in contradiction with it. It also includes water withdrawal through a water meter that does not record or records lower than the actual withdrawal values due to customer intervention or through a water meter the customer did not adequately protect against damage.

Unauthorized discharge of wastewater into the sewerage system is a discharge without a written contract for the discharge of wastewater or in contradiction with it. A discharge of wastewater in contravention of the conditions established for the customer by the sewerage system, or through a measuring device not approved by the operator, or through a measuring device that does not record the amount of discharged wastewater or records a lower amount than the actual one as a result of an intervention of the customer.

These facts constitute the merits of an offense of a natural person and an administrative offense of a legal person.

9.16 Sanctions

The administrative punishment in the Water Act is elaborated according to the Principles of Legal Regulation of Offenses and Other Administrative Offenses in the acts regulating the performance of public administration. The Water Act [1] contains both the offenses of natural persons (Part 1, § 116 to § 125) and administrative offenses of legal and self-employed natural persons (Part 2, § 125 to § 125 k). The common provisions related to administrative offenses are listed at the end of Title XII (Part 3, § 125). The offense proceedings take advantage of ancillary offense law, which remains the general procedural definition of the offense proceedings, the proceedings of the administrative offenses of legal entities and entrepreneurs use the ancillary provisions of the Administrative Procedure Code. All administrative offenses under the Water Act [1] are contained in Title XII and relate to the violations of specific legal enactments. In the case of offenses, it is always necessary to be aware of their relation to criminal offenses; an offense is a cased act that violates or threatens the interest of the society and is expressly identified as an offense in the Water Act or another law, unless it is another administrative offense punishable according to special legal regulations.

Also, in the section of the Water Supply and Sewerage System Act, the administrative punishment is contained in Title 8 of the Water Supply and Sewerage System Act [2] and refers to a violation of specific legal enactments. The concept of punishment divides punishments into offenses committed only by natural persons and administrative offenses involving legal persons and self-employed natural persons. The main principle is that the same administrative offense should have the same punishment regardless of whether it was committed by a natural or legal person.

9.17 Conclusion

This chapter is focused on the explanation and extension of the knowledge from the area of design documentation for the field of sewerage systems, sewer connections, water supply systems, and water-service pipes both for the professional and the non-professional public. At the beginning of the chapter, the basic terms from the given field of study are explained. The following parts deal with:

- technical requirements for design documentation of the sewage system and water supply system constructions,
- legislative documentation,
- subject matter and local jurisdiction,
- design documentation and designing activity in construction,
- location of the construction,
- building permit,
- construction change before its completion,
- use of the construction,
- permit of construction removal,
- construction passport and the documentation of the actual execution,
- obligations of the owners (water supply systems, water-service pipes, sewerage systems, and sewerage connections),
- unauthorized withdrawals, and
- wastewater drainage and sanctions.

This chapter was based on the legislative and normative documents valid in the Czech Republic.

9.18 Recommendations

This chapter provides a complete basic overview of the preparation of design documentation for the construction of sewerage systems, sewer connections, water supply systems, and water-service pipes. This chapter explains the basic terminology used in the given fields for easier orientation and understanding of the matter. Attention is also focused on the responsibilities of the owners of sewerage systems, sewer connections, water supply systems, and water-service pipes. It also describes the issue of unauthorized water consumption and the possible sanctions for administrative offenses arising from the Water Act. This chapter is intended for both professional and non-professional public. It should be noted that this chapter was based on the legislative and normative documents valid in the Czech Republic.

References

1. Zákon č. 254/2001Sb., o vodách a o změně některých zákonů (Vodní zákon). In: Sbírka zákonů. Praha: Ministerstvo vnitra, ČR, 2001, roč.2001, částka 98, číslo 254
2. Zákon č. 274/2001 Sb., o vodovodech a kanalizacích pro veřejnou potřebu a o změně některých zákonů (Zákon o vodovodech a kanalizacích). In: Sbírka zákonů. Praha: Ministerstvo vnitra, ČR, 2001, roč.2001, částka 104, číslo 274
3. Vyhláška č. 428/2001 Sb., kterou se provádí zákon č. 274/2001 Sb., o vodovodech a kanalizacích pro veřejnou potřebu a o změně některých zákonů (Zákon o vodovodech a kanalizacích). In: Sbírka zákonů. Praha: Ministerstvo vnitra, ČR, 2001, roč.2001, částka 161, číslo 428
4. ČSN 75 6101 (2012) Sewer systems and house connections
5. ČSN EN 752 (2017) Drain and sewer systems outside buildings—sewer system management
6. ČSN EN 1671 (1998) Pressure sewerage systems outside buildings
7. ČSN EN 1610 (2017) Construction and testing of drains and sewers
8. ČSN 75 6909 (2004) Testing water-tightness of drains and sewers
9. ČSN 75 0905 (2014) Water supply and sewerage tanks—testing of water-tightness
10. ČSN 75 5401 (2007) Designing of water pipeline
11. TNV 75 5402 (2007) Výstavba vodovodních potrubí
12. Zákon č.500/2004 Sb., správní řád, ve znění pozdějších předpisů. In: Sbírka zákonů. Praha: Ministerstvo vnitra, ČR, 2004, roč. 2004, částka 174, číslo 500
13. Zákon č. 183/2006 Sb., o územním plánování a stavebním řádu (Stavební zákon). In: Sbírka zákonů. Praha: Ministerstvo vnitra, ČR, 2006, roč. 2006, částka 63, číslo 183
14. Vyhláška č. 499/2006 Sb., o dokumentaci staveb. In: Sbírka zákonů. Praha: Ministerstvo vnitra, ČR, 2006, roč. 2006, částka 163, číslo 499
15. Zákon č. 360/1992 Sb., o výkonu povolání autorizovaných architektů a o výkonu povolání autorizovaných inženýrů a techniků činných ve výstavbě. In: Sbírka zákonů. Praha: Ministerstvo vnitra, ČR, 1992, roč. 1992, částka 73, číslo 360
16. Zákon č. 50/1976 Sb., o územním plánování s stavebním řádu (Stavební zákon). In: Sbírka zákonů. Praha: 1976, roč.1976, částka 9, číslo 50

Chapter 10
Evaluation of Technical Condition of Sewerage Systems Operated by Municipalities in the Czech Republic

Petr Hluštík and Martina Zelenakova

Abstract This chapter deals with the comparison of the technical condition of sewerage systems in the Czech Republic concerning water companies operating water management infrastructure and operators—mostly municipalities themselves, i.e. the owners of sewerage systems. The aim of this chapter is to investigate the technical condition of sewerage systems operated by municipalities' owners. The paper presents comparison and classification of various defects in the sewerage systems. We used the methodology corresponding to the applicable EN 13508-2 standard used to assess the condition of the sewerage systems and house connections. The assessed localities are municipalities up to 2000 inhabitants in the Czech Republic, which themselves operate sewerage systems. In these localities, the sewerage systems were monitored throughout 15 years, with the age of the system ranging from 40 to 50 years with regard to construction and development of the municipality in the relevant locality. This paper offers a solution for assessing the structural and technical condition of sewer systems using a multi-criteria analysis based on the failure mode and effects analysis (FMEA) method while keeping the relative assessment scale and comparing the results. The conducted analysis has determined the average defect rate of sewerage systems for selected municipalities operating their sewerage networks themselves.

Keywords Sewerage systems · Technical condition · Defect categories

P. Hluštík (✉)
Faculty of Civil Engineering, Institute of Municipal Water Management, Brno University of Technology, AdMaS Centre, Purkyňova 139, 612 00 Brno, Czech Republic
e-mail: hlustik.p@fce.vutbr.cz

M. Zelenakova
Department of Environmental Engineering, Technical University of Košice, Vysokoškolská 4, 042 00 Košice, Slovakia

M. Zelenakova et al. (eds.), *Management of Water Quality and Quantity*, Springer Water, https://doi.org/10.1007/978-3-030-18359-2_10

10.1 Introduction

Currently, various evaluation methodologies are employed to assess the technical condition of the sewer system as well as engineering constructions along with mathematical models and software products that include risk analysis of the operating conditions of the sewers [1–5]. Besides these instruments, a number of European studies are drawn up to compare various types of pipe systems, i.e. various pipe materials for municipal sewer systems and their impact on the environment over their service life [6–9]. The developed studies usually deal with the specificities of the relevant country as defined by laws and regulations, requirements related to the system operators and control strategies. The public study by Stein and Partner GmbH [10] from Germany compares sewerage failures in European countries and highlights the specificities of each country and differences in the sewerage failures.

The methodology assessing the technical condition of sewerage can be divided into simple and advanced, addressing development strategy and analyses based on models. The mathematical models predicting the pipeline development are based on probabilistic conditions of the pipelines and stochastic approaches. The current trend is to simulate sewerage ageing using semi-Markov chain or to simulate system failures using Monte Carlo algorithms with a fixed time-step, for example, the Bietigheim model, KAIN and Pforzheim [9]. The advanced models (software) are used for pipe repair and replacement planning support in the medium to short term. The applied software systems supporting strategic planning are KANEW [11], PiReM—Pipe Rehabilitation Management (Fuchs-Hanusch et al. 2008), WARP—Water Main Rehabilitation Planner [12], PARMS—Pipeline Asset and Risk Management [13], FAST—Fichtner Asset Services and Technologies [14] and strategy development and analysis model STATUS Sewer [9] and other products, also that ones that are not publicly accessible and are used internally by various companies.

Most of the mentioned approaches used principles of risk analysis. Risk analysis is widely applied in many fields that touch our daily lives. These include decisions about risks due to chemical and physical stressors (natural disasters, climate change, contaminants in food and water, pollution, etc.), biological stressors (human, plant and animal pathogens; plant and animal pests; invasive species, invasive genetic material), social and economic stressors (unemployment, financial losses, public security, including risk of terrorism), construction and engineering (building safety, fire safety, military applications) and business (project operations, insurance, litigation, credit, etc.) [15]. Risk analysis is a pervasive but often unnoticed component of modern society that is used by governments, private sector and individuals in the political, scientific, business, financial, social sciences and other communities [16]. There are many studies on risk analysis. Comparison of the most common methods, techniques, methodologies, models, tools are listed in numerous publications, including [17–24].

The risk analysis and mathematical models of sewer systems are developed in many areas of water engineering, e.g. flood modelling in urban areas, modelling of groundwater infiltration into sewerage systems and other advanced models to support planning and rehabilitation of sewerage systems. It can be stated that a data analysis for risk modelling related to sewers is specific and complex. Its specificities consist of a lack of data, thus making model calibration for planning rehabilitation of medium and large systems in European countries difficult. Most of the data focus on local effects caused by secondary conditions. Data specificities, distortion, model failure in case of limited data and calibration with distorted predictions are described by Scheidegger et al. [25]. An appropriate way to meet the calculation requirements using variable grading systems (descriptive, numerical) and units is fuzzy logic. The result of the simulation and model data generation using mathematical models is to the definition of links between local secondary conditions and characteristic specific pipeline failure. The evaluation and comparisons of the technical condition of the sewerage systems within the operating companies in Czech Republic using risk analysis method are the subject of this article.

10.2 Materials and Methods

In the Czech Republic, the duty to inspect the structural and technical condition of the sewerage systems is laid down in Act No. 183/2006 Coll. [26] on town planning and the building code (the Building Code) as amended, under which the owners of buildings and water-related structures are obliged to use and operate them in accordance with decisions issued by construction and water authority and keep them in a good condition. The owner of water-related infrastructure does not carry out these obligations itself, but they are transferred to an operator under an operation contract. Their performance is only possible provided that the actual condition of the operated or owned property is established. Thus, the assessment of the structural and technical condition of the sewerage system is not usually the subject of the project assessing the sewerage system, such as a drainage master plan. Nevertheless, its results provide essential input data for planning the renovation of sewerage systems.

The occurrence and risks of sewerage system defects depend primarily on the quality of design documentation, quality of the applied building materials and the construction of the sewerage system itself, quality of operating the sewerage system and environmental effects (internal and external), which can fundamentally change over the lifetime of the sewerage systems. The defect is a condition when the sewerage system is permanently or temporarily unable to fulfil its function and requirements, with deteriorating reliability of the system compared to its original condition. In extreme situations, the consequences may have a form of emergency breakdowns of the sewerage systems, caved-in streets or tramline bodies, subsidence in the worst cases even loss of life.

The collection and assessment of all relevant information on the sewerage system and its inspection rank amongst the essential components of operating the sewerage system and form a necessary operation in drawing up sewerage system renewal plans. In the Czech Republic, the information about the sewerage system renewal plans is defined under Act No. 274/2001 Coll. [27] on water mains and sewers, as amended. These renewal financial plans force the operators (owners) of the sewerage systems to develop internal methodologies for assessing the technical condition of the infrastructure.

Assessing the priorities of renewing individual sections of the sewerage systems is usually performed through a categorisation on the basis of a multi-criteria analysis, based on the results of visual inspection and sewerage system monitoring. The choice of the multi-criteria evaluation of the sewerage system depends on the operator (owner) of the infrastructure, and it is an internal matter of such a company. However, the content and extent of the internal methodology are always defined by the project management and meet the needs of the relevant organisation. Internal assessment methodologies are developed by professionals and are directly applicable to the relevant structures and facilities operated by the company. It may be stated that internal methodologies of the companies are simplified and adequate tools to prevent from accidents and bursts, but they are not developed transparently (in particular, there is no emphasis placed on other aspects, for example, competitive facilities in the sewerage system).

According to the focus of the analysis, the methods are divided into qualitative and quantitative. However, there are also combined methods (semi-quantitative), allowing qualitative and also quantitative analyses [15]. Failure modes and effects analysis (FMEA) which are used in this study belong to qualitative risk analysis method.

10.2.1 Multi-criteria Analysis of Sewerage Systems

To classify the individual defects according to applicable standards EN 13508-2: Investigation and assessment of drain and sewer systems outside buildings—Part 2: Visual inspection coding system [28], the evaluation criteria are proposed for the structural condition of the sewerage system using the failure mode and effects analysis. This is a multi-criteria analysis method, which describes systems and subsystems already in the conceptual stage and focuses on interactions between systems and system elements. Potential sewer section failure rate is rated by the highest code (number) of the risk priority. The proposed methodology can be used to detect a potential failure of the system and allocate the necessary time and funds in advance.

The groups of defects that are selected as criteria for assessing the technical condition of the sewerage system include:

- C1—structural damage to pipes;
- C2—damage to internal surfaces;
- C3—defected house drain connection;
- C4—operating defects;
- C5—sewerage system leakage.

Table 10.1 Definition of categories

Code and definition of categories		Description of measures
CI	Satisfactory condition	No relevant measures to change the indicator are required
CII	Good condition	No immediate solution is required, renovation within the long-term 15-year plan
CIII	Unsatisfactory condition	Critical values calling for renovation within the mid-term 5–10 years
CIV	Emergency condition	Undesirable and malfunctioning condition requiring immediate renovation within 1 year

Determining the importance weight for the operator/owner of the sewerage systems is a correction value. Setting the defect importance weight and defect classification into CI–CIV corresponds to the national standards, and it is a recommended value. It is based on the expertise of the author and discussions with the operators of the sewerage systems. The proposed methodology for evaluating defects is easily modifiable for the operators of sewerage systems. The methodology includes a qualitative systematic list of defects enabling quantification and also includes an estimate of the worst type of consequences. The description of categories CI–CIV is provided in Table 10.1.

In mathematical terms, the assessment of the technical condition of the sewer section can be described as follows using Eq. 10.1.

$$u^i = \sum_{j=1}^{n} v_j \cdot C_j^i \tag{10.1}$$

where

i 1, 2,…, p,
u^i general assessment of i-criterion,
v_j rate of importance of j-criterion,
K_j^i sub-assessment of i-section based on j-criterion,
n number of criteria,
p number of sewer sections.

Sections with the highest point scores are proposed to be included in the renovation plans (short-term, mid-term and long-term) to renew the sewerage networks and are used as the basis for setting priorities for sewer section rehabilitation. Optimal selection of sections to be renovated is made on the basis of information collected about the structural condition of the sewer, comprehensive assessment of the hydraulic condition of the sewer section, assessment of the impact of the structure on the surrounding environment, etc. To evaluate the technical condition of the sewer sections using the multi-criteria analysis—see Table 10.2—and to classify the individual defects, it is always necessary to monitor the sewerage system.

Table 10.2 Proposed evaluation criteria of the structure condition of sewerage in the Czech Republic

Group of criteria 1.00	Defect	Classification at the level				Definition to EN 13508-2
		CI	CII	CIII	CIV	
		0.1	0.15	0.25	0.5	
C1—0.35 Structural damage to pipes	Missing pipe material or part of wall	–	–	≤25 cm^2	>25 cm^2	Broken pipe
	Breaking, destruction	1.0				
	Longitudinal and cross cracks	<1 mm	2–3 mm	4–5 mm	>6 mm	Fissure
	Deformation of plastic pipe profile	<10%	11–15%	16–20%	>20%	Deformation
	Visible soil	0.5				Soil visible
	Hollow space	0.5				Void visible through defect
C2—0.20 Damaged internal sewer surface	Damaged surface, lining/channel	–	–	≤25 cm^2	>25 cm^2	Pipe deformation with crack
	Pipe material corrosion, protruding pipe material, reinforcement, aggregate	<25% ts <20 mm	26–50% ts 20–30 mm	51–75% ts 30–40 mm	>76% ts 40 mm	Damaged masonry, binder
	Internal lining, Mechanical abrasion	<25%	26–50%	51–75%	>76%	Lining defect

(continued)

Table 10.2 (continued)

Group of criteria 1.00	Defect	Classification at the level				Definition to EN 13508-2
		CI	CII	CIII	CIV	
		0.1	0.15	0.25	0.5	
C3—0.10 Incorrect sewer connection	Missing pipe material at the point of connection	<15 cm^2	16–25 cm^2	26–50 cm^2	>51 cm^2	Effective connection
	Incorrect connection position	–	–	11-01 h	–	
	Connection set in the sewer profile	<10% of the profile	11–20% of the profile	21–30% of the profile	>31% of the profile	Intruding house connection
C4—0.15 Operating defects	Incrustation	<10% of the profile	11–20% of the profile	21–30% of the profile	>31% of the profile	Attached deposit
	Growing roots	<10% of the profile	11–20% of the profile	21–30% of the profile	>31% of the profile	Roots
	Obstacles in the outflow	<10% of the profile	11–20% of the profile	21–30% of the profile	>51% of the profile	Settled deposit
	Foreign pipes and cables	1.00				Other obstacles
C5—0.20 Leaking sewer system	Infiltration/exfiltration wastewater		–	drops	leaks	Infiltration Exfiltration
	Displaced joint at the angle	<10%	11–15%	16–20%	>20%	Displaced joint
	Displaced joint—longitudinally	<5% of the profile	6–10% of the profile	11–15% of the profile	>16% of the profile	
	Displaced joint—radially	<5% of the profile	6–10% of the profile	11–15% of the profile	>16% of the profile	
	Intruding sealant	0.25				Intruding sealant

Note Wall thickness

Table 10.3 Sewer system defect rate in the company Veolia Czech Republic, a.s.

Sewer system	Number of defects/10 km/year	Material	Number of defects/10 km/year
Sewers	1.21	Stoneware	1.9
Connections	1.91	Concrete	7.1
Structures	1.17	Plastics	1.8
Total	4.29	Others	1.2

The final assessment values u^i are dimensionless figures in an interval of <0 = trouble less network, 1 = necessary rehabilitation>. The methodology recommends sections with the highest ranking as sewers to be reconstructed.

10.2.2 Study of Czech Republic

As of 2016, there are 6259 municipalities in the Czech Republic with a total of 10.56 million inhabitants, and the total length of the sewerage system was 47,070 km. The number of inhabitants connected to the sewage system is at 84%, and the amount of treated wastewater with respect to the total volume exceeds 97%. The percentage of municipalities that operate themselves in the sewerage systems is 23.94%, including the capital of Prague, which is a separate region, and the sewerage network is operated by a single company. In the Czech Republic, there is a slight increase in the number of registered owners with the 2015/2014 index = 1.04 and operators with the 2015/2014 index = 1.02.

Defect rate of sewerage systems in the Czech Republic—systems operated by Veolia Czech Republic, a.s.

The company Veolia Czech Republic, a.s., [29] a leading supplier of water-related services and providing sewerage services to more than 3.2 million inhabitants, has published statistical information on the average defect rate of the sewerage networks and facilities of its subsidiaries—see Table 10.3.

Defect rate of sewerage systems in the Czech Republic—systems operated by owners—municipalities

The monitoring and evaluation of the technical condition of the sewerage system according to the proposed selection criteria were conducted in ten municipalities in the South Moravian Region and at five municipalities in the Zlín Region in the Czech Republic over 2010–2015. These municipalities have been owners as well as operators of the infrastructure since the sewerage construction in the municipality. The major part of the sewerage network was probably built in the 1970s—it is combined sewerage consisting mainly of concrete or reinforced concrete pipes. In total, the monitoring was conducted over 110 km of the entire sewerage system

Table 10.4 Statistical data of selected municipalities in the Czech Republic

Statistical data of the evaluated municipalities in the Czech Republic operated by the municipalities					
Material	km	%	Profile (mm)	km	%
Stoneware	23.4	21.3	<300	51.8	47.1
Concrete	50.5	45.9	301–500	35.6	32.4
Plastics	29.8	27.1	501–800	15.2	13.8
Others	6.3	5.7	>800	7.4	6.7
Total	110	100	Total	72.4	100

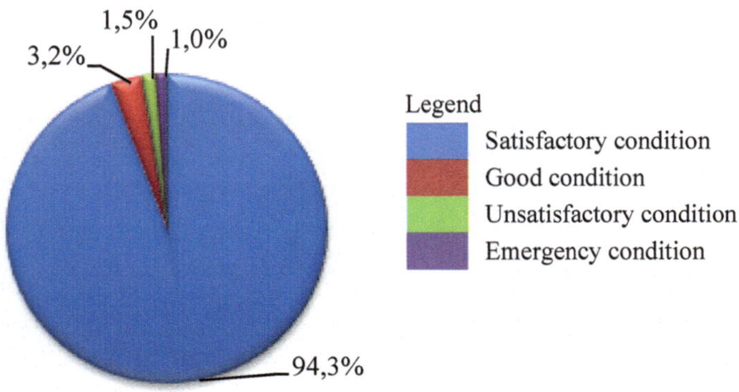

Fig. 10.1 C1—Structural damage to pipes: breaking, destruction

in all the municipalities. Statistical data on the sewerage networks in the selected municipalities are given in Table 10.4.

Based on the size of the assessed municipalities, the sewerage systems can be categorised as 500–2000 population equivalent. The evaluated data are unique. There is currently no information on the technical condition and defect rate of the sewerage systems operated by the owners—municipalities themselves.

10.2.3 Statistical Evaluation of Data of the Technical Condition of the Sewerage System

Based on the individual technical indicators, the most interesting sewerage failures were selected for graphic representation as shown in Figs. 10.1, 10.2, 10.3, 10.4, 10.5 and 10.6. Data were evaluated on the basis of camera surveys of the sewer network.

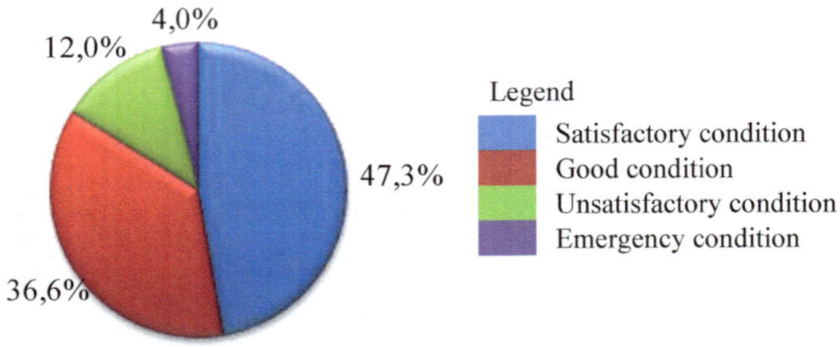

Fig. 10.2 C1—Structural damage to pipes: longitudinal and cross cracks

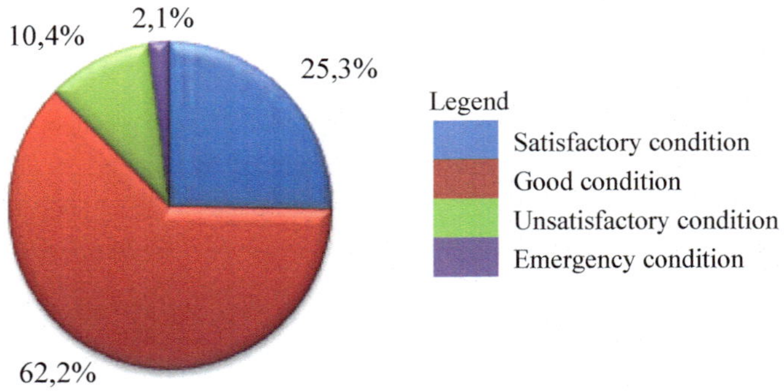

Fig. 10.3 C2—Damage internal sewer surface: pipe material corrosion, protruding pipe material, reinforcement, aggregate

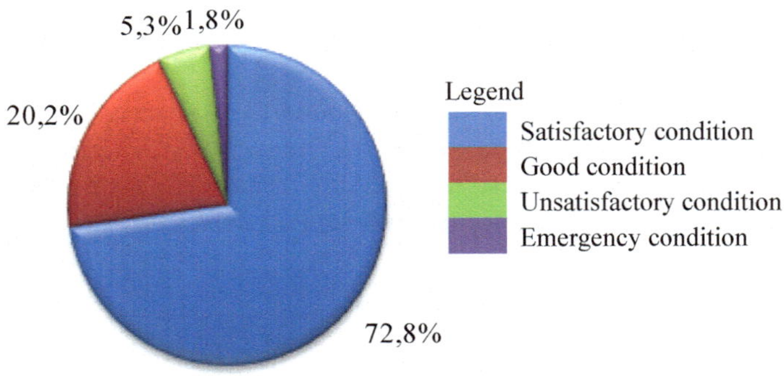

Fig. 10.4 C3—Incorrect sewer connection: connection set in the sewer profile

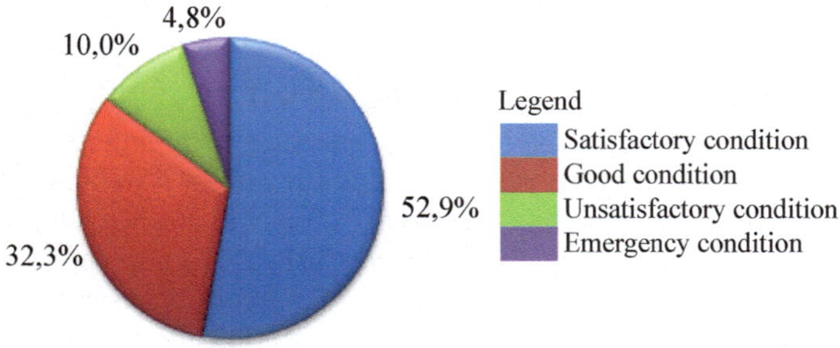

Fig. 10.5 C4—Operating defect: obstacles in the outflow

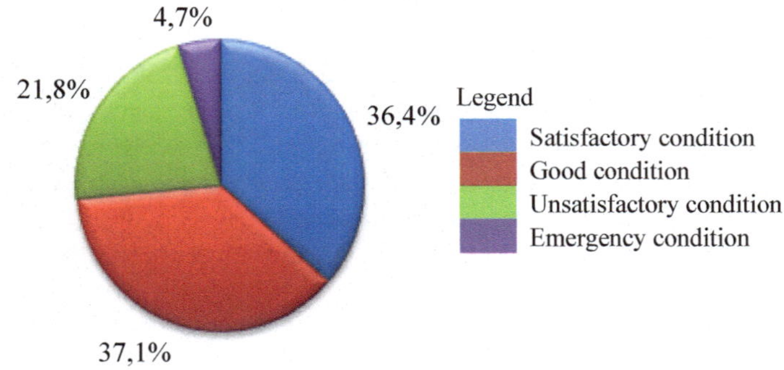

Fig. 10.6 C5—Leaking sewer system: displaced joint—longitudinally

Statistical expression of failures

The resulting deformation of the pipe (analysis of Fig. 10.1) usually indicates an emergency condition that needs to be addressed immediately. Missing parts of sewer pipes cause clogging, and wastewater seeps into groundwater.

The cracks (evaluation of Fig. 10.2) may range from hairline cracks that usually need not be repaired up to large cracks and cracks along the entire pipe or even the entire pipe profile. Cracks result in sewage leakage out of the pipes, water flow reduction and sewer blockages.

One of the common causes of degradation of sewer pipelines made of materials bonded with cement is biogenic sulphuric acid corrosion (evaluation of Fig. 10.3). This corrosion accelerates the degradation of the walls, in particular the dry parts of the pipes of sewerage networks which shortens the sewer service life. Salt solutions react with concrete components, particularly with hydrated cement. Biogenic sulphuric acid corrosion of concrete is caused by the products of the life processes of higher animals (urine, excrements) as well as bacteria occurring in biological processes.

Due to limited funds available to the operators (owners), camera surveys of the sewerage systems to detect the technical condition are not conducted. The sewerage system is not operated fully in lien with the sewerage regulations and without proper inspections and regular maintenance.

Roots grow into the sewer largely in the area of leaking sockets, damaged pipes or manholes. The roots then catch impurities, and the pipes get clogged over time. Roots growing through the sewer were rated as a satisfactory condition in 80.2%, in 18.5% as good condition, as unsatisfactory in 1.1% and as an emergency in 0.2%.

Unprofessional connection pipes (evaluation of Fig. 10.4) do not usually cause sewer clogging unless the branch pipe is displaced in the sewer to such an extent that it prevents from wastewater outflow. In places of unprofessional connections, there may be erosion in spaces surrounding the leaking pipes resulting in the formation of cavities. Roads may cave in, and pipes may collapse in the area of the cavity.

The sewer system profiles must be designed to ensure minimum driving speed at which no clogging of pipes occurs (evaluation of Fig. 10.5). When designing lower gradients, it is always necessary to make an individual assessment; the egg-shaped or circular profile is hydraulically more favourable. The frequency of inspections is governed by local conditions, it is carried out at least once every 5 years, and this cycle is later shortened in proportion to the age of the sewerage system and its technical condition. Deposits removed from sewers with inadequate gradients should not exceed the sewer connection bottom level, and they must not significantly reduce the flow capacity of the sewer.

The joint displacement may be either vertical or horizontal or axial (evaluation of Fig. 10.6). A dent is formed here which then catches impurities which start amassing here until the pipe becomes clogged. In the area of the damaged joint, there is also water leakage from the pipe and the pipe then loses the driving force needed to carry wastewater to the main sewer.

Evaluation of the technical condition of sewerage systems operated by their owners

It can be stated that from the structural point of view the most common problem is damaged or displaced pipe socket joint. Other common problems are hairline cracks line cracks, fissure lines and open cracks. The displaced connection or connections blockage is caused by improper connection and poorly designed connection gradient to the sewerage network. The connection was often made without the consent of the owner/operator or individually without any special equipment.

In operational terms, the most severe problem is related to the surface damage caused by sulphuric acid corrosion and the related loss of material on the upper surface of the body. Based on the obtained technical indicator values in all the evaluated municipalities for the same period, we have drawn up the percentage of defects in the sewer system as shown in Fig. 10.7.

The sewer manholes on the sewer networks are often located in front gardens of detached houses where they are covered and hidden which helps develop anaerobic conditions in the sewers. Deterioration of sewer pipe section surface when the sewer manholes are covered over long sections has been observed.

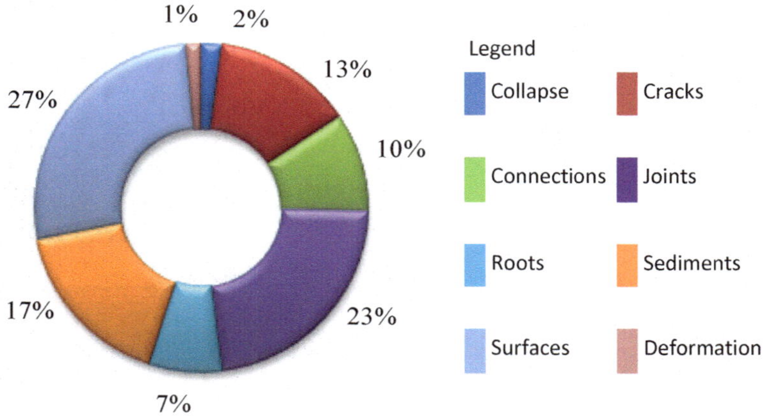

Fig. 10.7 Types of defects and their occurrence in sewerage

10.3 Results

The unnamed subsidiaries of Veolia Czech Republic, a.s., are represented by companies in Graph A to O. Operators are represented in a range of small municipalities up to large cities. Operators A, B, F are cities over 100,000 inhabitants, smaller towns and rural areas. Operators C, D, E, H are cities with 50–100 thousand inhabitants, smaller towns and rural areas. Operators G, I, J are cities up to 50 thousand inhabitants and rural areas. The letter Z marks selected municipalities that operate sewerage systems themselves. The resulting Fig. 10.8 shows the overall comparison of the number of defects per 10 km of sewers of the individual companies and operators (individual municipalities). Figure 10.8 shows the resulting comparison of the number of defects per 10 km of sewers of the individual companies and independent operators of the sewer systems. Currently, no statistical data providing a comprehensive comparison of the number of defects and their percentage are available in the Czech Republic.

As regards concrete pipes, this concerns smaller-size sewers made of plain concrete that is used in the past for the construction of sewerage systems in small municipalities which are presently at the end of their service life. The defect rate of stoneware and plastic pipes is only apparent, cecause stoneware pipes have been used in the Czech Republic for more than 100 years while plastic pipes have been used for a much shorter period.

When comparing the defect rates of the individual types of piping materials in company Z, we also take into account the time factor, i.e. comparison of the defect rate of sewers of the same age. Defects of the sewerage system were caused by lack of qualification in constructing the sewerage systems involving local residents, the age of the sewerage system and use of lower quality materials during construction conducted in the past. Experience shows that the assessed selected municipalities do not carry out regular inspections and maintenance of sewers, i.e. CCTV inspections

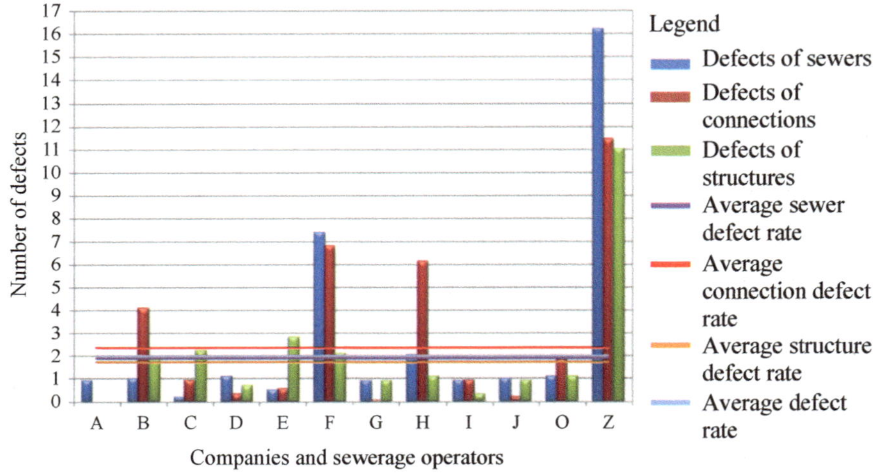

Fig. 10.8 Average number of defects per 10 km

nor physical visual inspection of the structural and technical condition. The average defect rate of the sewerage systems in comparison with the subsidiaries of Veolia Czech Republic, a.s., is up to 5 times higher totalling an average of 13 defects per 10 km. The evaluated municipalities do not develop water infrastructure renewal plans, they have no overview of the sewerage facility records, and they do not generate financial reserves or funds for maintenance, repairs, renovation or reconstruction of the sewerage network and sewerage-related structures. The reasons above have made the municipal councils in 40% of the assessed municipalities change the operators, being usually replaced with regional water utilities in the relevant regions of the Czech Republic.

10.4 Conclusions

The reason for postponing repairs today is presently often the lack of funds available to the operator (owner) for repairs, which is a frequent problem of a vast majority of the operators. The sewerage system repairs necessitated by the defects form a significant cost item of every operator and are a factor affecting the sewage tariff level. Each operator makes surveys of the sewerage system in their own interest. They keep records of defects and their monitoring. The survey results are an invaluable source of information on the types and characteristic defects related to the individual sewer material groups, civil structures in the system, causes of the defects and frequency of defects in the system.

This paper offers an alternative solution for assessing the structural and technical condition of sewer systems using a multi-criteria analysis based on the failure mode and effects analysis method while keeping the relative assessment scale and comparing the results. This method has been implemented in selected municipalities of the Czech Republic, which themselves operate their sewerage systems and know their technical condition. The results of assessing the technical conditions classified into technical indicators of defects are unique both in the Czech Republic and abroad. Currently, water companies operating water infrastructure do not develop and classify any detailed statistical data of the sewer system defects. Internal databases of these companies provide information about sewerage CCTV inspections indicating defects; however, these defects are not further analysed or classified.

The conducted analysis has determined the average defect rate of sewerage systems for selected municipalities operating their sewerage networks themselves, and the data were compared with Veolia Czech Republic, a.s. The empirical experience is not recorded and is usually included in internal company's documents.

The CCTV inspections in selected municipalities can be comprehensively compared, the sewerage construction took place over the same years, and the material used is of the same age just like the construction of the sewerage. The CCTV inspections were conducted in all parts of the sewerage system of the selected municipalities, and therefore, no random selection was made. For the duration of the CCTV inspections, we can ignore the newly formed defects following the sewerage monitoring which could affect the statistical data. The final analysis of the evaluated municipalities did not reveal any link between a defected pipe and local ecological and environmental conditions of the landscape.

There is a need for a general and thorough approach to justifying, explaining, demonstrating, working, sampling, using and creating real skills in risk analysis in any area of the human society. Such a need is embedded in human evolution and it gains importance every time one needs to develop forecasts of future courses of action [18]. All management decisions are concerned with the future, and everything about the future is to a greater or lesser extent uncertain. Nowhere is this more true than in project planning, where ignoring uncertainty can have serious consequences on the quality of the plans [30].

The achieved results from the presented research can be integrated into the pipe ageing model and the forecasting model.

10.5 Recommendations

Based on the outputs from the present article, financial plans for the reconstruction of the water management infrastructure and the planning of the reconstruction of the sewerage networks can be prepared.

This article is a document for future studies in sewerage systems. This is a document that can be a basis for water companies when planning investments to the sewerage infrastructure.

Acknowledgements This chapter has been worked out under the project No. LO1408 "AdMaS UP—Advanced Materials, Structures and Technologies", supported by Ministry of Education, Youth and Sports under the "National Sustainability Programme I".

References

1. Frey DD, Herder PM, Wijnia Y, Subrahmanian E, Katsikopoulos K, de Neufville R, Oye K, Clausing DP (2010) Research in engineering design: the role of mathematical theory and empirical evidence. Res Eng Design 21:145–151
2. Eisenbart B, Gericke K, Blessing LTM, McAloone TM (2017) A DSM-based framework for integrated function modelling: concept, application and evaluation. Res Eng Des 28:25–51
3. Ic YT (2016) Development of a new multi-criteria optimization method for engineering design problems. Res Eng Des 27:413–436
4. Liang R, Dong Z, Sheng Y, Wang X, Wu C (2017) Case study of selecting decision-making schemes in large-scale infrastructure projects. J Infrastruct Syst 23(4):06017001
5. Samuel T, Ariaratnam AE, Yang Y (2001) Assessment of infrastructure inspection needs using logistic models. J Infrastruct Syst 7(4):160–165
6. Hochstrate K (1999) Substanzwertorientierte Zustandsklassifizierung von Kanälen. Das Modell Bietigheimer, Korrespondenz Abwasser, Abfall (KA) 46 http://www.sintef.no/projectweb/care-s/. Accessed 12 Nov 2016
7. Waldmann KH, Stocker UM (2003) Stochastische modelle. Springer Verlag, Karlsruhe
8. Schmidt T (2009) Modellierung von Kanalalterungsprozessen auf der Basis von Zustandsdaten. Dissertation. Technischen Universität Dresden, Fakultät für Bauingenieurwesen, Institut für Stadtbauwesen und Straßenbau
9. Stein R (2015) Evaluation models for the assessment of the structural and operational condition of drain and sewer systems. http://www.unitracc.com/aktuelles/artikel/evaluation-models-for-the-assessment-of-the-structural-and-operational-condition-of-drain-and-sewer-systems-2015-part-iii. Accessed 12 Nov 2016
10. Stein D, Partner GmbH (2005) Final Report: European study of the performance of various pipe systems, respectively pipe materials for municipal sewage systems under special consideration of the ecological range of effects during the service life. Germany Bochum
11. Fuchs-Hanusch D, Gangl G, Kornberger B, Kölbl J, Hofrichter J, Kainz H (2008) PiReM—pipe rehabilitation management developing a decision support system for rehabilitation planning of water mains. Water Pract Technol 3(1):1–9
12. Cardoso MA, Silva M Santos, Coelho ST, Almeida MC, Covas DIC (2012) Urban water infrastructure asset management—a structured approach in four water utilities. Water Sci Technol 66(12):2702–2711
13. Burn S, Tucker S, Rahilly M, Davis P, Jarrett R, Po M (2003) Asset planning for water reticulation systems—The PARMS model. Water Sci Technol Water Supply 3(1–2):55–62
14. Scholten L, Scheideggera A, Reicherta P, Maurera M, Lienert J (2014) Strategic rehabilitation planning of piped water networks using multicriteria decision analysis. Water Res 49:124–143
15. Zeleňáková M, Zvijáková L (2017) Using risk analysis for flood protection assessment. Springer International Publishing, Switzerland. Available online 15 Oct 2018. https://www.springer.com/gp/book/9783319521497
16. Arthur JR, Bondad-Reantaso MG (2012) Introductory training course on risk analysis for movements of live aquatic animals. FAO SAP, Samoa, p 167. Available online 15 Oct 2018, http://www.fao.org/docrep/015/i2571e/i2571e.pdf
17. Alverbro K, Nevhage B, Erdeniz R (2010) Methods for risk analysis. TRITA-INFRA-FMS, Stockholm, Sweden
18. Bujoreanu IN (2012) Risk analysis series part one—why risk analysis? J Defense Resour Manag 3(1):139–144. Available online 15 Oct 2018, http://journal.dresmara.ro/issues/volume3_issue1/15_bujoreanu.pdf

19. Marhavilas PK, Koulouriotis D, Gemeni V (2011) Risk analysis and assessment methodologies in the work sites: on a review, classification and comparative study of the scientific literature of the period 2000–2009. J Loss Prevent Proc Ind 24(5):477–523
20. Rasche T (2001) Risk analysis methods—a brief review. The University of Queensland. Mineral Industry Safety and Health Center. http://www.mishc.uq.edu.au/Publications/Risk_Analysis_Methods_a_Brief_Review.pdf. Accessed 12 Nov 2016
21. Raspotnig Ch, Opdahla A (2013) Comparing risk identification techniques for safety and security requirements. J Syst Softw 86(4):1124–1151
22. Reniers GLL, Dullaert W, Ale BJM, Soudan K (2005) Developing an external domino prevention framework: Hazwim. J Loss Prev Process Ind 18(3):127–138
23. Rouvroye JL, van den Bliek EG (2002) Comparing safety analysis techniques. Reliab Eng Syst Saf 75(3):289–294
24. Segudovic H (2006) Qualitative risk analysis method comparison. Infigo IS d.o.o., Zagreb
25. Scheidegger A, Scholten L, Maurer M, Reichert P (2013) Extension of pipe failure models to consider the absence of data from replaced pipes. Water Res 47(11):3696–3705
26. Act No. 183/2006 Sb. on town planning and the building code (the Building Code). AION, The Czech Republic, c 2010–2016, vol 2006, no 183
27. Act No. 274/2001 Sb. on Water Mains and Sewers. AION, The Czech Republic, c 2010–2016, vol 2001, no 274
28. EN 13508-2 Evaluation of condition of outside sewer network system and sewer hook-ups. Part 2. Coding system for visual inspection. CEN, c2012
29. Veolia (2016) Veolia, Czech Republic. http://www.veolia.cz/cs. Accessed 8 Aug 2016
30. Welsh T (2012) Full monte—the better approach to schedule risk analysis. User guide. Available online 15 Oct 2018. http://www.barbecana.com/events/fullmonte-UserGuide.pdf

Part V
Hydrodynamics Modelling

Chapter 11
Numerical Modelling of the Fluid Flow at the Outlet from Narrowed Space for a Better Water Management

V. Michalcová and K. Kotrasová

Abstract In the water management, events of significant changes of the running fluid flow space occur very frequently. Such change influences significantly the characteristics of the flow field and thus its effects on the environment. These might be the effects on adjacent objects in the close proximity such as walls of the designated space, or effects on objects bypassed with the running fluid. Such a situation frequently influences the surrounding area. The flow run may affect the terrain of the land or even the ambient climate in some cases. This chapter is dedicated to problems of numerical modelling of Newtonian fluid flows in changing flow space. Showing a particular task, which is solved numerically using computational fluid dynamics (CFD) codes in ANSYS Fluent software, the reader becomes familiar with both problems of the mathematical specification of fluid movements and principles of the correct choice of the numerical model. To resolve this task, four numerical models are selected. Basic turbulent characteristics of the flow field are monitored. Outputs of individual models are evaluated and compared with each other. Selected results are also verified by experimental measuring.

Keywords Channel · Fluid · CFD · Finite volume method · Turbulence · Reynolds number

List of Symbols

A	Turbulent viscosity function (SST k-ω) (-)
C_{ij}	Convection member (RSM) (kg m^{-1}s^{-3})

V. Michalcová (✉)
Department of Structural Mechanics, Faculty of Civil Engineering, VŠB—Technical University of Ostrava, L. Podéště 1875/17, 708 33 Ostrava-Poruba, Czech Republic
e-mail: vladimira.michalcova@vsb.cz

K. Kotrasová
Department of Structural Mechanics, Faculty of Civil Engineering, Institute of Structural Engineering, The Technical University of Kosice, Kosice, Slovak Republic
e-mail: kamila.kotrasova@tuke.sk

M. Zelenakova et al. (eds.), *Management of Water Quality and Quantity*, Springer Water, https://doi.org/10.1007/978-3-030-18359-2_11

C_μ	Constant (-)
C_μ	Model function for μ_t calculation (Realisable k-ε) (-)
$C_{1\varepsilon}$	Empiric constant (-)
$C_{2\varepsilon}$	Empiric constant (-)
$D_{L,ij}$	Molecular diffusion member (RSM) (kg $m^{-1}s^{-3}$)
$D^*_{T,ij}$	Turbulent diffusion member (RSM) (kg $m^{-1}s^{-3}$)
D_ω	Mixing member (SST k-ω) (kg $m^{-3}s^{-2}$)
f_i, F_i	Force components (N)
G_k	Generation member of turbulent kinetic energy (kg $m^{-1}s^{-3}$)
G_ω	Generation member of dissipation (SST k-ω) (kg $m^{-3}s^{-2}$)
k	Turbulent kinetic energy (m^2s^{-2})
I	Linear scale factor of turbulence (m)
p	Compression (Pa)
P_{ij}	Compression production member (RSM) (kg $m^{-1}s^{-3}$)
t	Time (s)
u	Momentary velocity (m s^{-1})
u_i	*I*-fold component of momentary velocity (m s^{-1})
u_i	Wind velocity at fall in the *i*-fold building point (m s^{-1})
$\overline{u}$	Mean velocity (time averaged) (ms^{-1})
$\overline{u_i}$	Mean wind velocity at fall in the *i*-fold building point (m s^{-1})
$\overline{u_{i,j,k}}$	*i, j, k*-fold component of mean velocity (m s^{-1})
$u'_{i,jk}$	*i, j, k*-fold component of fluctuation velocity (m s^{-1})
x_i	Cartesian coordinate system $[x_1, x_2, x_3]$ or $[x, y, z]$ (m)
s	Tensor module of the mean deformation velocity (k-ε) (s^{-1})
$S_{i,j}$	Deformation velocity tensor (k-ε) (s^{-1})
Y_k	Dissipation member for k (SST k-ω) (kg $m^{-1}s^{-3}$)
Y_ω	Dissipation member for ω (SST k-ω) (kg $m^{-3}s^{-2}$)
δ_{ij}	Kronecker delta symbol (-)
ε	Velocity of kinetic energy dissipation (m^2s^{-3})
ε^*_{ij}, ε_{ij}	Dissipation member (RSM) (kg $m^{-1}s^{-3}$)
Φ^*_{ij}, Φ_{ij}	Compressive stress member (RSM) (kg $m^{-1}s^{-3}$)
κ	Von Karman constant (-)
μ	Dynamic viscosity (Pa s), (kg $m^{-1}s^{-1}$)
μ_t	Dynamic turbulent viscosity (Pa s), (kg $m^{-1}s^{-1}$)
ν	Kinematic viscosity (m^2s^{-1})
ν_t	Kinematic turbulent viscosity (m^2s^{-1})
ρ	Density (kg m^{-3})
σ_k	Turbulent Prandtl number for k (-)
σ_ε	Turbulent Prandtl number for ε (-)
σ_{υ_i}	Turbulent Prandtl number for ν_t (-)
τ	Viscous stress (Pa)
ω	Dissipation per turbulent kinetic energy (s^{-1})

Indexes

i	Velocity component index, iteration index, point index (-)
i	Summary index (-)
j, k, l	Summary Einstein index (-)

11.1 Introduction

If the flow space of the running medium is modified, significant changes of flow field characteristics occur [1–4]. This also changes the effects of the flow on the immediate and more distant surroundings of the specific running process to a great extent. The basis for the examination of effects of this flow field change on the surroundings is the assessment of its turbulent characteristics [5, 6]. These problems are inspiring for both the water management and other industrial sectors [7–9]. Numerical modelling is considered one of the possibilities of its solution [10, 11], which this chapter is dedicated to.

Numerical modelling of many physical effects relates closely to modelling of a particular motion form using mathematical means. Resolutions of a whole series of technical, meteorological, and ecological problems relate closely to fluid motions [12–14].

The task is solved using CFD codes (Computational Fluid Dynamics) in the software complex called ANSYS Fluent. It is a state-of-the-art computing method that makes calculations using a finite volume method [15–17]. It enables modelling of a fluid flow [18, 19] (Newtonian and non-Newtonian) and full heat exchange [20] (by conduct, circulation, and radiation). It is capable of solving polyphase streaming and chemical reactions [21]. CFD is a field which joints immediately in several disciplines. Applying CFD tools requires an understanding of the principle of the simulated problem, theoretical knowledge in the area of streaming and numerical modelling, and last but not least experimenting in the development of computing area, network generation, selection of boundary conditions, calculation parameters, etc.

As in any numerical solution, it is necessary to create a model for CFD. Subsequently, mathematical procedures are applied to it, so that selected data on processes running in the whole examined area are acquired from the specified boundary and initial conditions. It is at the same time necessary to respect physical principles. Without the knowledge of turbulence physics, the researcher cannot be sure about the right model selection. For this reason, separate chapters in this thesis are dedicated to methods of mathematical modelling and specification of selected numerical models.

11.2 Task Specification

In the task presented, air flow through a confined narrowing space of rectangular cross section is simulated (Fig. 11.1), and subsequent outflow to the space has identical dimensions with the original one (Fig. 11.2). Such specific simulation was selected because control results from measuring in an aerodynamic channel are available within cooperation with an experimental research site of ITAM CAS, v. v. i., Centre of Excellence Telč [22, 23], (Fig. 11.3). The objective is to assess the flow field change when passing through a nozzle using numerical calculations.

The area of calculation for the numerical solution is 4.9 m long, while 0.5 m is formed by onset, 1.4 is the nozzle length where the cross section changes according to the diagram in Fig. 11.1. Behind the narrowed cross section, there is a space of 3 m according to Fig. 11.4. The calculation grid uses 9×10^5 hexa cells. In the cross section at the nozzle output, there is distinct strong densification around the

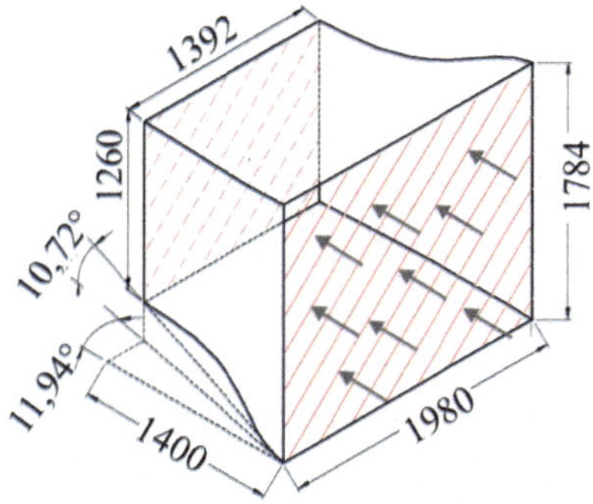

Fig. 11.1 Contraction diagram (length dimensions in mm) [22]

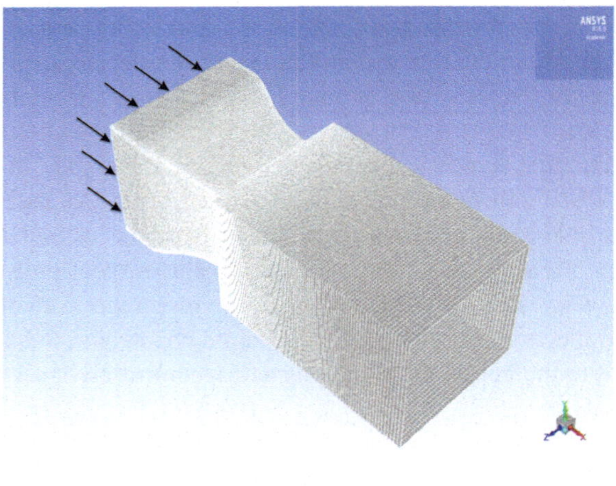

Fig. 11.2 Whole area diagram (grid from numerical simulation)

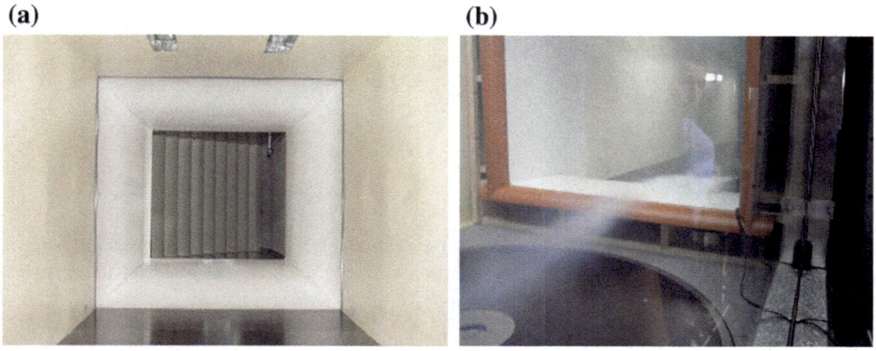

Fig. 11.3 Contraction in tunnel: **a** downstream, **b** upstream view [23]

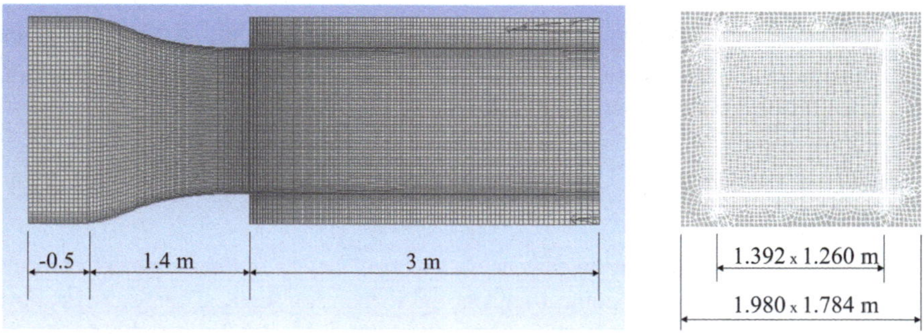

Fig. 11.4 Grid for numeric computation

connection of both areas for more precise data transmission on the interface. Also, the steady increase of cell sizes farther from boundary areas of the narrowed space is worthy of mention. The reason is the correct calculation at the walls of this space [24, 25].

Values of input variables have been set to correspond with experimental measuring: turbulence intensity $i_u = 1\%$ and inlet velocity $u_x = 10$ m s^{-1}. To verify results, turbulent characteristics (u_x, i_u) from the physical experiment are available at the distance of 0.36 m behind the nozzle outlet (Fig. 11.5). Velocity distribution in the cross section monitored is defined by non-dimensional coefficient given by actual velocity value/mean velocity value ratio. Individual scanning points of the experiment are 20 mm from each other in both the horizontal and vertical directions.

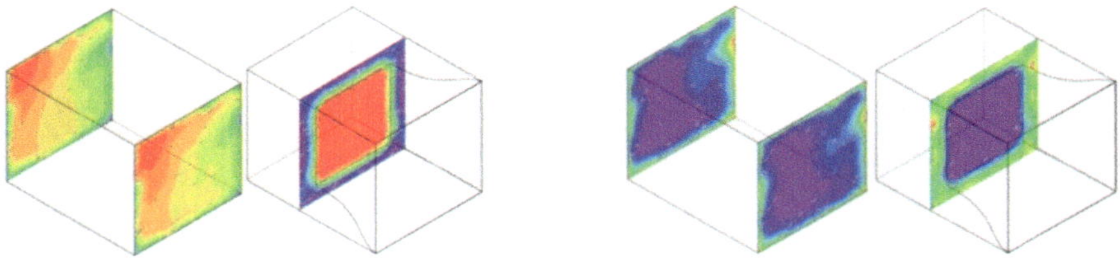

Fig. 11.5 Experiment: Diagram of velocity field distribution (left) and turbulence intensity (right) when passing through the space without cross-sectional change and in the outflow from the narrowed space [23]

11.3 Methods of Numerical Modelling of Turbulent Flow

The type of this flow is defined by non-dimensional variable, Reynolds number *Re* [26]. It indicates the ratio between inertial and friction forces influencing the element, and it is given by the well-known relation:

$$Re = \frac{ul}{v} \tag{11.1}$$

where u (m s^{-1}) represents the flow velocity, ν(m^{-2} s^{-1}) kinematical viscosity of the running fluid, which is equal to 1.5×10^{-5} m^{-2} s^{-1} here. Characteristic flow dimension l (m) is in this case defined by the hydraulic diameter of the narrowed space, $l = 1.3$ m.

The flow of the described task is $Re = 8.6 \times 10^5$, which represents fully developed turbulence. It is characteristic for the turbulent flow that its variables show chaotic fluctuations in space and time. It is a flow with the rotary motion of eddies. Important features of turbulence include diffusivity dissipation ε. Diffusion represents the mixing of components and thus quick and intensive transfer of mobility. Dissipative nature of the flow represents a transfer of kinetic energy of the medium into its internal thermal energy. The transfer proceeds in cascades, when the energy is gradually transferred to eddy motions of smaller and smaller dimensions. Eddies are thus deformed, and the kinetic energy is withdrawn from the basic time medium flow. Deformation thus leads to increased eddy energy withdrawn from the velocity field, which caused the deformation. The power transfer cascade ends with eddies so small that their movement is absorbed effectively by medium viscosity.

Mathematics specification of the flow is provided by two balance equations [27, 28]:

- Motion Navier–Stokes (NS) equations—weight balance. It ensures a balance of surface, weight, and inertial forces.
- Continuity equation—flow continuity equation.

Balance equations are described in more details in Sect. 11.5. There are three general methods of numerical modelling that are explained below.

11.3.1 Method of Direct Numerical Simulation—DNS Method

Direct numerical simulation (DNS) method [29, 30] is used under certain limiting conditions only, given by high demands for the computer capacity. The reason is the necessity of a very fine network. The number of nodal points of this network increases rapidly with the Reynolds number, which leads to technical non-viability of calculations with the existing computer technology, and it is not dealt with in this thesis.

11.3.2 Large Eddy Method—LES Method

Large eddy simulation (LES) method [31, 32] is based on direct modelling of large eddies that may be captured by the network. These turbulent structures of large scale demand kinetic energy from the main flow and are very dependent on the position in the flow field and on time. They are thus simulated directly in a 3D and time-dependent form. Energy transfer from large eddies, turbulent eddies of small scales are formed. Turbulent eddies of small scales are generally isotropic and participate little in transport events. Through them, however, dissipation of kinetic turbulent energy occurs due to the viscosity of the running medium. These small eddies are parameterized by so-called sub-grid models that are included in each model based on the LES method. For mathematical specification of individual sub-grid models, see the manual of ANSYS Fluent [27].

11.3.3 Statistic Turbulence Models—RANS Models

Reynolds-averaged Navier–Stokes equations (RANS) models [33] remain the most frequently used tool for the majority of engineering tasks of turbulent flows. They are based on the method of time (Reynolds) averaging of variables of the turbulent flow and the following procedure of time averaging of balance equations. In Navier–Stokes equations, new variables emerge, of the so-called Reynolds (*Re*) stress. Their presence is a fundamental problem in the calculation of the turbulent shear flow in equations describing the mean fluid motion. The system of motion equations is thus not closed as in the event of the laminar flow. The set of additional equations and empiric relations that together with motion equations form a soluble system is called the turbulence model. Turbulence models can be divided into several groups. Statistical models either solve *Re* stress using differential equations [Reynolds stress model (RSM)], or using the so-called Boussinesq hypothesis, seeking for simplified expression of these stresses.

11.4 Balance Flow Equation in ANSYS Fluent

Mathematical specification of the turbulent flow and individual numerical models include a considerable amount of frequently repeated variables. For more comfortable orientation in the text, an alphabetic index of all variables used in equations is inserted at the end of this thesis (before bibliography).

The task in this thesis represents a non-stationary isothermal flow. Balance equations describing this type of flow transport can be expressed for averaged variables in the differential form:

Continuity equation:

$$\frac{\partial \rho}{\partial t} + \frac{\partial}{\partial x_j}(\rho \overline{u_j}) = 0 \tag{11.2}$$

Motion equation:

$$\frac{\partial}{\partial t}(\rho \overline{u_\iota}) + \frac{\partial}{\partial x_j}(\rho \overline{u_\iota u_j}) + \frac{\partial}{\partial x_j}\left(\rho \overline{\acute{u}_\iota \acute{u}_j}\right) = -\frac{\partial p}{\partial x_i} + \mu \frac{\partial^2 \overline{u_\iota}}{\partial x_j^2} + f_i \rho, \tag{11.3}$$

where u_i (m s^{-1}) represents velocity components, ρ (kg m^{-3}) the density of running medium, t (s) the time, p (Pa) the pressure, μ (Pa s), (kg m^{-1}s^{-1}) dynamic viscosity of the medium and f_i components of the volumetric force, i is a component index, and j is a summary index. Equations (11.2) and (11.3) thus represent one continuity equation and three motion equations for three directions of the system of coordinates.

11.4.1 Boussinesq Hypothesis

Equations (11.2) and (11.3) include the mean value of the product of velocity fluctuation components $-\overline{\acute{u}_\iota \acute{u}_j}$, which is proportional to the so-called Reynolds stresses $-\rho\overline{\acute{u}_\iota \acute{u}_j}$ Transport equations can also be derived for this stress. Turbulent flow of motion thus acts as a stress (fluid moving upwards $(\acute{u}_3)$ is mixed with the fluid moving horizontally $(\acute{u}_1)$. The result is a component *Re* stress $-\rho\overline{\acute{u}_\iota \acute{u}_j}$). There are nine of these stresses, or six, since they form a tensor, and symmetry applies as in viscose stresses.

Equations for Reynolds stresses are as follows:

$$\frac{\partial}{\partial t}\left(\rho\overline{\acute{u}_\iota \acute{u}_j}\right) + \overline{u_k}\frac{\partial}{\partial x_k}\left(\rho\overline{\acute{u}_\iota \acute{u}_j}\right)$$
$$= -\frac{\partial}{\partial x_k}\left(\rho\overline{\acute{u}_\iota \acute{u}_j \acute{u}_k}\right) + p\left(\overline{\delta_{kj}\acute{u}_\iota + \delta_{k\iota}\acute{u}_j}\right) - \mu\frac{\partial}{\partial x_k}\left(\overline{\acute{u}_\iota \acute{u}_j}\right)$$

$$-\rho\left(\overline{\acute{u}_\iota\acute{u}_k\frac{\partial\acute{u}_j}{\partial x_k}}+\overline{\acute{u}_j\acute{u}_k\frac{\partial\acute{u}_\iota}{\partial x_k}}\right)+p\left(\overline{\frac{\partial\acute{u}_\iota}{\partial x_j}+\frac{\partial\acute{u}_j}{\partial x_\iota}}\right)-2\mu\left(\overline{\frac{\partial\acute{u}_\iota}{\partial x_k}+\frac{\partial\acute{u}_j}{\partial x_k}}\right) \tag{11.4}$$

Re stresses exist in the turbulent flow only and have identical characteristics as the viscose stress τ used in the mechanics of fluids. They form an extensive system of hardly soluble differential equations, in which i and j are component indexes, and k is a summary index. Therefore, the Boussinesq theory [34] was created, dealing with the more straightforward expression of Reynolds stresses in the Eq. (11.3). It uses the analogy with Newton's law of viscous flow:

$$\tau_{ij}=-\mu\frac{\partial\overline{u}_j}{\partial x_{xi}} \tag{11.5}$$

Moreover, we express the relation of *Re* stress to averaged velocity gradients as follows:

$$\rho\overline{\acute{u}_\iota\acute{u}_j}=\mu_t\left(\frac{\partial\overline{u_\iota}}{\partial x_j}+\frac{\partial\overline{u_j}}{\partial x_i}\right)-\frac{2}{3}\left(\rho k+\mu_t\frac{\partial\overline{u_\iota}}{\partial x_i}\right)\delta_{ij}, \tag{11.6}$$

where $\overline{\acute{u}_\iota\acute{u}_j}$ (m^2s^{-2}) is a mean value of the product of velocity fluctuation components, k(m^2s^{-2}) turbulent kinetic energy, δ_{ij}(-) Kronecker delta symbol, τ(Pa) shear stress, μ_t(Pa s), (kg $m^{-1}s^{-1}$) turbulent dynamic viscosity, and ν_t(m^2s^{-1}) (see below) turbulent kinematical viscosity.

In the Eq. (11.6), and after insertion subsequently also in the basic Navier–Stokes Eq. (11.3), a new variable emerges—the so-called turbulent dynamic viscosity μ_t, or kinematical ν_t, which the following applies to:

$$\mu_t=\rho\upsilon_t \tag{11.7}$$

It expresses complex functional dependencies of the running fluid state and the considered point position. It is thus not a physical constant (feature) of the fluid, but the function specifying given turbulent flow.

The difference between individual models consists just in the method of defining μ_t. The disadvantage of the Boussinesq hypothesis consists in the assumption that μ_t is an isotropic scalar variable.

11.5 Selected Mathematical Models of Turbulence

From the extensive offer of models of the ANSYS Fluent complex, this thesis describes and evaluates models, the approach of which to the solution of balance Eqs. (11.2) and (11.3) is suitable for mathematical expression of the process described herein only.

Following models have been selected and tested for the task solution:

- Statistical models of turbulence based on the Boussinesq theory (models for isotropic turbulence): Realisable k-ε model and SST k-ω model.
- A statistical model of turbulence resolving differential equations for the turbulent stress directly (model for anisotropic turbulence): RSM model. It enables stationary task resolution; it is, however, very sensitive and frequent problems with convergence occur if used.
- Turbulence model based on the LES method of simulations: DES model. This model is based on the principle of anisotropic turbulence modelling, which better corresponds with the examined process. It is, however, more demanding for the quality and density of network; tasks need to be resolved non-stationary; and computing times are incomparably longer.

Details for the specification of mathematical models stated in this and the following chapters are drawn from the extensive manual of ANSYS Fluent [27]. In this thesis, they are revised as possible to the briefest form, so that they describe the above process only. It means, for example, skipping all members in equations, which do not relate to the isothermal flow, or defining specific values of certain empiric constants. It means, for example also, or description of certain variable functions in the form suitable for isothermal process only, with the constant density of the running medium and the flow with high *Re*.

It applies in general that some constants used in equations are represented in individual models in Fluent and determined based on examination and measuring under laboratory conditions, where the turbulent flow corresponding with the overwhelming majority of technical problems is generated. It is thus not appropriate to change their values. Incorrect determination of these constants would disrupt calculation balance just as, for example, incorrect specification of boundary conditions.

11.5.1 Realisable k-ε Model

Realisable k-ε model is the two-equation model, which the dynamic viscosity μ_t is presented in two transport equations. These equations enable to define the turbulent velocity (turbulent kinetic energy k (m^2s^{-2})) and linear scale factor (ε (m^2s^{-3})—dissipation of turbulent energy).

The model of realisable k-ε counts on the turbulent viscosity in the relation:

$$\mu_t = \rho C_\mu \frac{k^2}{\varepsilon}, \tag{11.8}$$

where C_μ(-) is the function of the mean stress, rotation, and turbulent field (k and ε). The major difference of this model from other k-ε models is that C_μ is not a constant. Its value changes from $C_\mu = 0.09$ in the equilibrium boundary layer up to $C_\mu = 0.05$ in the strongly shear flow, which is in the eventual consequence shown in the calculation of turbulent viscosity.

The transport equation for the transfer of turbulent kinetic energy k is as follows:

$$\frac{\partial}{\partial t}(\rho k) + \frac{\partial}{\partial x_i}(\rho k \overline{u_\iota}) = \frac{\partial}{\partial x_j}\left[\left(\mu + \frac{\mu_t}{\sigma_k}\right)\frac{\partial k}{\partial x_j}\right] + G_k - \rho\varepsilon \tag{11.9}$$

σ_k is the turbulent Prandtl number for k and its value is $\sigma_k = 1.0$ (-).
G_k (kg m^{-1}s^{-3}) is the production member of the turbulent kinetic energy. It expresses the change of turbulent kinetic energy due to a change in the velocity gradient:

$$G_k = -\rho\overline{\acute{u}_\iota \acute{u}_j}\frac{\partial \overline{u_j}}{\partial x_k} \tag{11.10}$$

On the condition of the Boussinesq hypothesis, following relations apply:

$$G_k = \mu_t S^2, \quad S \equiv \sqrt{2S_{ij}S_{ij}} \quad \text{and} \quad S_{ij} = \frac{1}{2}\left(\frac{\partial \overline{u_j}}{\partial x_i} + \frac{\partial \overline{u_\iota}}{\partial x_j}\right). \tag{11.11}$$

σ_ε is the turbulent Prandtl number for ε, its value is $\sigma_\varepsilon = 1.0$ (-).
S (s^{-1}) is a module of the mean velocity deformation tensor (k-ε).

Transport equation for the transfer of dissipation velocity of turbulent kinetic energy ε is as follows:

$$\frac{\partial}{\partial t}(\rho\varepsilon) + \frac{\partial}{\partial x_i}(\rho\varepsilon\overline{u_\iota}) = \frac{\partial}{\partial x_j}\left[\left(\mu + \frac{\mu_t}{\sigma_\varepsilon}\right)\frac{\partial \varepsilon}{\partial x_j}\right] + \rho C_1 S\varepsilon - \rho C_2 \frac{\varepsilon^2}{k + \sqrt{v\varepsilon}} \tag{11.12}$$

The product of $\rho C_1 S\varepsilon$ (kg m^{-1}s^{-4}) is a production member of dissipation of the turbulent kinetic energy.

C_1, C_2 (-) are present empiric constants.

11.5.2 SST k-ω *Model*

Model k-ω differs from k-ε in another approach to modelling of turbulent viscosity in the Eq. (11.6), which is here more dependent on the distance from the wall. It thus changes its values in the proximity of walls. The variable k is as usually the turbulent kinetic energy, and ω ($m^{-1}s^{-1}$) is dissipation per turbulent kinetic energy unit, it is calculated as follows:

$$\omega = \frac{\varepsilon}{kC_\mu} \tag{11.13}$$

It is also more beneficial to express the function of turbulent viscosity in a simplified fashion for better understanding:

$$\mu_t = A\rho\frac{k}{\varepsilon} \tag{11.14}$$

A represents here a complicated function that is described in details in the manual and depends on many variables. For large *Re* and greater distances from the wall, it approximates the constant of the value $A = 1.0$ (-). Then by inserting (11.13) to (11.14), the value of μ_t will be identical with the turbulent viscosity as in the standard k-ε model (11.8).

The transport equation for the transfer of turbulent kinetic energy k is as follows:

$$\frac{\partial}{\partial t}(\rho k) + \frac{\partial}{\partial x_i}(\rho k\overline{u_i}) = \frac{\partial}{\partial x_j}\left[\left(\mu + \frac{\mu_t}{\sigma_k}\right)\frac{\partial k}{\partial x_j}\right] + G_k - Y_k \tag{11.15}$$

The transport equation for the transfer of dissipation velocity of turbulent kinetic energy ω is as follows:

$$\frac{\partial}{\partial t}(\rho\omega) + \frac{\partial}{\partial x_i}(\rho\omega\overline{u_i}) = \frac{\partial}{\partial x_j}\left[\left(\mu + \frac{\mu_t}{\sigma_\varepsilon}\right)\frac{\partial \omega}{\partial x_j}\right] + G_\omega - Y_\omega + D_\omega \tag{11.16}$$

G_k (kg $m^{-1}s^{-3}$), G_ω (kg $m^{-3}s^{-2}$) represent generation of the turbulent kinetic energy k, or dissipation velocity ω.
Y_k (kg $m^{-1}s^{-3}$), Y_ω (kg $m^{-3}s^{-2}$) specify dissipation of variables due to turbulence.
D_ω (kg $m^{-3}s^{-2}$) represents a mixing member connecting standard k-ε with k-ω models. All these members depend on many factors, and they are explained by extensive relations in the ANSYS Fluent manual [27].

11.5.3 RSM Model

Reynolds stress model includes complicated calculation of *Re* stresses using approximation through six differential transport Eqs. (11.4).

Calculated *Re* stresses are then inserted in the RANS motion Eq. (11.3). These are thus solutions of:

- Six transport Eqs. (11.4) with modification of individual members (11.17) to (11.23),
- Three motion equations for mean velocity components (11.3) and continuity Eq. (11.2),
- One transport equation for dissipation transfer ε,
- One transport equation for turbulent kinetic energy in cells at the wall.

To clarify the approach to the transport equation, it is necessary to indicate individual members of the equation:

Convection member (kg m^{-1}s^{-3}):

$$C_{ij} = \overline{u_k}\frac{\partial \rho \overline{\acute{u}_\iota \acute{u}_j}}{\partial x_k} \tag{11.17}$$

Turbulent diffusion member (kg m^{-1}s^{-3}):

$$D^*_{T,ij} = \frac{\partial}{\partial x_k}\left[\left(\rho \overline{\acute{u}_\iota \acute{u}_j \acute{u}_k}\right) + p\left(\overline{\delta_{kj}\acute{u}_\iota + \delta_{k\iota}\acute{u}_j}\right)\right] \tag{11.18}$$

Molecular diffusion member (kg m^{-1}s^{-3}):

$$D_{L,ij} = \frac{\partial}{\partial x_k}\left(\mu \frac{\partial}{\partial x_k}\left(\overline{\acute{u}_\iota \acute{u}_j}\right)\right) \tag{11.19}$$

Pressure production member (kg m^{-1}s^{-3}):

$$P_{ij} = -\rho\left(\overline{\acute{u}_\iota \acute{u}_k}\frac{\partial \overline{u_j}}{\partial x_k} + \overline{\acute{u}_j \acute{u}_k}\frac{\partial \overline{u_\iota}}{\partial x_k}\right) \tag{11.20}$$

Compressive stress member (kg m^{-1}s^{-3}):

$$\emptyset^*_{ij} = p\overline{\left(\frac{\partial \acute{u}_\iota}{\partial x_j} + \frac{\partial \acute{u}_j}{\partial x_i}\right)} \tag{11.21}$$

Dissipation member (kg m^{-1}s^{-3}):

$$\varepsilon^*_{ij} = 2\mu\overline{\left(\frac{\partial \acute{u}_\iota}{\partial x_k} + \frac{\partial \acute{u}_j}{\partial x_k}\right)} \tag{11.22}$$

Members of convection (11.17), molecular diffusion (11.19), and pressure production (11.20) are solved directly; other members must be simplified due to numerical stability.

Dissipation member (11.22) is resolved for the incompressible flow as follows:

$$\varepsilon_{ij} = \frac{2}{3}\delta_{ij}\rho\varepsilon \tag{11.23}$$

Turbulent diffusion member (11.19) is modified to the relation (kg m^{-1}s^{-3}) as follows:

$$D_{T,ij} = \frac{\partial}{\partial x_k}\left(\frac{\mu_t}{\sigma_k}\frac{\partial}{\partial x_k}\left(\overline{\acute{u}_\iota \acute{u}_j}\right)\right) \tag{11.24}$$

Turbulent viscosity μ_t is solved just as in k-ε models (8) including $C_\mu = 0.09$ (-).

An essential role in the transport equation is played by a member of the compressive stress (11.22). There are more options for modification of this member. In the presented task, the so-called linear model of compressive stress was used, with the format (11.25) for transport equation solution, where $\emptyset_{ij}$(kg m^{-1}s^{-3}). Individual partial members are solved under (11.26) to (11.28).

$$\emptyset_{ij} = \emptyset_{ij,1} + \emptyset_{ij,2} + \emptyset_{ij,w} \tag{11.25}$$

$$\emptyset_{ij,1} = -C_1\rho\frac{\varepsilon}{k}\left(\overline{\acute{u}_\iota \acute{u}_j} - \frac{2}{3}\delta_{ij}k\right) \tag{11.26}$$

$$\emptyset_{ij,2} = -C_2\left[\left(P_{ij} - C_{ij}\right) - \frac{2}{3}\delta_{ij}(P - C)\right] \tag{11.27}$$

$$\begin{aligned} \emptyset_{ij,w} = & -\acute{C}_1\rho\frac{\varepsilon}{k}\left(\overline{\acute{u}_k \acute{u}_m n_k n_m \delta_{ij}} - \frac{3}{2}\overline{\acute{u}_\iota \acute{u}_k n_j n_k} - \frac{3}{2}\overline{\acute{u}_j \acute{u}_k n_i n_k}\right)\frac{k^{\frac{3}{2}}}{C_1\varepsilon d} \\ & + \acute{C}_2\frac{\varepsilon}{k}\left(\emptyset_{km,2} n_k n_m \delta_{ij} - \frac{3}{2}\emptyset_{ik,2} n_j n_k - \frac{3}{2}\emptyset_{jk,2} n_i n_k\right)\frac{k^{\frac{3}{2}}}{C_1\varepsilon d} \end{aligned} \tag{11.28}$$

where compression production is $P = 0.5P_{kk}$ (kg m^{-1}s^{-3}),
convection $C = 0.5C_{kk}$ (kg m^{-1}s^{-3}),
n_i represent unit directional vectors, κ (-) von Karman constant, d (m) the distance from the wall and $C_1 = \frac{C_\mu^{\frac{3}{4}}}{\kappa}$ (-).

Non-dimensional constants are of the following values:
$C_1 = 1.8$; $C_2 = 0.6$; $\acute{C}_1 = 0.5$; $\acute{C}_2 = 0.3$.

Just the relation (11.28) ensures distribution of *Re* normal stresses in the proximity of the wall. It tends to absorption $-\rho\overline{\acute{u}_\iota \acute{u}_j}$ (perpendicularly to the wall), increases parallel stress with the wall, and thus eliminates lack of isotropic condition of the Boussinesq hypothesis.

11.5.4 DES Model

This hybrid model attempts to treat near-wall regions in a RANS-like manner (in this task, the SST *k-ω* is used), while the rest of the flow is treated in an LES-like manner. Switching between both modes is based on the turbulent length scale and grid dimension. As the turbulent length scale exceeds the grid dimension, the regions are solved using the LES mode. This task is solved using the DES SST *k-ω* model.

11.6 Task Solution and Results

Models described in the previous chapter were selected for the task solution. In all models, the flow field was monitored from the velocity u_x and turbulence point of view, or turbulence intensity i_u point of view. In all calculations, identical types of boundary conditions were set. It is the so-called velocity-inlet at the input and the pressure condition pressure-outlet at the output. Boundary conditions on both sides and the lower and upper surface of the calculation area were defined so that they correspond with the restricted space of the aerodynamic tunnel.

For DES model, a time step of 0.0005 s was selected; the calculation simulated the flow for 4.3 s, provided that averaging of variables was set after 1.6 s of the simulated process. Then, the flow field could be considered stabilized. The resulting flow time at time averaging of variables represented approximately quadruple air exchange in the calculation area. Within single calculation, about $8.6 \cdot 10^4$ iterations were carried out. Other models admittedly enable to solve the task stationary, but also these calculations had to be carried out non-stationary due to bad convergence. For RSM model, the time step was set to 0.001 s, and a total of 5.1×10^3 iterations was carried out.

11.6.1 Results

The course of velocity u_x and turbulence intensity i_u in the longitudinal axis of the monitored area are apparent in Fig. 11.6.

Values of the velocity u_x and turbulence intensity i_u including their distribution inside the nozzle are apparent in Figs. 11.7 and 11.8.

At the distance of 0.36 m behind the outlet from the narrowed space back to the expanded space (total of 1.76 m), isolines of the velocity u_x are shown in Fig. 11.9, and isolines of turbulence intensity i_u in Fig. 11.10, from numerical calculations and the experiment. As stated above, velocity distribution in the monitored cross section is defined by non-dimensional coefficient given by actual velocity value/mean velocity value ratio.

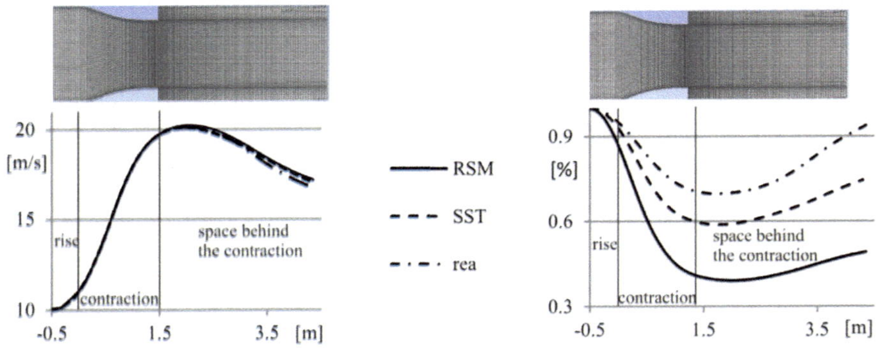

Fig. 11.6 Course of velocity (left) and turbulence intensity (right) in the longitudinal axis of the monitored area

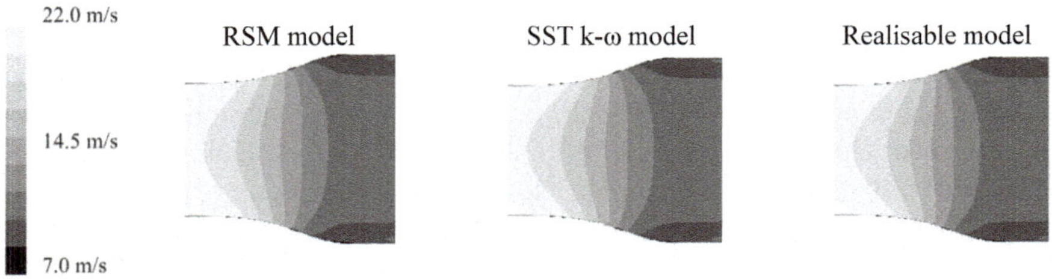

Fig. 11.7 Velocity field of inner nozzle space in the vertical longitudinal section

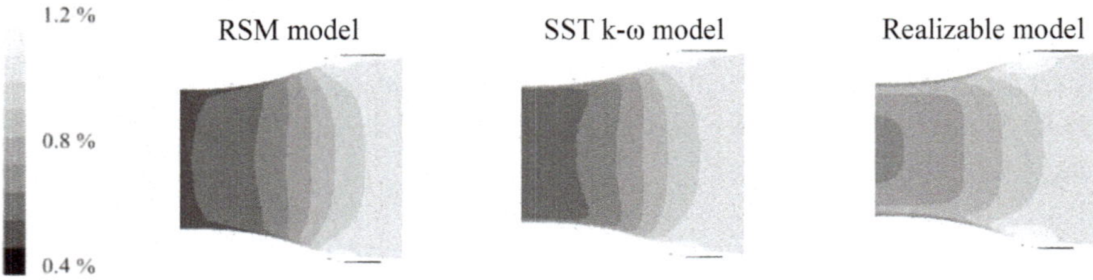

Fig. 11.8 Turbulence intensity of inner nozzle space in the vertical longitudinal section

In DES model, turbulence intensity can be assessed for large eddies only. Its records u_x and i_u are therefore shown separately in Fig. 11.11.

Distribution of the velocity field including values of velocity u_x inside the narrowed space in three different sections is apparent in Fig. 11.12. Input section corresponds with dimensions 1.98 × 1.784 m under Fig. 11.1.

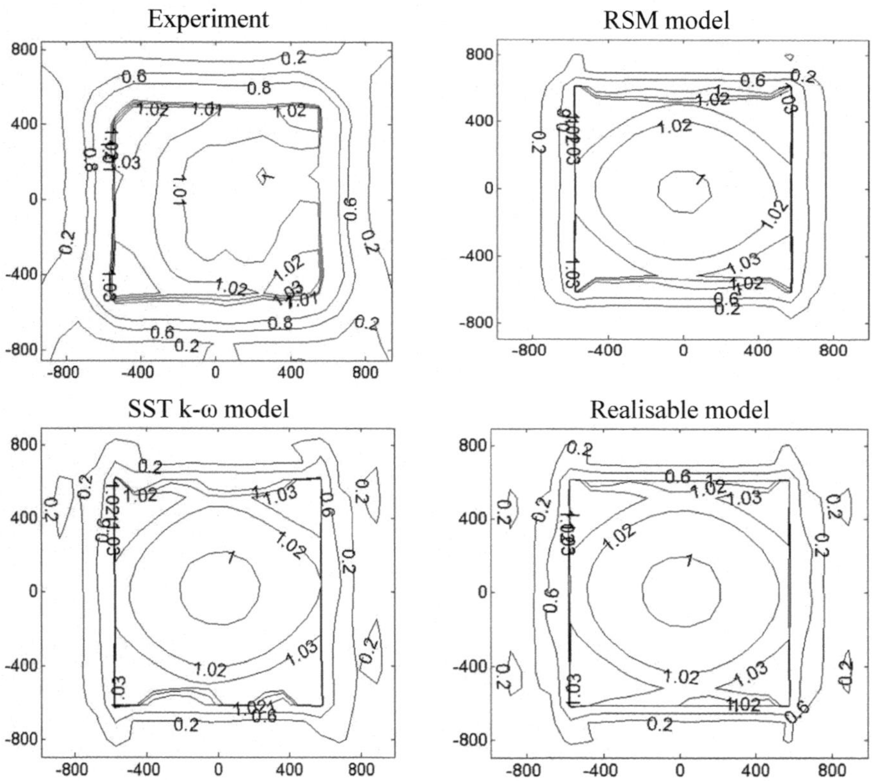

Fig. 11.9 Velocity coefficient isoline (-) 0.36 m behind the narrowed space

11.6.2 Illustrative Practical Examples

The findings of the described task can be used in many cases in the field of water management. Typical examples are the reconstructions of water dams. Figure 11.13 shows shots from the reconstruction of the Šance waterworks in the years 2015–2018.

Valuable knowledge from numerical modelling of a given issue can be used, for example, in solving culvert (Fig. 11.14).

11.7 Conclusion

In this chapter, the reader has become more familiar with the possibilities of numerical modelling of the fluid flow in changing flow space. The presented task is solved using a finite volume method in ANSYS Fluent software. Total of four numerical models were selected, suitable for the solution of described problems. Their mathematical

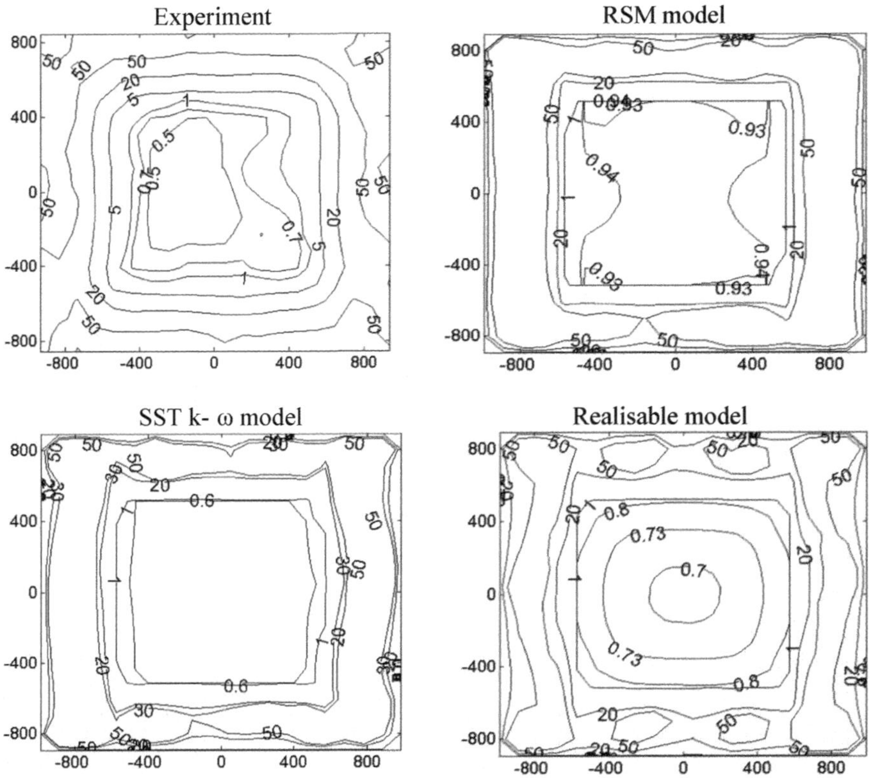

Fig. 11.10 Turbulence intensity isoline (%) 0.36 m behind the narrowed space

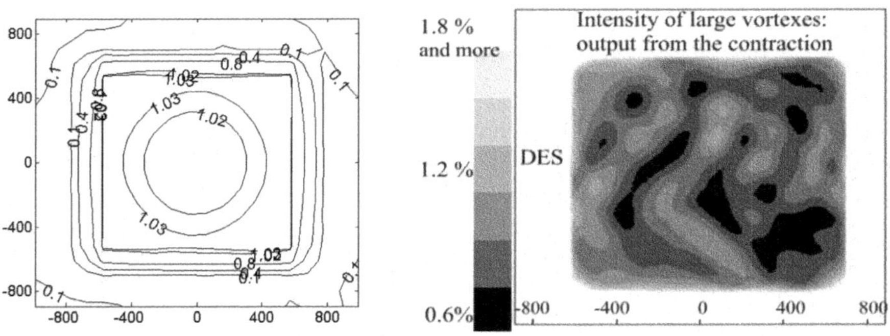

Fig. 11.11 Velocity coefficient isoline (left) and turbulence intensity (right) 0.36 m behind the narrowed space from the numerical calculation using DES model

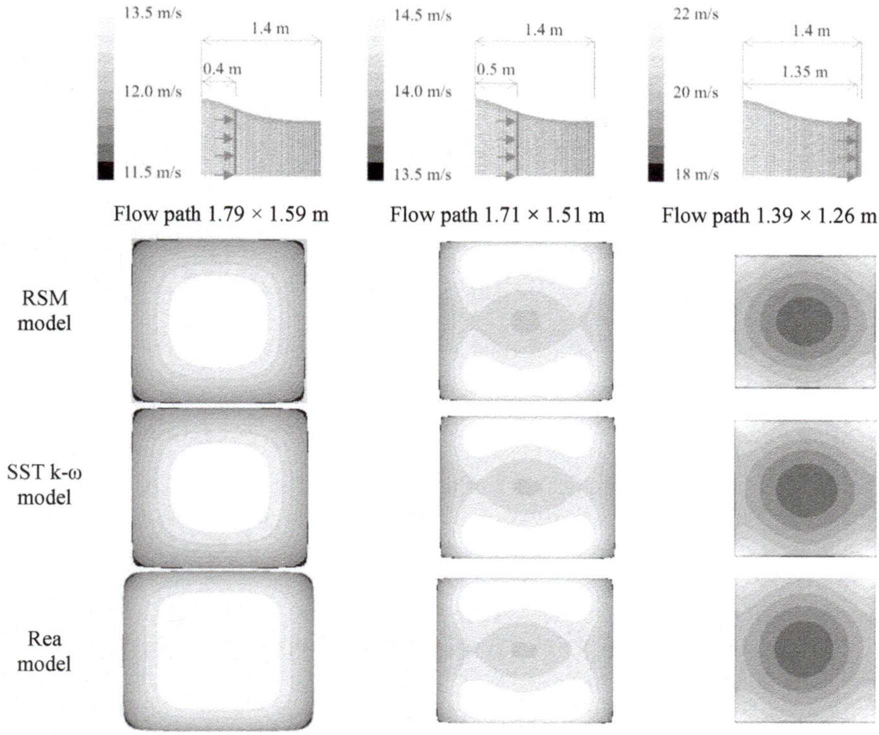

Fig. 11.12 Velocity field in cross sections in inner nozzle space

Fig. 11.13 Rekonstrukce vodní nádrže Šance (https://www.msstavby.cz/, https://fm.denik.cz/)

Fig. 11.14 Culverts (a typical and non-environmental solution) [35]

approach to the solution and differences between individual models are explained in this thesis. The flow field change was monitored in the changing flow space. Results related to flow velocity and turbulence intensity are presented. Some results could be verified by experimental measuring performed in an aerodynamic tunnel of the experimental research site of ITAM CAS, v. v. i., Centre of Excellence Telč.

Calculations and the experiment confirmed an assumption of accelerated flow field through a narrowed space (Fig. 11.6). The change in the velocity field shape is, however, impressive. In the narrowed space, the flow accelerates in the proximity of peripheral walls of the nozzle (Figs. 11.7 and 11.12). At the distance of 0.4 m from the entry to the narrowed space, the flow field shape is unchanged yet, the first acceleration in the proximity of walls appeared in 0.5 metres. At the outlet (1.35 m), acceleration I the area of walls is marked already. The flow in the proximity of walls is faster by ca 18% than in the cross-sectional axis.

Numerical models describe the shape of velocity field with a very satisfactory result. All models show almost identical shapes of the velocity field in the monitored narrowed space of the area (Figs. 11.7 and 11.12). Velocity values in the whole calculation area (even outside the narrowed space) are almost identical for all calculations. Insignificant variation can be observed about 1.5 m behind the nozzle outlet (see Fig. 11.6).

On the contrary, turbulence intensity i_u decreases in the narrowed space, which is confirmed by numerical calculations and the experiment (Figs. 11.8 and 11.10). The great increase of i_u in the proximity of walls right behind the exit to the more open space is worthy of mention (Fig. 11.10). In the main flow closer to the centre, i_u is, however, lowered temporarily (Fig. 11.6). Numerical description of turbulence intensity is very problematic, which shows first in the different course of values in the longitudinal axis of the monitored area (Fig. 11.6), especially at a greater distance from the outlet from the narrowed space.

The subject matter of further examination will be a more detailed assessment of turbulence intensity as compared with the prepared physical measuring of the flow field inside the nozzle and at a greater distance from the outlet. Based on this, it would be possible to define the most suitable model that should serve for detail description of the simulated process. Based on this, it would be possible to monitor the impacts of the changed flow on the land and selected neighbouring objects.

Acknowledgements Financial support from VŠB-Technical University of Ostrava by means of the Czech Ministry of Education, Youth and Sports through the Institutional support for conceptual development of science, research, and innovations for the year 2018 and by the Scientific Grant Agency of the Ministry of Education of Slovak Republic and the Slovak Academy of Sciences the project VEGA 1/0477/15 and VEGA 1/0374/19 is gratefully acknowledged.

References

1. Sargison JE, Walker GJ, Rosi R (2004) Design and calibration of a wind tunnel with a two dimensional contraction. In: Proceeding of the 15th Australian fluid mechanic conference, Sydney, Australia, pp 143–146. ISBN 1-864-87695-6
2. Fang FM, Chen JC, Hong YT (2001) Experimental and analytical evaluation of flow in a square-to square wind tunnel contraction. J Wind Eng Ind Aerodyn 89(1):247–262. ISSN: 0167-6105
3. Fang FM (1997) A design method for contractions with square end sections. J Fluids Eng 119(2):454–458. ISSN: 1555-1415
4. Wolf T (1995) Design of a variable contraction for a full-scale automotive wind tunnel. J Wind Eng Ind Aerodyn 56(1):1–21. ISSN: 0167-6105
5. Michalcová V, Kuznětsov S, Pospíšil S, Brožovský J (2014) Numerical and experimental investigations of air flow turbulence characteristics in the wind tunnel. Appl Mech Mater 617:275–279
6. Vijiapurapu S, Cui J (2010) Performance of turbulence models for flows through rough pipes. Appl Math Model 34(6):1458–1466. ISSN: 0307-904X
7. Michalcová V, Kuznětsov S, Pospíšil S (2013) Models of load on buildings from the effects of the flow field. Trans VŠB Tech Univ Ostrava Constr Ser 13(2):91–97. ISSN: 1804-4824 (Online); ISSN: 1213-1962 (Print)
8. Rodríguez M, Lastra J, Oro MF, Galdo M, Vega E, Marigorta B, Morros CS (2013) Novel design and experimental validation of a contraction nozzle for aerodynamic measurements in a subsonic wind tunnel. J Wind Eng Ind Aerodyn 118:35–43. ISSN: 0167-6105
9. Liu P, Duan H, Zhao W (2009) Numerical investigation of hot air recirculation of air-cooled condensers at a large power plant. Appl Therm Eng 29(10):1927–1934. ISSN: 1359-4311
10. Tropea C, Yarin AL, Foss JF (eds) (2007) Measurement of turbulent flows. Springer (Chapter 10)
11. Toda HB, Cabrit O, Truffin K, Bruneaux G, Nicoud F (2014) Assessment of subgrid-scale models with an LES-dedicated experimental database: the pulsatile impinging jet in turbulent cross flow. Phys Fluids Am Inst Phys 26:075108
12. Kotrasová K, Grajciar I (2014) Dynamic analysis of liquid storage cylindrical tanks due to earthquake. Adv Mater Res 969:119–124
13. Kotrasová K (2014) Sloshing of liquid in rectangular tank. Adv Mater Res 969:320–323
14. Kotrasová K (2017) Study of hydrodynamic pressure on wall of tank. Procedia Eng 190:2–6. ISSN: 1877-7058
15. Michalcová V, Kotrasová K (2012) Influence of numerical diffusion on exactness of calculation in software FLUENT. Bull Transilvania Univ'Brasov 5(54):99–106
16. Eymard R, Gallouët T, Herbin R (2000) Finite volume methods. In: Handbook of numerical analysis, vol 7. Elsevier, pp 713-1018. ISSN: 1570-8659
17. Santis DD, Geraci G, Guardone A (2012) Finite volume and finite element schemes for the euler equation in cylindrical and spherical coordinates. J Comput Appl Math ISSN: 0377-0427

18. Mehta DR, Bell JH (1989) Boundary-layer predictions for small low-speed contractions. AIAA J 27(3):372–374
19. Michalcová V, Kuznětsov S, Pospíšil S (2014) Numerical and experimental study of the load of an object due to the effects of a flow field in the atmospheric boundary layer. Int J Math Comput Simul 8:135–140. ISSN: 0378-4754
20. Irrenfried C, Steiner H (2017) DNS based analytical P-function model for RANS with heat transfer at high Prandtl numbers. Int J Heat Fluid Flow 66:217–225. ISSN: 0142-727X
21. Taraba B, Michalec Z, Michalcová V, Blejchař T, Bojko, Kozubková M (2014) CFD simulations of the effect of wind on the spontaneous heating of coal stockpiles. Fuel 118:107–112. ISSN: 0016-2361
22. Michalcová V, Kuznětsov S, Pospíšil S (2014) Numerical modelling of air flow attributes in a contractions chamber. Trans VŠB Tech Univ Ostrava Civil Eng Ser 14(2):11–16. ISSN: 1804-4824 (Online); ISSN: 1213-1962 (Print)
23. Michalcová V, Kuznětsov S, Pospíšil S, Brožovský J (2014) Numerical and experimental investigations of air flow turbulence characteristics in the wind tunnel contraction. Appl Mech Mater 617:275–279
24. Kocich R, Bojko M, Macháčková A, Klečková Z (2012) Numerical analysis of the tubular heat exchanger designed for co-generating units on the basis of microturbines. J Heat Mass Transfer 55(19/20):5336–5342. ISSN: 00179310
25. Bojko M, Kozdera M, Kozubkova M (2014) Investigation of viscous fluid flow in an eccentrically deposited annulus using CFD methods. In: Proceeding of the 8th international conference on experimental fluid mechanics, Kutna Hora, Czech Republic. ISSN: 21016275, ISBN: 978-802605375-0
26. Balbus S (2017) When is high Reynolds number shear flow not turbulent? J Fluid Mech 824:1–4
27. ANSYS® Fluent Release 18.2, Fluent documentation, Theory Guide
28. Khadra K, Angot P, Parneix S, Caltagirone JP (2000) Fictitious domain approach for numerical modelling of Navier-Stokes equations. Int J Numer Meth Fluids 34:651–684
29. Coleman G, Sandberg R (2010) A primer on direct numerical simulation of turbulence—methods, procedures and guidelines. Aerodynamics and Flight Mechanics Research Group
30. Spalart PR, Coleman GN, Johnstone R (2008) Direct numerical simulation of the Ekman layer: a step in Reynolds number, and cautious support for a log law with a shifted origin. Phys Fluids 20:101507
31. Zhiyin Y (2015) Large-eddy simulation: past, present and the future. Chin J Aeronaut 28(1):11–24. ISSN: 1000-9361
32. Xie ZT, Castro IP (2008) Efficient generation of inflow conditions for large-eddy simulation of street-scale flows. Flow Turbul Combust 81:449–470. ISSN: 1386-6184
33. Hertwig D, Efthimiou GC, Bartzis JG, Leitl B (2012) CFD-RANS model validation of turbulent flow in a semi-idealized urban canopy. J Wind Eng Ind Aerodyn 111:61–72. ISSN: 0167-6105
34. Schmitt FG (2007) About Boussinesq's turbulent viscosity hypothesis: historical remarks and a direct evaluation of its validity. Comptes Rendus Mécanique, Elsevier Masson 335(9–10):617–627
35. Sedláček D (2017) Modelový výzkum proudění na vtoku do propustku. Thesis, ČVUT, Praha

Chapter 12
Numerical Modelling of Fluid Domain Flow in Open Top Channel

K. Kotrasová and V. Michalcová

Abstract Ground-supported liquid-transporting structures are used to store and transport of fluid. Water flow in open channel deals with the equilibrium forces in the fluid. The fluid develops pressure on the channel. During an earthquake, the fluid exerts hydrodynamic impulsive and hydrodynamic convective pressures together with hydrostatic pressure on tank walls and bottom of the ground-supported fluid filling endlessly long channel. This paper provides a theoretical background for the specification of fluid effect on the liquid-filled rectangular channel during ground motion. The numerical model of the seismic response of the fluid filling of endlessly long shipping channel was obtained by using finite element method (FEM) in software ADINA. The results from the numerical parametrization: mesh density of the 2D fluid region, mesh parameter 'PATTERN' of the 2D fluid region or 3D fluid region have to influence for finally the model of numerical simulation of water-filled channel grounded on the soil.

Keywords Fluid · Earthquake · Finite element method · Pressure · Numerical analysis

12.1 Introduction

Water flow in open channels, throughs of flowing water is the secondary discipline that deals with the equilibrium forces in the fluid, whether is in the stay or in the move. The concept of the elementary volume was gradually introducing, which is,

K. Kotrasová (✉)
Department of Structural Mechanics, Faculty of Civil Engineering, Institute of Structural Engineering, The Technical University of Kosice, Vysokoskolska 4, 042 00 Kosice, Slovak Republic
e-mail: kamila.kotrasova@tuke.sk

V. Michalcová
Department of Structural Mechanics, VŠB—Technical University of Ostrava, Faculty of Civil Engineering, L. Podéště 1875/17, 708 33 Ostrava-Poruba, Czech Republic
e-mail: vladimira.michalcova@vsb.cz

M. Zelenakova et al. (eds.), *Management of Water Quality and Quantity*, Springer Water, https://doi.org/10.1007/978-3-030-18359-2_12

on the one hand, a very small flow area, on the other hand, it is a large enough volume to compare individual molecular pathways [1].

The explaining of the movement of liquids as such is useless. A more appropriate way is to consider the flowing fluid as a continuous environment, and at each point in this environment, we assign the velocity of the vector, $v = v(r, t)$. In this case, we can divide the flow into hydrostatic ($v = 0$) and hydrodynamic ($v \neq 0$) [2].

The properties describing the movement and flow of liquids derived from the basic volume of liquid can be expressed by:

- weight conservation law—continuity equation,
- act of preservation of momentum,
- energy conservation law,
- entropy change law.

Fluid movement can be generally defined as unstable and three-dimensional. The role of hydrodynamics is to determine the speed division into sections of a given flow and to determine the pressure when water moves in the area under consideration [3]. The results can be presented through physical and mathematical modelling. It is more advantageous to use mathematical modelling, because physical research is usually more demanding, both financially and temporally [4]. The use of modelling results in real conditions, i.e. in larger dimensions, is conditional on the fluctuation of velocities describing fluid dynamics [5].

By solving the open flow mathematical modelling is necessary to adopt certain assumptions that apply the general solution procedure. A complete description of the model is practically inalienable [6].

The solution of mathematical modelling is the numerical solution of the physical equation describing the flow time of the current model section. From this point of view, we can divide the individual events that are going on in time into constant ones, where the values are a function of time and spatial coordinates and on fixed ones, where the values of the variables are functions of spatial coordinates only and do not change over time [7]. Determining the known boundary conditions for the steady flow model addresses unknowns that are the result of the mathematical model. In addition, each model is also defined by the initial conditions, which are the values of all the variables determined at the beginning of the solution (e.g. at time $t = 0$). They are, therefore, the basis for determining the compromise depending on the level of available data with the time or financial possibilities of the contracting entity [8].

Implementation can be divided into three parts:

- Preprocessing—preparing and entering the necessary inputs into the model and defining outputs from the model;
- processing—calculation;
- postprocessing—processing and exporting outputs from the model for further data processing.

Outputs from flow patterns are of great importance to specialties that focus their interests on waterway construction, sailing operations, trade unions using water, water transport as a hydropotential carrier (water—energy) [8].

Issues of sustainable and integrated water supply in the country and the urbanized area, water management, melioration and water supply systems, communal and industrial sector requires theoretical and practical knowledge of water protection and management are based on mathematics, physics, and other theoretical and natural sciences disciplines [9].

The knowledge of subjects of the theoretical, natural science base and mathematics—physical basis, and using computer support for engineering work helps with modelling and simulations of these issues [10].

One of the most problems is the analysis of supply and wastes channel, which could be loaded by different kinds of loadings [11].

Ground-supported liquid-containing or liquid-transporting structures are used to store and transport of fluid, mainly water, between two points in a safe and cost-effective manner, includes economic, safety, and aesthetic aspects. The fluid develops pressure on the walls and bottom of the channel [12].

When subjected to external excitation like an earthquake, the fluid exerts hydrodynamic impulsive and hydrodynamic convective pressures together with hydrostatic pressure on walls and bottom of ground-supported liquid-containing or liquid-transporting structures [13].

The seismic analysis and design of liquid-containing or liquid-transporting structures is a complicated task due to the high complexity of the problem. A number of particular aspects should be taken into consideration. For example, dynamic interaction between contained fluid and structure, sloshing motion of the contained fluid; and dynamic interaction between structure and sub-soil [14]. Dynamic interaction between contained fluid and structure and sloshing motion of the contained fluid belong to the full range of so-called fluid-structure interactions (FSI). The tank–soil interaction could under specific conditions have a significant effect on the seismic response of the liquid-containing and liquid-transporting structures [15].

The knowledge of pressures and effects acting onto walls and the bottom of liquid-containing or liquid-transporting structures and dynamic response of their structures during an earthquake plays an essential role in the reliable and durable design of earthquake resistance structure/facility [16].

12.2 Seismic Effect of Fluid on Channel

The complete solution of the Laplace equation for the motion of fluid contained in rigid channels can be expressed as the sum of two separate contributions, called ‘impulsive rigid’, and ‘convective’, respectively. The ‘impulsive rigid’ component of the solution satisfies exactly the boundary conditions at the walls and the bottom of the channel. There is the compatibility between the velocities of the fluid and the tank. However, the ‘impulsive rigid’ component gives incorrectly zero pressure at the original position of the free surface of the fluid in the static situation. This is caused due to the presence of the waves in the dynamic response.

The 'convective' component does not alter those boundary conditions that are already satisfied while fulfilling the correct equilibrium condition at the free surface [17].

We consider the ground supported endlessly rectangular channel with the total rigid walls having the uniform thickness whom the interior width is $2L$. We consider fluid filling of the rectangular channel to the height H that is excited by a time-dependence horizontal acceleration $A_g(t)$ in the direction of channel width $2L$.

The solution of the Laplace equation of fluid excited by the horizontal seismic excitation can be obtained in the form (12.1). The total hydrodynamic pressure is given by the sum of an impulsive and a convective contribution by using the absolute summation rule [18]

$$p(z,t) = p_i(z,t) + p_c(z,t) \tag{12.1}$$

The impulsive rigid pressure component along the height of the channel wall in (Pa) is given by

$$p_i(z,t) = C_i(z)\rho H A_g(t) \tag{12.2}$$

where

L is the half-width of the channel in the direction of the seismic action in (m),
ρ is the mass density of the liquid in (kgm^{-3}),
$C_i(z)$ is a function that presents the distribution of the impulsive hydrodynamic pressure along with the height of the channel wall as a function of the slenderness parameter γ.

The time-dependence the pressure $p_i(z, t)$ in Eq. (12.2) is given by the function $A_g(t)$, while pressure $p_i(z, t)$ is constant in the direction that is orthogonal to the seismic action [19]. $A_g(t)$ is the time-history-dependent ground acceleration in free—field motion of the ground in (ms^{-2}).

$$C_i(z) = \frac{\sqrt{3}}{2}\left(1 - z^2\right)\tanh\sqrt{3}\frac{L}{H}. \tag{12.3}$$

Figure 12.1 presents function $C_i(z)$ independence of the slenderness parameter γ. The slenderness parameter is given by the relation $\gamma = H/R$, where H is the height of the fluid-filled rectangular channel and L is half-width of the channel.

The convective pressure component, called as 'sloshing', is given by the summation of modal terms (sloshing modes), where each one has different variations in time. The dominant contribution for excited fluid filling is the fundamental—first mode. The other modes have extraordinary influence. Usually, they have a fractional value of the first mode. The distribution of the hydrodynamic convective pressure $p_{c1}(z, t)$ is given

$$p_{c1}(z,t) = Q_{c1}(z)\rho L Ag(t) \tag{12.4}$$

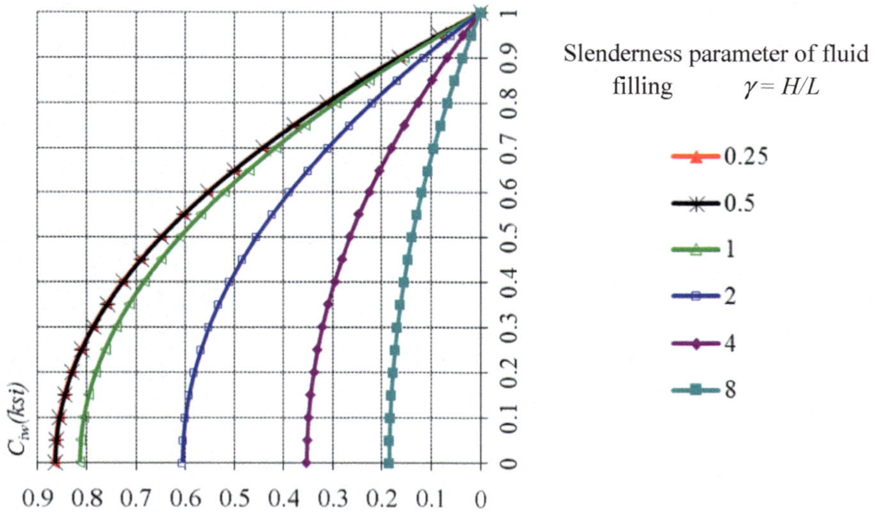

Fig. 12.1 Functions $C_i(z)$ present the distribution of the impulsive hydrodynamic pressure as a function of the slenderness parameter γ

The function $Q_{c1}(z)$ represents the distribution of hydrodynamic convective pressure distribution along with the height of the channel wall as a function of the slenderness parameter γ.

$$Q_{c1}(z) = 0.833\left(\cosh\left(z\frac{1}{2}\sqrt{\frac{5}{2}\frac{H}{L}}\right)/\cosh\left(\frac{1}{2}\sqrt{\frac{5}{2}\frac{H}{L}}\right)\right). \tag{12.5}$$

Figure 12.2 shows functions $Q_{c1}(z)$ independence of the slenderness parameter γ. $A_1(t)$ in (ms^{-2}) is the acceleration response function of the simple oscillator having the frequency of the first mode, the appropriate value of the damping and subjected to an input acceleration $A_g(t)$ (ms^{-2}) [20].

The period of oscillation of the first 'sloshing' mode is

$$T_{c1} = \sqrt{(L/g)/\left(\frac{\pi}{2}\tanh\left(\frac{\pi H}{2L}\right)\right)}. \tag{12.6}$$

12.2.1 Earthquake as Extremely Loading

The data that can describe the effects of the earthquake in a form suitable for calculating the seismic response of the medium (structure, fluid) and which allow the determination of the extreme forces or deformation effects of the earthquake on the

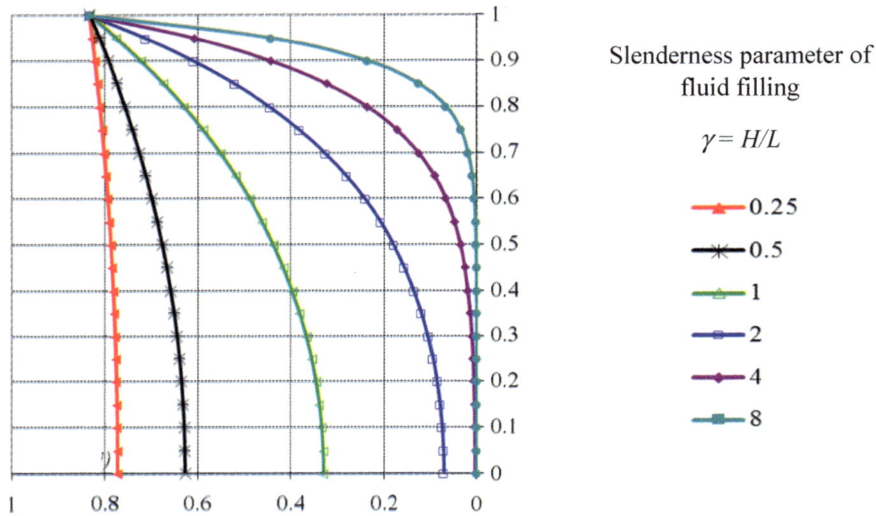

Fig. 12.2 Functions $Q_{c1}(z)$ present the distribution of the hydrodynamic convective pressure as a function of the slenderness parameter γ

structure is called a seismic load. In physical terms, the seismic load is the oscillation of the subsoil. A seismic motion can be defined by accelerograms—time courses of acceleration of motion of the subsoil, or corresponding displacement and alternatively velocity. Seismic motion consists of three components, two horizontal and one vertical. Excitation in horizontal directions is usually more pronounced compared to that in the vertical direction. All three courses act simultaneously and represent a complex spatial movement.

It consists of two ways of the solution:

- the best is considering the effects of all three components at the same time when designing and designing the structure,
- very often an individual component is applied individually, and then an effect is additionally superimposed.

The characteristics used to describe the complex seismic movement of the earth's surface with its properties can be subdivided into characteristics in the time domain and frequency domain and response spectra.

The most commonly used in engineering practice are the spectra of seismic responses that they give the basic information about the characteristic of the seismic response structure on the seismic motion. The seismic response spectrum is defined as a set of maxis (maximum displacement, velocity or acceleration) of the one pendulum system, depending on its frequency, in the effect of the analysed seismic loading. The elastic response of medium (structure, fluid) on the seismic motion of the soil can be approximated by the response of a simple oscillator whose own frequency is identical to the actual frequency of the structure.

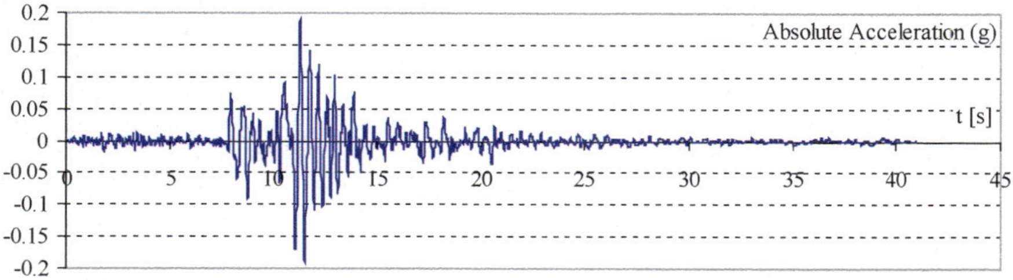

Fig. 12.3 Absolute acceleration of earthquake Loma Prieta, California (18.10.1989)

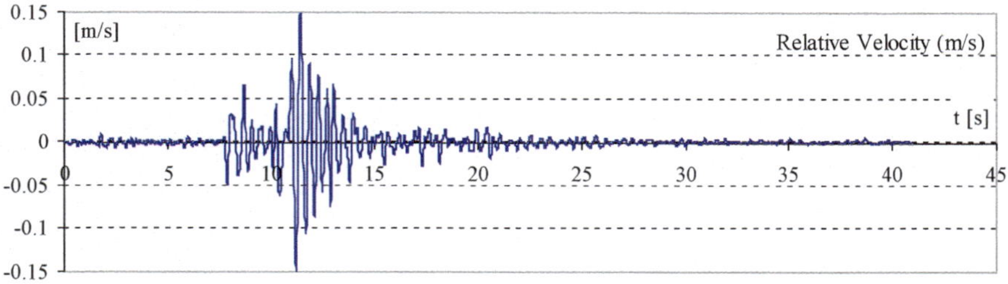

Fig. 12.4 Relative velocity of earthquake Loma Prieta

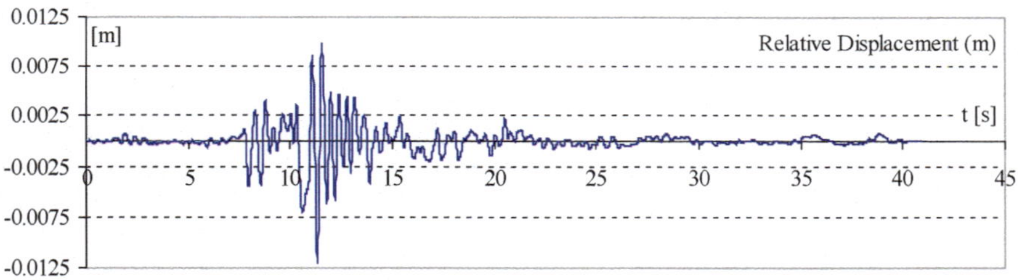

Fig. 12.5 Relative displacement of earthquake Loma Prieta

The recordings of the absolute acceleration, the relative velocity, the relative displacement of earthquake Loma Prieta, California (October 18, 1989) are documented in Figs. 12.3, 12.4 and 12.6. The elastic spectrums absolute acceleration of earthquake Loma Prieta presented in Fig. 12.3 are shown in Fig. 12.6 for 0.5, 2, 4, 5 and 7% damping (Fig. 12.5).

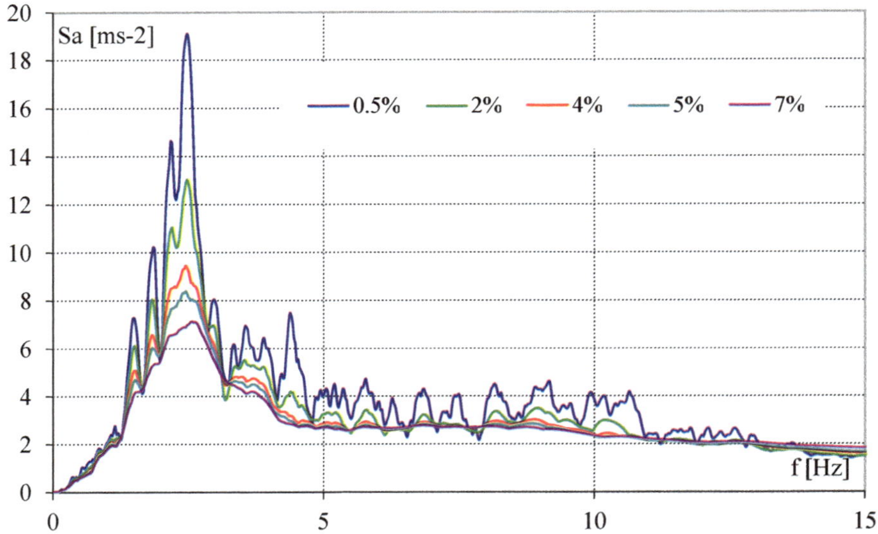

Fig. 12.6 Elastic spectrums of absolute acceleration of earthquake Loma Prieta, California for 0.5, 2, 4, 5 and 7% damping

12.3 Finite Element Method

The finite element method (FEM) is well established for the analysis of complex engineering problems involving structures, fluids and fluid–structure interaction [21].

12.3.1 Problem Formulation of Water-Filled Channel Grounded on Soil

For the fluid-structure interaction analysis are three different finite element approaches to represent fluid motion [22]:

- Eulerian is used to describe the behaviour of the fluid region by velocity and pressure,
- Lagrangian is used to describe the behaviour of the fluid region by displacement,
- mixed methods are used both: the pressure and displacement fields are included in the element formulation.

In fluid–structure interaction analyses, the following steps are repeated:

- fluid forces are applied to the solid,
- the solid deformation changes the fluid domain.

For most interaction problems, the computational domain is divided into two separated models:

- the fluid domain,
- solid domain.

A fluid model and a solid model are defined, respectively, through their material data and boundary conditions. The interaction occurs along with the interface of the two domains. Having the two models coupled, we can perform simulations and predictions of many physical phenomena [23].

12.3.1.1 Fluid Flow Equations

The motion of a continuous fluid medium is governed by the principles of theories classical mechanics and thermodynamics [24]. These theories can be expressed in conservative forms for mass, momentum and energy

$$\begin{aligned} \frac{\partial \rho}{\partial t} + \nabla \cdot (\rho \mathbf{v}) &= 0 \\ \frac{\partial \rho \mathbf{v}}{\partial t} + \nabla \cdot (\rho \mathbf{v}\mathbf{v} - \boldsymbol{\tau}) &= \mathbf{f}^B \\ \frac{\partial \rho E}{\partial t} + \nabla \cdot (\rho \mathbf{v} E - \boldsymbol{\tau} \cdot \mathbf{v}) &= \mathbf{f}^B \cdot \mathbf{v} + q^B \end{aligned} \tag{12.7}$$

where

the stress tensor $\boldsymbol{\tau}$ is given by $-(p + \lambda \nabla \cdot \mathbf{v})\mathbf{I} + 2\mu\mathbf{e}$ in (Pa),
the specific total energy E is given by $e + (\mathbf{v} \cdot \mathbf{v})/2$ in (m^2s^{-2}),
the heat flux $\mathbf{q}$ is given by $-k\nabla\theta$ in (Wm^{-3}),
t is the time in (s),
ρ is the fluid density in (kg m^{-3}),
$\mathbf{v}$ is the velocity vector (m s^{-1}),
$\mathbf{f}^B$ is the body force per unit volume (kN m^{-3}),
q^B is the specific rate of heat generation per unit volume (Wm^{-3}).

The specific total energy E in (m^2s^{-2}) is defined as and

$$E = \frac{1}{2}\mathbf{v} \cdot \mathbf{v} + e \equiv b + e \tag{12.8}$$

The fluid stress is defined by

$$\tau = (-p + \lambda \nabla \cdot \mathbf{v})\mathbf{I} + 2\mu\mathbf{e} \tag{12.9}$$

where

e is the specific internal energy (kgm^{-3}),
b is the specific kinetic energy (m^2 s^{-2}),
p is the fluid pressure (Pa),
μ is fluid viscosity (kg/m-s),

λ, is second fluid viscosity (.),
tensor $\mathbf{e}$ is the velocity strain in (1/s) given by

$$\mathbf{e} = \frac{1}{2}\left(\nabla\mathbf{v} + \nabla\mathbf{v}^T\right). \tag{12.10}$$

The heat flux vector in (Wm^{-2}) is assumed to obey Fourier law of heat conduction

$$\mathbf{q} = -k\nabla\theta \tag{12.11}$$

where

k is the heat conductivity coefficient in (W/m-K),
θ is the temperature in (K).

The gravitational force as one of the most important body forces is included in $\mathbf{f}^B$, where the gravitational acceleration vector in (ms^{-2}) is $\mathbf{g}$ [25].

$$\mathbf{f}_g^B = \rho\mathbf{g}. \tag{12.12}$$

The equations for the description of continuum movement are based on the Eulerian approach. The based characteristic properties of the fluid domain are considered as function time and area. The primitive variables p, $\mathbf{v}$, θ or the conservative variables ρ, $\rho\mathbf{v}$, ρE are the solution of the governing equations [26].

12.3.2 Free Surface of Fluid Domain, Boundary Conditions

The free surface is the moving boundary. In the case of moving boundaries, the conditions that must be satisfied are $\hat{\mathbf{u}}\cdot\mathbf{n} = \hat{u}_s$ and $\hat{\mathbf{u}}\cdot\mathbf{t} = \hat{u}_t$, where $S_{\hat{u}}.S_{\hat{u}}$ corresponds to the part of the surface with imposed displacements $\hat{u}_s$ and $\hat{u}_t$ in the normal and tangential directions, respectively. $\mathbf{n}$ and $\mathbf{t}$ are unit normal and tangent vectors to the boundary, and $\hat{\mathbf{u}}$ is the boundary displacement [27].

The effect of air, in the case of a free surface, is considered usually included only as the pressure p_0

$$-p_0\mathbf{n}\cdot\tau\cdot\mathbf{n}\tau = \alpha\left(\frac{1}{R_t} + \frac{1}{R_s}\right)\mathbf{n}, \tag{12.13}$$

where $\mathbf{n}$ is a unit normal vector to the interface surface pointing outwards of the free surface, τ is stress tensor, α is the coefficient of surface tension between the fluid and air and R_t and R_s are the principal radii of curvatures of the interface surface [16].

The surface in a reference time t_0 is represented by the function $S\left({}^0\mathbf{x}, \mathbf{t}_0\right) = 0$, where ${}^0\mathbf{x}$ is the vector of coordinates of the particles that are located on the free surface at time t_0. The following condition has to be satisfied

$$\frac{\delta \mathbf{S}}{\delta \mathbf{t}} + \left(\mathbf{v} - \hat{\mathbf{v}}\right) \cdot \nabla S = \mathbf{0}, \tag{12.14}$$

which ensures that the particles that are at the free surface at time t0 will remain on that surface for all times.

The balance of forces between interactive forces of liquid and gas expresses the dynamic boundary condition for free surface

$$\begin{aligned} \mathbf{f}_l.\mathbf{n} + \sigma\, K &= -\mathbf{f}_g\ \mathbf{n}, \\ \mathbf{f}_l.\,\mathbf{t}\ + \sigma\, K &= -\mathbf{f}_g\ \mathbf{t}, \\ \mathbf{f}_l.\,\mathbf{s} + \sigma\, K &= -\mathbf{f}_g\ \mathbf{s}, \end{aligned} \tag{12.15}$$

where $\mathbf{f}_l$, respectively, $\mathbf{f}_g$ are forces exerted by a *liquid and gas*, respectively, **t** a **n** tangent and normal to free surface and **s** is surface tension (if present).

Kinematic boundary condition declares that the velocity at a point of free surface moves together with the point of FE-mesh

$$(\mathbf{v} - \mathbf{v}_b) \cdot \mathbf{n} = 0. \tag{12.16}$$

The conditions in normal direction **n** and in both two tangent directions **t** and **s** to the surface of the free surface are given for every point on the free fluid surface.

12.4 Numerical Experiments

12.4.1 Effect of Mesh Density of the 2D Fluid Region

The fluid filling of an infinitely long rectangular tank in the x-axis direction with the fluid-filling width of 2 m and the fluid height of 2 m is considered, see Fig. 12.7. The material characteristics of fluid filling (water, H_2O) are bulk modulus $B = 2.1 \times 10^9$ N/m^2 and mass density $\rho_w = 1000$ kgm^{-3}.

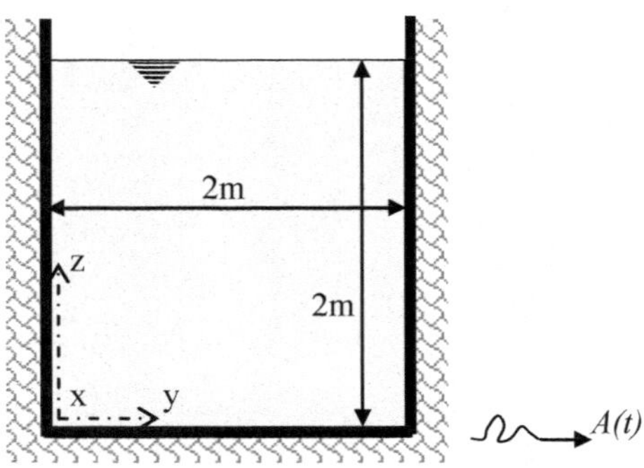

Fig. 12.7 Fluid domain

The tank is considered as total rigid solid, situated on a total rigid subsoil and excited in the indicated direction of the y-axis. The seismic load was modelled by recording the horizontal shifts recorded during the Loma Prieta earthquake. The ground-supported total rigid rectangular endlessly long channel considered in this numerical experiment, shown in Fig. 12.7, is without roof slab structure covered the channel.

The fluid domain of H_2O with volume modulus $B = 2.1 \times 10^9$ N/m^2, density ρ_w $= 1000$ kg/m^3, represents an incompressible fluid, defined separately considering total interaction with the structure. The fluid velocity and pressure were obtained by the Navier-Stokes equations solution, considering the assumption of the boundary conditions for the fluid domain: zero pressure on the surface, zero velocity of the fluid domain, volume in the fluid corresponding to its gravity. The boundary conditions were considered as total rigid: zero displacement of the structure, time-dependent horizontal displacement of the bottom of the vessel according to the selected accelerogram (rigid subsoil was considered). The liquid domain (water) was modelled in the ADINA program as a plane problem using finite element method (MKP) 3-nodes 2D FLUID finite elements ('plane strain') with the choice of PATTERN 1 mesh parameter by the considered mesh.

The fluid region was meshed uniform in both directions of the fluid plane domain. The uniform mesh 64×64 of fluid region 2×2 m for the "f4096" model gives 16,384 2D fluid finite elements, the uniform mesh 32×32 gives 4096 2D fluid finite

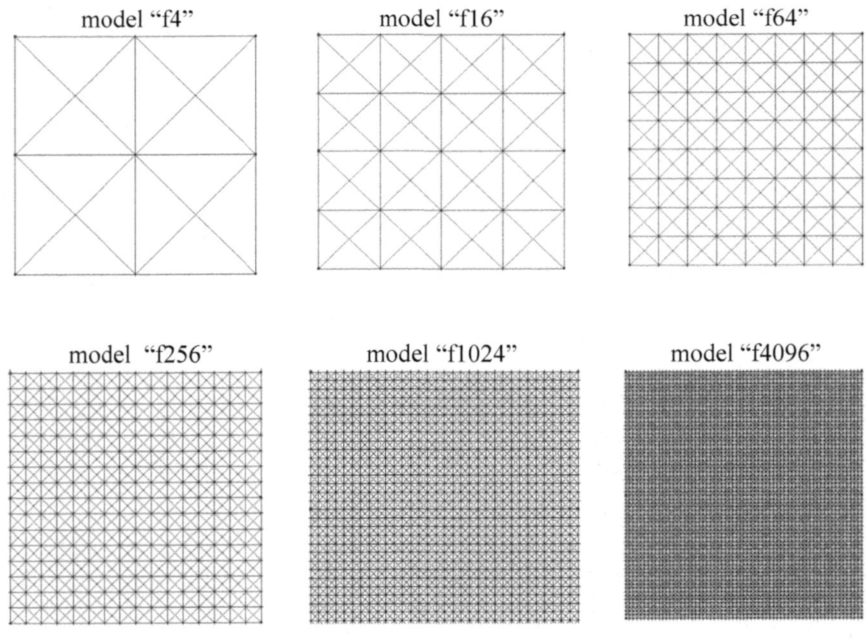

Fig. 12.8 Mesh models

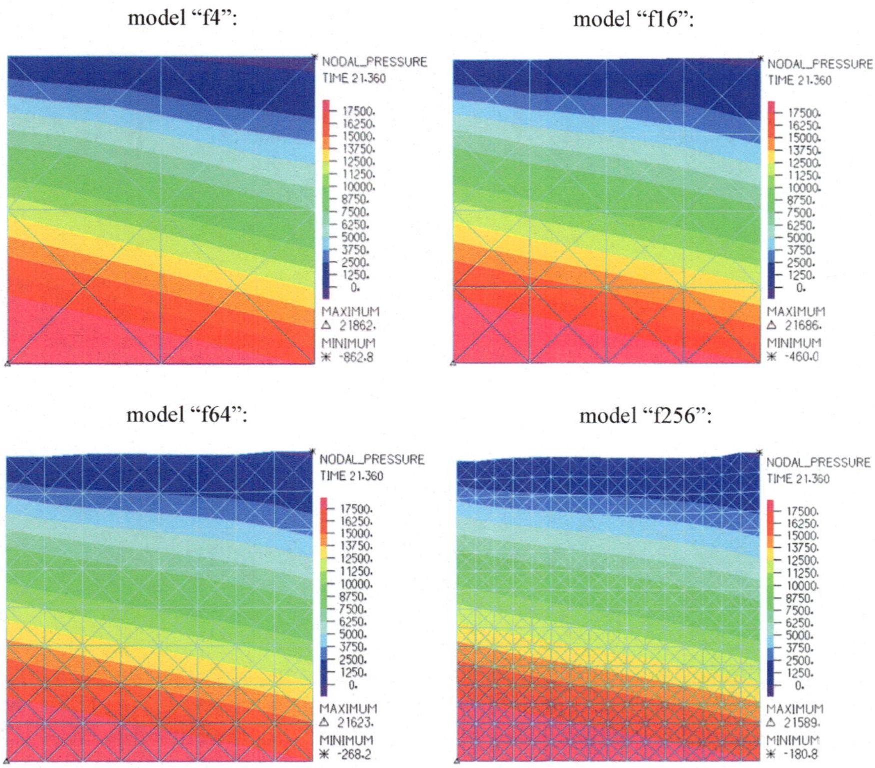

Fig. 12.9 Fluid region shapes and pressure distributions in time 21.36 s for models "f4", "f16", "f64", "f256"

elements for the "f1024" model, the uniform mesh 16 × 16 gives 1024 2D fluid finite elements for the "f256", the uniform mesh 8 × 8 gives 256 2D fluid finite elements for the "f64" model, the uniform mesh 4 × 4 gives 64 2D fluid finite elements for the "f16" model and the uniform mesh 2 × 2 gives 16 2D fluid finite elements for the "f4" model.

The peak values of pressure are generated at the bottom in the lower left corner of the fluid domain in time 21.36 s, Fig. 12.9, Table 12.1. The maximum wave of liquid is generated in the left corner of the free surface, it is in the upper left corner in times 21.12, 21.32 or 21.56 s. The offered mesh models of the fluid domain were presented in Fig. 12.8. The mesh models, shapes and the pressure distributions of the fluid domain at the time 21.36 s for the models "f4", "f16", "f64" and "f256" were documented in Fig. 12.9.

Table 12.1 summarizes the obtained results: the models, the number of finite elements in the considered meshing of fluid domain, the peak fluid area pressure and the maximum wave height obtained by numerical simulation of FEM for considered models. It is seen from Table 12.1 that the influence of the mesh density does not

Table 12.1 Model, number of finite elements, peak pressure and peak wave of fluid domain for the effect of mesh density of 2D fluid region

Model	Number of finite element	Peak pressure (kPa)	Peak wave (in time) (mm)
f4	16	21.862	3.03 (21.12 s)
f16	64	21.686	19.76 (21.12 s)
f64	256	21.623	28.31 (21,12 s)
f256	1024	21.589	34.27 (21.32 s)
f1024	4096	21.613	38.14 (21.56 s)
f4096	16,384	21.528	40.24 (21.12 s)

have a significant on the peak values of pressure, but its effects on the maximum wave are obvious. With increasing mesh density, the height of the maximum liquid wave also increases.

12.4.2 *Effect of Mesh Parameter 'PATTERN' of 2D Fluid Region*

The fluid filling of the infinitely long rectangular tank in the *x*-axis direction with fluid-filling width 2 m and a fluid-filling height 2 m is considered, Fig. 12.7. The material characteristics of fluid filling (water, H_2O) are bulk modulus $B = 2.1 \times 10^9$ N/m^2, mass density $\rho_w = 1000$ kg/m^3.

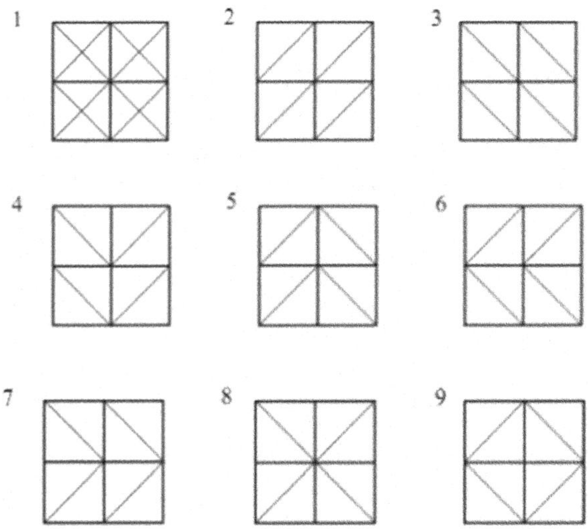

Fig. 12.10 Types of meshing (parameter PATTERN) of the 2D fluid of finite elements in the ADINA [26]

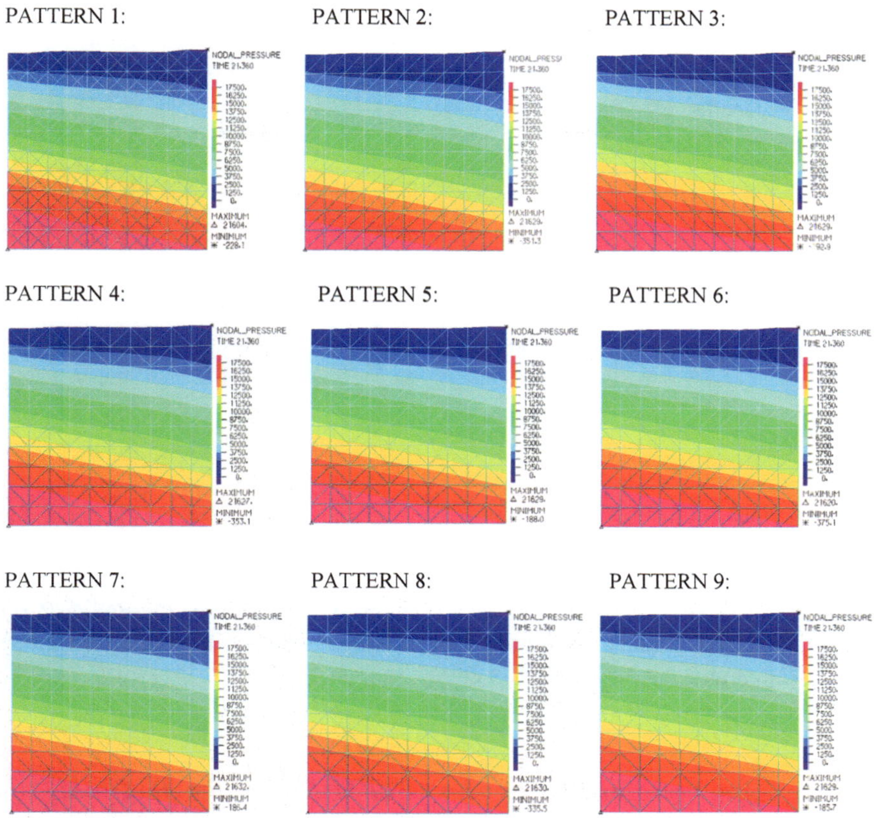

Fig. 12.11 Fluid domain shapes and pressure distributions in time 21.36 s for different types of meshing

The tank is considered as total rigid, situated on a total rigid subsoil and excited in the indicated direction of the y-axis. The seismic load was modelled by recording the horizontal displacement recorded during the Loma Prieta earthquake.

The fluid domain was modelled as a plane problem by using a uniform mesh density of 10×10 with PATTERN 1–9 mesh selection, and these dividing got 400 or 200 2D 3-nodal fluid finite elements 'plane strain'.

The peak values of pressure are generated at the bottom of the reservoir at the lower left corner of the fluid area at 21.36 s. The fluid domain shapes of width 2 m and height 2 m are shown in Fig. 12.11, their fluid domain shapes, the pressures distribution from software ADINA at 21.36 s. The model, the number of finite elements of the fluid domain, peak values of pressure in time 21.36 s and peak values of the vertical wave with a time of uprise were documented in Table 12.2 (Fig. 12.10).

The results for each PATTERN mesh: the models, the number of finite elements for the used fluid domain of the uniform meshing of 10×10, the maximum pressure in (kPa) and the maximum wave in (m) obtained by simulating the considered models

Table 12.2 Model, number of finite elements, peak pressure and peak wave of fluid domain for effect of mesh parameter "PATTERN" of 2D fluid region

Model	Number of finite element	Peak pressure (kPa)	Peak wave (in time) (mm)
PATTERN 1	400	21.604	28.87
PATTERN 2	200	21.629	26.63
PATTERN 3	200	21.629	26.99
PATTERN 4	200	21.627	27.12
PATTERN 5	200	21.629	26.40
PATTERN 6	200	21.620	26.96
PATTERN 7	200	21.632	27.16
PATTERN 8	200	21.630	26.59
PATTERN 9	200	21.629	25.99

using FEM were summarized in Table 12.2. The maximum pressure is generated at the bottom of the reservoir in the lower left corner of the fluid area at 21.36 s, and the maximum wave of liquid is generated in the left corner of the free surface, in the upper left corner at 21.08 s. Table 12.2 shows that mesh type the PATTERN 1 creates, the double number of 3-node finite elements with comparing with another PATTERN parameter, i.e. PATTERN 2–9.

12.4.3 Effect of Mesh Parameter 'PATTERN' of the 3D Fluid Region

The fluid filling of the endlessly long rectangular tank in the x-axis direction with the fluid-filling width 2 m and the fluid height of 2 m is considered. The material characteristics are bulk modulus $B = 2.1 \times 10^9$ N/m^2 and mass density $\rho_w = 1000$ kg/m^3. The tank is rigid, situated on a total rigid subsoil and excited by recording the horizontal displacement recorded during the earthquake Loma Prieta.

The numerical model is presented fluid domain in plane yz: 2 m $\times$ 2 m and in direction x cut out 0.4 m with up-described boundary condition. The uniform mesh density of 10 $\times$ 10 for different types of meshing with PATTERN 1–5 mesh parameter was used, and this division got 4800 or 1200 3D 8-nodal fluid finite elements.

The maximum possible pressure is generated at the bottom of the reservoir at the lower left corner of the fluid area arose in time 21.36 s. The fluid domain shapes (original dimensions 2 m $\times$ 2 m $\times$ 0.4 m), its final element's distributions and pressures distribution in time 21.36 s are shown in Fig. 12.12. The type of model, peak values of pressure, peak values of vertical waves and time of their meshing were documented in Fig. 12.12.

The results for each PATTERN mesh: the models, the number of finite elements for the used fluid domain of the uniform meshing of 10 $\times$ 10 $\times$ 2, the maximum

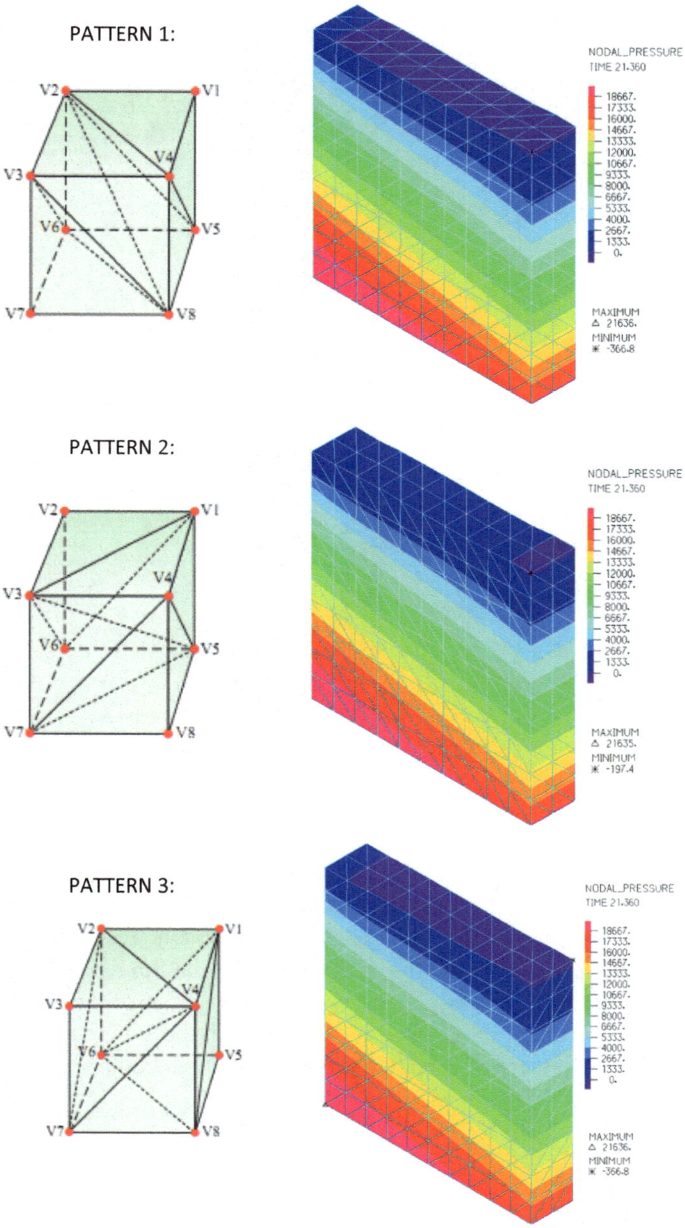

Fig. 12.12 Fluid domain shapes, 3D fluid finite elements and pressure distributions in time 21.36 s [26]

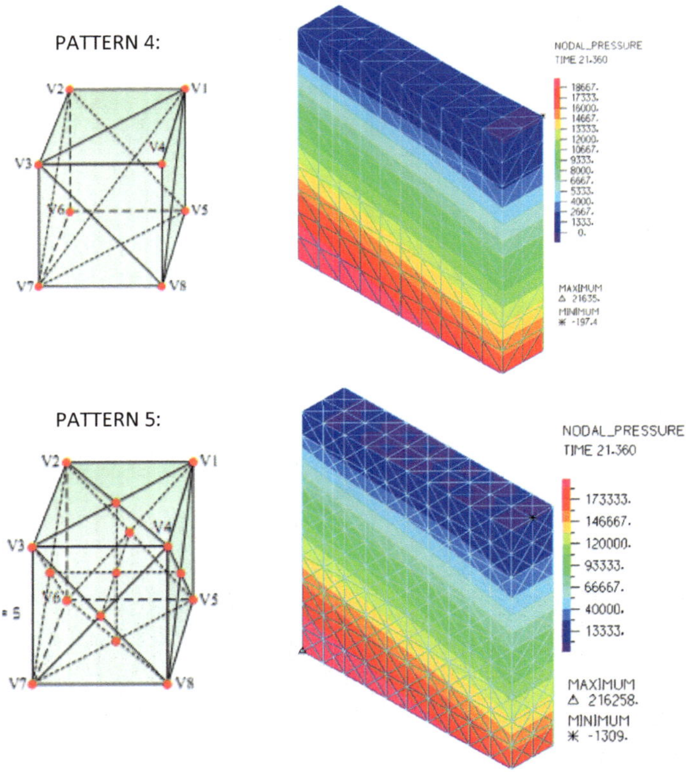

Fig. 12.12 (continued)

pressure in (kPa) and the maximum wave in (m) obtained by the considered models using FEM were summarized in Table 12.3. The maximum pressure is generated at the bottom of the reservoir in the lower left corner of the fluid area at 21.36 s, and the maximum wave of liquid is generated in the left corner of the free surface, in the upper left corner at 21.08 s. Table 12.3 shows that mesh type the PATTERN 5 creates, the four time more of 8-node finite elements with comparing with another PATTERN, PATTERN 1–4.

12.5 Concluding Remarks

The fluid domain analysed using by the FEM. The fluid domain in the total rigid channel was excited by recording the horizontal displacement measured during the earthquake Loma Prieta in California. Basic responses of the interest were the pressure of fluid domain.

Table 12.3 Model, number of finite elements, peak pressure and peak wave of fluid domain for effect of mesh parameter "PATTERN" of 3D fluid

Model	Number of finite element	Peak pressure (kPa)	Peak wave (in time) (mm)
PATTERN 1	1200	21.6356	28.0020 (21.08 s)
PATTERN 2	1200	21.6346	28.2792 (21.08 s)
PATTERN 3	1200	21.6356	28.0020 (21.08 s)
PATTERN 4	1200	21.6346	28.2792 (21.08 s)
PATTERN 5	4800	21.6258	29.1660 (21.16 s)

- The peak values of hydrostatic pressure are located at the bottom of the fluid domain. The maximum value using numeric simulation FEM ALE FSI is 19.597 kPa, it correlates with the analytical result of hydrostatic pressure 19.62 kPa, given by $\rho \cdot g \cdot h = 1000 \times 9.81 \times 2.0$.
- The peak pressure values in the lower left corner of the fluid area and peak waves in the left corner of the free surface of the fluid domain given by numerical simulation FEM are presented in Tables 12.1, 12.2 and 12.3 for different mesh densities and mesh parameters PATTERN.
- The hydrodynamic pressure of fluid domain along the left side was similar, asymmetric and got little higher values in comparison with the right side.
- The type of mesh parameters does not have a significant influence on the solution of the fluid domain with considered boundary conditions, but it influences the number of finite elements.
- The mesh density does not have a significant influence on the peak values of pressure. However, its effects on the maximum wave are obvious. With increasing mesh density, the height of the liquid maximum wave increases.

Acknowledgements This work was supported by the Scientific Grant Agency of the Ministry of Education of Slovak Republic and the Slovak Academy of Sciences the project VEGA 1/0374/19, and the financial support from VŠB-Technical University of Ostrava by means of the Czech Ministry of Education, Youth and Sports through the Institutional support for conceptual development of science, research and innovations for the year 2018 is gratefully acknowledged.

References

1. Zeleňáková M, Zvijáková L (2017) Using risk analysis for flood protection assessment. Springer Nature, 128 p. ISBN 978-3-319-52149-7
2. Batchelor GK (1967) An introduction to fluid dynamics. Cambridge University Press, Cambridge

3. Bathe KJ, Zhang H, Ji S (1999) Finite element analysis of fluid flows with structural interactions. Comput Struct 72(2/2):1–16
4. Bathe KJ, Zhang H, Wang MH (1995) Finite element analysis of incompressible and compressible fluid flows with free surfaces and structural interaction. Comput Struct 56(2/2):193–213
5. Bathe KJ, Zhang H (2004) Finite element developments for general fluid flows with structural interaction. Int J Numer Methods Eng 60(1):213–232
6. Di Carluccio A, Fabbrocino G, Salzano R, Manfredi G (2008) Analysis of pressurized horizontal vessels under seismic excitation. In: icsv18: 18th the world conference on earthquake engineering, Beijing, China, 12–17 Oct 2008
7. Helou SH (2014) Seismic induced forces on rigid water storage tanks.Asian J Eng Technol 2(4):310–315. ISSN: 2321 - 2462
8. Jacobs GB, Don WS, Dittmann T (2012) High-order resolution Eulerian-Lagrangian simulations of particle dispersion in the accelerated flow behind a moving shock. Theor Comput Fluid Dyn 26(1):37–50. ISSN 0935-4964
9. Kock E, Olson L (1991) Fluid-structure interaction analysis by the finite element method a variational approach. Int J Numer Methods Eng 31(3):463–491. John Wiley & sons, Ltd
10. Krejsa M, Janas P, Krejsa V (2014) Software application of the DOProC method. Int J Math Comput Simul 8(1):121–126
11. Kotrasova K (2017) Study of hydrodynamic pressure on wall of tank. In: Procedia engineering, vol 190. Amsterdam, Elsevier, pp 2–6. ISSN 1877-7058
12. Kotrasova K, Harabinova S, Hegedusova I, Kormanikova E, Panulinova E (2017) Numerical experiment of fluid—structure—soil interaction. In: Procedia engineering, vol 190. Amsterdam, Elsevier, pp 291–295. ISSN 1877-7058
13. Kotrasova K, Grajciar I, Kormanikova E (2014) Dynamic time-history response of cylindrical tank considering fluid—structure interaction due to earthquake. Transport structures and wind engineering. In: Applied mechanics and materials, vol 617, pp. 66–69. ISSN 1660-9336
14. Kotrasova K, Grajciar I, Kormanikova E (2015) A case study on the seismic behavior of tanks considering soil-structure-fluid interaction. J Vibr Eng Technol 3(3):315–330. ISSN 2321-3558
15. Kotrasová K, Kormaníková E (2017) The study of seismic response on accelerated contained fluid. In: Advances in mathematical physics, vol 2017, pp 1–9. ISSN 1687-9139
16. Kotrasová K, Kormaníková E (2016) A case study on seismic behavior of rectangular tanks considering fluid-structure interaction. Int J Mech 10:242–252. ISSN 1998-4448
17. Kotrasová K, Kormaníková E (2017) Response of endlessly long shipping channel due to earthquake. In: MATEC web of conferences, pp 1–6. ISSN 2261-236X
18. Housner GW (1955) Dynamic pressures on accelerated fluid containers. Division of Engineering, California institute of technology, Pasadena California
19. Eurocode 8 (2006) Design of structure for earthquake resistance—Part 4: silos, tanks and pipelines
20. IITK-GSDMA (2005) Guidelines for seismic design of liquid storage tanks—provisions with commentary and explanatory examples. Kanpur, Indian Institute of Technology Kanpur
21. Melcer J (2015) Dynamic response of a bridge due to moving loads. J Vib Eng Technol Vibr Eng Technol 3(2):199–209. ISSN 2321-3558
22. Sucharda O, Brozovsky J (2013) Bearing capacity analysis of reinforced concrete beams. Int J Mech 7(3):192–200
23. Michalcová M, Lausová L (2015) Numerical modelling of the aerodynamic roughness of trapezoidal and corrugated sheets. In: Civil-Comp proceedings. ISSN 1759-3433
24. Wang X, Fundamentals of fluid—solid interactions. In: Analytical and computational approaches. Elsevier, Linacre house, Oxford. ISSN 157-6917
25. Zmindak M, Grajciar I (1997) Simulation of the aquaplane problem. Comput Struct 64(5–6):1155–1164
26. Manual ADINA (2005) 71 Elton Ave, Watertown, USA, Adina R&D, Inc.
27. Rugonyi R, Bathe KJ (2001) On finite element analysis of fluid coupled with structural interaction. Comput Model Eng Sci 2(2):195–212

Chapter 13
Monitoring of Changes in Water Content in Soil Pores of Earth-Fill Dams

J. Pařílková, Z. Zachoval, M. Gomboš, D. Pavelková, B. Gjunsburgs, J. Gints, Y. Yanev, D. Toneva-Zheynova, T. Zydroń and A. Gruchot

Abstract The Czech Republic (CR) lies on the watershed divide of three seas—The North, Baltic and Black Seas. The watershed of these seas divides the territory of the CR into three international catchment areas (Labe, Odra and Danube). In conjunction with the fulfilment of obligations arising from the Convention of the European Economic Commission of the United Nations Organisation on the protection and use of transboundary watercourses and international lakes, an "International Commission" was founded in 1990–1998 for each international catchment area. The CR is a member of the International Commissions for the protection of these catchment areas. The purpose of the activity of these institutions is to support cooperation in water management at the level of international catchment areas. It is obvious that practically all significant watercourses in the CR drain water beyond its boundary on the territories of the neighbouring states, resulting in the fact that water sources in the CR are practically dependent on the amount and distribution of atmospheric precipitation. In the context of discussing changing climatic conditions, atmospheric precipitation of very different intensities increases the likelihood of the occurrence of extreme floods and dry episodes. The theme of protection from the consequences of floods and drought is part of joint debates of all three international commissions. The approach to the solution of the issues above, however, can be different because

J. Pařílková (✉) · Z. Zachoval
Laboratory of Water Management Research, Institute of Water Structures, Faculty of Civil Engineering, Brno University of Technology, Veveří 331/95, 602 00 Brno, Czech Republic
e-mail: parilkova.j@fce.vutbr.cz

M. Gomboš · D. Pavelková
Institute of Hydrology, Slovak Academy of Science, Bratislava, Slovak Republic

B. Gjunsburgs · J. Gints
Department of Water Technology, Institute of Heat, Gas and Water Technology, Water Engineering and Technology, Riga Technical University, Riga, Latvia

Y. Yanev · D. Toneva-Zheynova
Department Automation of Manufacturing, Faculty of Automation and Computing, Technical University of Varna, Varna, Bulgaria

T. Zydroń · A. Gruchot
Department of Hydraulic Engineering and Geotechnics, Faculty of Environmental Engineering and Land Surveying, University of Agriculture in Krakow, Krakow, Poland

M. Zelenakova et al. (eds.), *Management of Water Quality and Quantity*,
Springer Water, https://doi.org/10.1007/978-3-030-18359-2_13

it is based on the current conditions of the given international catchment area. Many fields and areas give attention to the theme of threats posed by natural disasters created by flood situations and drought not only from the view of science. It is not possible to prevent floods or drought by the present scientific knowledge and technical options. In case of floods, it is possible to reduce their impact on the lives and property of inhabitants by constructing efficient flood control works. However, it is necessary to have in mind that even the best flood control measure will only be as reliable, efficient and safe as its weakest element will be reliable, efficient and safe. In many cases, this element is earth-fill dams. Therefore, it is necessary to have continuous and long-term information on their condition. When addressing the issue of drought in the CR, it is necessary to realise that the absolute majority of water sources depends on the retention and accumulation of water in the territory of the CR. At present, the impacts of drought and lack of water in the conditions of the CR are significantly mitigated by the existing water management infrastructure. However, it can be expected for the future that the existing water sources will not be sufficient. Not only the aspect of the potentially diminishing available amount of surface water and subsurface water in the CR will be problematic, but also the aspect of the unsuitable quality of water. Also from this aspect, it is possible to see the earth-fill dam as an element that helps retain and accumulate water in the landscape. The reality is that every inhabitant of the CR is a significant user of its water sources, and it is necessary to bear the responsibility for them. Unfortunately, people usually realise neither the significance nor the value of the discussed irreplaceable nature wealth for everyday life, nor the scope of activities and financial costs that are associated with this “matter of course”. In this chapter, the team of the authors provides a view of international cooperation and its achievements when dealing with a project of applied research in the EUREKA programme, which was focused on the development and construction of monitoring technology enabling changes in water content to be monitored in a porous medium. The selected results are documented in a link to remedial measures applied to earth-fill dams of water reservoirs in the conditions of the CR, emphasising the use of the method of electrical impedance spectrometry. The authors are aware of the fact that it is only a fragment of the solution of the complex of the outlined issue. However, it is obvious that without the safe operation of water management works, the irreplaceable part of which is their monitoring, it is impossible neither to ensure water sources nor to manage them.

Keywords Water content · Soil · Earth-fill dams · Electrical impedance spectrometry · EUREKA programme

13.1 Introduction

Across the history of the planet Earth, the enlightened personalities [1, 2] knew that the greatest real mineral wealth of every state was not oil or noble ores, but land and water. Therefore, man has learned and is still learning how to manage land and water. To impound water, they built a dam [3–5]; to make a living, they began to cultivate land [6]. Perhaps it is a little paradox in today's modern world of advanced technologies and largely consumer societies that both the substances become ever more significant. Many scientific fields and areas of human activity deal with the interaction of both substances. Here, attention is given to earth-fill dams (EFDs) designed for water management.

EFDs are among the oldest building structures [7–9] which man has created, and its significance is not lost at present either. In hydraulic engineering, EFDs constructed inland serve for protection from floods, for water transfer, water retention and the like. In relation to their safe operation, attention is given to the issues of resistance and stability of the structure of EFDs and the basement and to the relating risk analysis [10]. EFDs are operated with permanent (fishpond dams, reservoir dams, dams of small water reservoirs) or occasional (mainly protective dikes and dams of detention basins) water load. Therefore, they can be exposed to irregular hydrodynamic load when it is necessary to assume that the soil can be saturated with water in a certain part, and, conversely, unsaturated in another part. It is experimentally proved that the hydraulic conductivity of unsaturated soil is by up to three to four order of magnitude lower than that of saturated soil. Therefore, all processes that take place in the soil when loaded will be different in time and space. Extreme demands for measuring technology monitoring selected processes [11–18] arise from the complexity of the whole group of changes that take place in soil with its hydrodynamic load [19–23].

The role is also complicated in terms of materials used for the structure of the body of EFDs and technologies used for its construction, which causes the heterogeneity and anisotropy of the environment. Mainly for economic and ecological reasons EFDs are built of suitable, locally available materials. If the available material is sufficiently permeable and, at the same time, stable (gravel with a high proportion of loam, loam, loamy-sandy soils, etc.) and the dam is relatively low, then EFDs are built of a single material—homogeneous dams. Otherwise, dams are built of several materials—zoned dams, which are always composed of the stabilisation part (gravelly soils, riprap) and the sealing part (clayey and loamy materials). The materials of EFDs are very heterogeneous, which causes great difficulties in the use of methods monitoring their water saturation.

When considering the aspects above, it is obvious that to solve the evolution problem of the behaviour of EFDs as a heterogeneous anisotropic medium usually of three components (air, water, soil particles) using the methods of conventional deterministic continuum mechanics is difficult, from both the theoretical and the practical point of view. The initial information on the saturation of the medium is usually unknown, or it can be estimated only with considerable uncertainty. The subsequent development of changes in saturation is variable in time and space. The knowledge of the saturation of the soil of EFDs, however, is an essential prerequisite for solving the state of stress, stability and inversion problems of determining the hydraulic conductivity of soils. Therefore, when obtaining the required information, the technology of measurement and long-term monitoring has an irreplaceable role.

This paper gives two applications of a monitoring system working with the method of electrical impedance spectrometry (EIS). These are an EFD of a reservoir and an EFD of a small water reservoir, in which water inrush areas appeared on the downstream side. As a consequence, the safety of their operation could have been threatened, and thus, it was decided on the necessity of their remediation. In case of the EFD of a reservoir, the method of a clay–cement diaphragm wall installed into the sealing core of the EFD was used. In the other case, the method of BioSealing was used, based on the life cycle of microorganisms naturally occurring in the soil. Both the EFDs were monitored using the EIS method before and after the remedial action. Some of the achieved results and conclusions are given below.

13.2 Specification of the Porous Soil Environment and Monitoring of Water in the CR

The porous soil environment, in this part, means earth and soil, including water content. It is the regulator of the material cycle. It can function not only as a repository of potentially risky substances but also as their source. It forms not only a complicated open, dynamic system but also a relatively independent system by its capability of self-regulating internal processes [24–27]. While earth can be defined as an independent natural feature formed by surficial solid products of weathering of the Earth's crust and by organic residues under the action of soil-forming factors, the term soil is used in the engineering-geological classification of rocks and is based on their structural cohesion. The soil is designated as unconsolidated rocks that are subdivided into cohesive (clay), non-cohesive (sand, gravel), organic (peat) and artificial rocks (material of dump sites, fills, etc.).

In terms of monitoring, it is a very complicated three-component environment (Fig. 13.1), the composition of which differs by the place of sampling and by the depth from which the sample was collected (e.g. humous, loamy, sandy, clayey earth, etc.). The physical, chemical and biological properties of the given medium are measured and monitored.

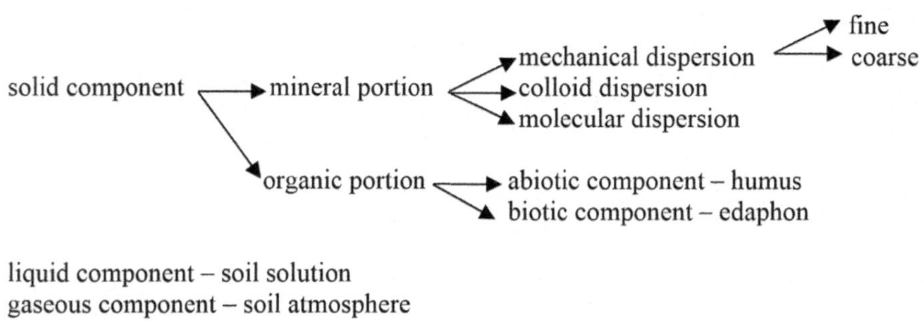

Fig. 13.1 Schematic composition of the porous soil environment

The issue of earth and soil monitoring is closely tied to water monitoring [28, 29]. Two fundamental organisations work in the section of water stage monitoring in the CR, the Czech Hydrometeorological Institute (CHMI) [30] and the T.G. Masaryk Water Research Institute (WRI) [31]. The CHMI mainly collects data and creates hydrological information, and the WRI chiefly collects data from the catchment basins and creates water resource balances for water management purposes. Both the organisations together have built a hydro-ecological information system (HEIS), the main objective of which is to take inventory of water resources, information about their regime, balancing of data from monitoring the hydrosphere, publishing of information for the needs of decision-making of the state administration bodies, etc.

Water monitoring is divided into the observation of surface and subsurface waters (particularly groundwater). It comprises data on the physical and chemical properties of water and their spatial and temporal occurrence.

13.3 Monitoring of Earth-Fill Dams (EFDs)

EFDs, including their basement, are exposed to effects that can threaten their stability. These are the effects of dead weight, water pressure, flowing water, the action of ice, pounding by waves, weather effects and other natural effects. To ensure the problem-free operation of EFDs, it is necessary to carry out monitoring and evaluation of their technical condition in terms of safety, operational reliability, causes of potential failures and their consequences, i.e. dam safety supervision (DSS) [20, 32]. It also includes a proposal of safety measures, if and when necessary.

Considering the basic factors determining the goals of monitoring EFDs, these can be the formation of fractures of different characters and the growth of pore pressures in the body of the dam, originating as a consequence of uneven subsidence of different parts of the dam, filled with materials of different deformation properties. This mainly concerns the difference in subsidence of the stabilisation part and the sealing part of the dam. Moreover, the problems are caused by water seeping through the dam and by fluctuations in water level in the reservoir. Other factors influencing the methods and procedures of monitoring are the shape of the valley, the geological structure of the surrounding rock environment and the co-action of construction materials. Ensuring the reliability of EFDs, therefore, requires comprehensive monitoring (Fig. 13.2) that is based on measurement:

- Before construction (an engineering-geological survey—EGS)—particular measurement of seismic activity and evaluation of long-term changes in the water regime in the given area (piezometric measurements). Piezometers used before construction itself can also become part of permanent control monitoring during the operation of the facility after being connected to remote read-out;

- During construction—monitoring is chiefly focused on the development of pore pressures in the body of the dam and in the basement (piezometers). Moreover, monitoring is focused on deformation (extensometers, inclinometers), subsidence, pressures of the dam on the basement (pressure cells) and the state of stress in the dam body;
- With the first fluctuations of water levels (first filling/emptying)—during the first filling, monitoring is particularly focused on deformation (extensometers, inclinometers), subsidence and seepage through the body of the dam and in the basement. During the first rapid emptying, monitoring is focused on changes of pore pressures and the formation of shear planes, if any, in the slopes of the reservoir;
- During the operation—monitoring focuses on the deformation of the dam body, its basement and the lateral slopes of the reservoir (geodetically, extensometers, inclinometers) because it takes place even several years after the completion of the dam. The development of pore pressures in the dam body (piezometers); stress in the dam body (pressure cushions); seismic activity (seismographs); in case of anchored lateral slopes, axial forces in anchors are measured too.

The gauges are installed both on the surface and also in the dam body, and the monitoring system of the engineering-geological survey (EGS) is retained, if and when necessary. These gauges currently constitute a standard monitoring system. In many cases is usually supplemented, e.g. by a camera system for "online" monitoring of EFDs, automatic detectors of emergency situations, or special measurements requested in response to a suspicion of undesirable changes. Then can be used such as thermovision, radar, infiltrometric, resistance and impedance measurements, use of optic cables and other contemporary modern methods and technology.

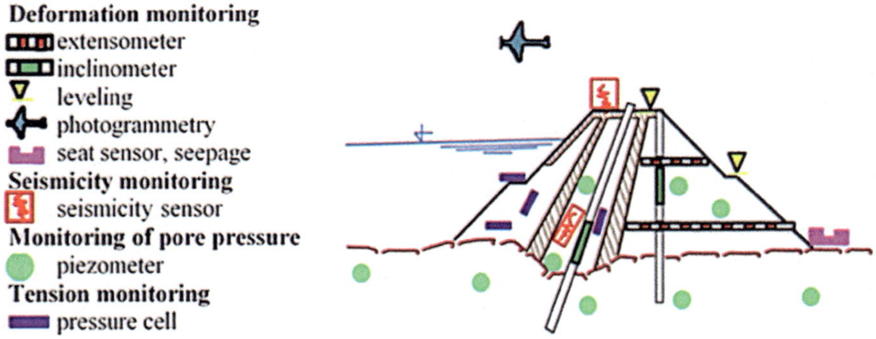

Fig. 13.2 Scheme of design for hydrogeotechnical monitoring of the earth-fill dam

13.4 EUREKA Programme, the Method of Electrical Impedance Spectrometry and the Developed Device Z-Meter IV

13.4.1 EUREKA Programme

The EUREKA programme [33] is the first European framework that opened the door for cooperation with Central and Eastern European countries in the early 1990s. Moreover, this chance was entirely taken by both research organisations and industry for initiating real European research and development cooperation. EUREKA Network Projects are transnational, market-driven innovative research and development projects, labelled by EUREKA and supported by the public administrations and public funding agencies that represent EUREKA in each of its 40 + member countries. The EUREKA "bottom-up" approach to project creation continues to be a characteristic which differentiates EUREKA from other such initiatives. This approach allows the project consortia to define the nature of the technologies to be developed and how the project comes together, to agree upon the intellectual property rights and build partnerships, to share expertise and ease access to international markets with the results of their research. EUREKA Network Projects are market-driven international R&D projects. They aim to develop marketable products, services or processes. Participation in international cooperation projects through EUREKA offers businesses, research institutes and higher education institutions a range of advantages. EUREKA's flexible and decentralised approach allows for a short and concise application procedure. The types of the EUREKA project are clusters, umbrellas and individual projects; special parts are EUROSTARS projects. Each of these types of project has specific rules and plays a key role in different areas of the market. Also, the research areas are very different. It is possible to classify ten main technological areas. There are electronics, microelectronics and telecoms technology; industrial manufacturing, material and transport; other industrial technologies; and energy technology. Next are chemistry, physical and exact sciences; biological sciences; agriculture and marine resources; and agro-food technology. Special areas are measurements and standards; and technology for protecting humankind and the environment.

The goal of the EUREKA programme in the Czech Republic [34] is to steadily enhance the competitiveness of the Czech economy and to increase its technological potential by supporting European cooperation of enterprises and research organisations in the area of progressive technologies. The output of the project solution is new, top-quality products, technologies and services with a high degree of innovation, capable of prevailing on the world market. The EUREKA programme does not preset thematic tasks and does not centralise financing or selection of projects. It follows the rule that proposals and initiatives must come out from bottom to top (the so-called bottom-up principle), from individual industrial enterprises and companies, research institutes and universities which are spontaneously interested in cooperation. Key criteria continue to be the innovative content, market needs and the quality of the team.

Each applicant in the Czech Republic can apply for special-purpose support in the Czech Republic. Special-purpose support is provided to the recipient, which is a legal entity. It is in the form of grants after having carried out a public tender in research and development pursuant to Act No. 130/2002 Coll. on the Support of Research and Development from Public Funds and on the Amendment to Certain Related Acts (the Act on the Support of Research and Development) as subsequently amended. In projects of research and development in the EUREKA programme in the conditions of the Czech market, the portion of special-purpose support shall not exceed 50% of the eligible costs of research and development project solution and must meet certain described conditions.

Selected results of earth-fill dam monitoring were obtained by carrying out project E!7614 entitled "A System of Monitoring of Selected Parameters of Porous Substances sing the EIS method in a Wide Range of Applications" (a part of the Czech researcher of the project bears the designation LF13019). The main leader of these projects (at international and national levels) is the company GEOtest, a.s. in cooperation with the Institute of Water Structures at the Faculty of Civil Engineering of Brno University of Technology, where the apparatus was tested in the Laboratory of Water Management Research [35, 36].

13.4.2 The Method of Electrical Impedance Spectrometry

Electrical impedance spectrometry (EIS) enables the detection of the distribution of electrical impedance $\mathbf{Z}$ or other electrical variables arising from it (admittance $\mathbf{Y}$, conductance G, susceptance B, phase φ, resistivity ρ, conductivity σ, etc.) [37], inside a monitored (homogeneous and heterogeneous) object, and thus the observation of its inner structure and its changes. Due to the electrodes used and the way of their location in a monitored object, it is possible to use the method as non-destructive invasive or non-destructive non-invasive if it does not interfere with the object being measured and does not affect the monitored process. The method ranks among indirect, contact electrical methods [38–40] and is used in measuring the properties of organic and inorganic substances. The method is based on the ability of charged particles to migrate into an electric field that is commonly found in the natural environment and contributes, for example, to weathering minerals. The artificially created electric field supplies electrons to the natural environment. This enhances geo-electro-physico-chemical processes that can be monitored. The carrier of the entire system of natural exogenous processes is water. Water has a number of unique properties that determine the behaviour of the earth matrix and biogenic components. The content of solutes in water essentially determines the electrical resistance of the environment, that is, the amount of electrons (current density) that can be transmitted to the soil at a certain voltage. It constitutes a very sensitive tool for monitoring phenomena that

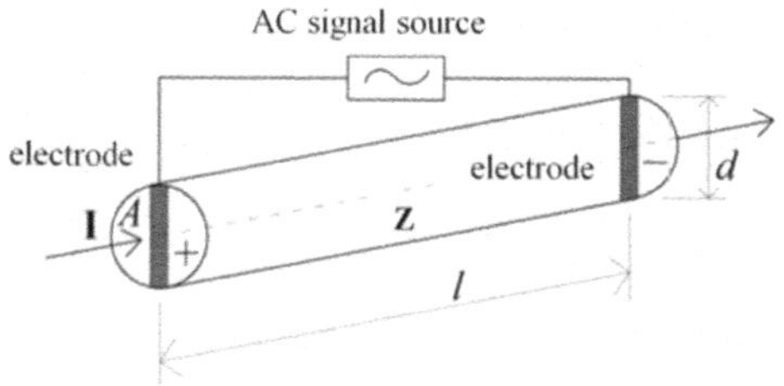

Fig. 13.3 Soil as an electrical conductor

take place in objects. There are changes occurring in earth dams when loaded by water, in wet masonry when being dried, in sewerage systems during the transport of water with sediments. Next areas are electrokinetic phenomena at boundaries (e.g. electrode/soil grain, between soil grains) or for describing basic ideas about the structure of an inter-phase boundary (e.g. electrode/water).

The EIS method is based on the alternating periodic driving signal. If the low amplitude of the alternating signal is used, concentration changes of charge are minimal at the surface of an electrode connected with measurement, which is very important in systems sensitive to so-called concentration polarisation. The range of frequencies used of the driving signal enables the characterisation of systems comprising more interconnected processes with different kinetics. The analytical EIS method was newly applied for monitoring the boundary between saturated and unsaturated soil when being loaded with water, for measuring changes in the electrical conductivity of soils in various areas of bodies of earth dikes [41, 42] and for monitoring the water/sediment boundary in water reservoirs or streams.

The basic principle of the method is the measurement of the frequency characteristic of electrical impedance **Z** of the measured porous environment (Fig. 13.3), as the case may be. The frequency characteristic can generally be expressed as the function of a complex variable in the following formula

$$\mathbf{Z} = R + jX,$$

where R is resistance forming the real part of **Z** theoretically independent of frequency f, X is reactance, the imaginary component of **Z**, the magnitude of which changes with frequency f of the alternating signal.

13.4.3 The Z-Meter IV Device and Measuring Probes

Within the solution of the E!7614 project, the measuring apparatus with a Z-meter device has been developed, tested and constructed. This apparatus was verified by laboratory experiments and measurements on objects in situ. Phenomena identifiable by changes in electrical impedance, which take place in a porous environment (soil, stone masonry, woods), are investigated.

Table 13.1 Technical specification of Z-meter IV device [43]

Parameter	Z-meter IV device
Impedance range	10 Ω–1 MΩ
Frequency range	100 Hz–200 kHz
Measuring voltage	0.2 V; 1.0 V
Accuracy of module Z	±2% of the range
Accuracy of phase	±2°
Communication interface	USB, SD card, Ethernet, Bluetooth
Number of channels	1, 8, 16, 32, 64, 128, 256
Switch	Internal to count 16 channels, external
Power	Li-Ion battery (a concept with a mains power source or a solar panel)
Hours of continuous operation	10 h
Weight	496 g
Width × length × height	100 mm × 212 mm × 35 mm

EIS techniques developed through the project within the EUREKA programme are very modular [43]. It is possible to change monitoring frequency (detection of environment characteristics), the excitation amplitude of the signal (sensitivity, reduction errors caused by the measurement technique), but also a number of sensors including the geometry of monitoring (Table 13.1, Fig. 13.4) and the connection of sensors in the probes.

From the physical point of view, the probes of the EIS method can be characterised as passive sensors that change their characteristic property under the action of the measured variable. If the distance between the sensors and the Z-meter device is long, then the sensors or probes must be constructed as active. A change in the property is thus a degree of the value of the measured non-electrical variable. In terms of connection, two- or three-terminal connection was used in measurement, while adhering to the principle of the close contact of the probe sensor with the surrounding environment. The two-terminal connection does not eliminate the effect of voltage loss on the resistance of the supply cables and on the contact resistance between the sensors and the measured environment. It is, however, sufficiently sensitive and accurate for changes in observed electrical impedance due to changes in the water content of the monitored soil.

During the whole time of monitoring, the sensors are in the same position including the cables (their length and position). Thus, it is possible to assume that parasitic impedance will always be the same.

The probes are mounted on the crest and downstream slopes of the embankment dam. The measuring probes are formed by one (three-terminal connection of an electrode) or two tubes (two-terminal connection of an electrode) fitted with passive/active sensors (depending on the transmission distance of the measured signal).

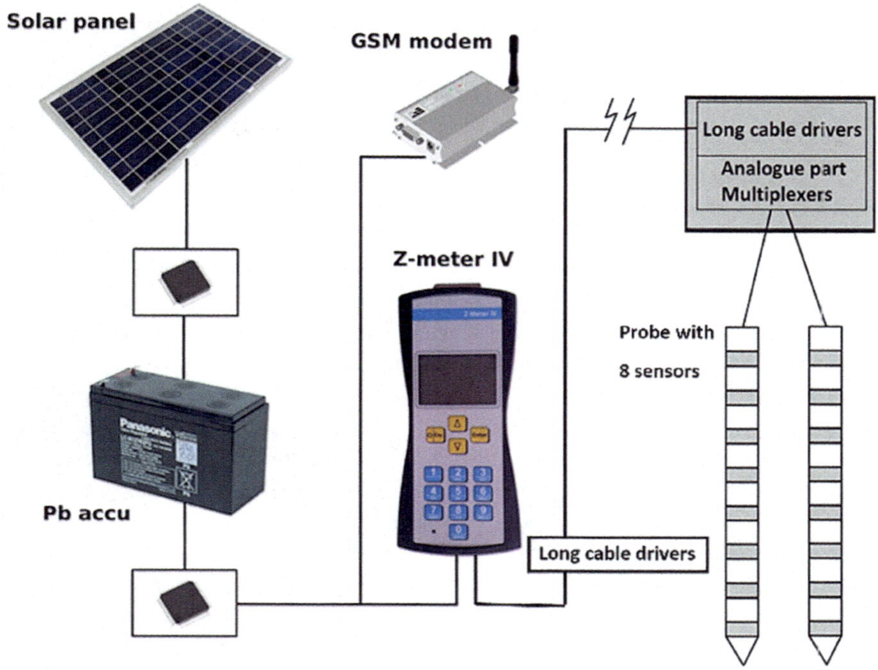

Fig. 13.4 Block diagram of the system

A conductive material alternates with a non-conductive material (an insulant) regularly. The measuring electrodes are made of stainless tubes; a polyamide tube of the same outer parameters was used as an insulant. The total length of the probes and its connection corresponds to the measurement and installation requirements. The spacing between the electrodes is maximally $l = 2.0$ m because for higher distances, it may affect the measured values of the Earth's geo-electro-magnetic field. The conductors ensuring the transmission of the exciting signal (electric current) and the measuring signal (electric voltage) are conveyed through the centre of the tubes to the individual measuring electrodes. To avoid the penetration of moisture to the area of conductors (parasitic electrical resistance), the individual parts of the probe are connected with a silicone non-conductive sealing material which also seals the probes in their surface layer. The conductors are terminated with a connector.

13.5 Field Testing

13.5.1 Karolinka Earth-Fill Dam

The Karolinka dam and water reservoir [44] is located in the CR on the Stanovnice stream at the river kilometre 0.75 above the town of Karolinka in the Vsetin District (Fig. 13.5). The total volume of the reservoir is 7.644 mil. m^3, and the area of the catchment is 23.1 km^2. It was built from 1977 to 1985. It is a straight (plan view), inhomogeneous earth-fill dam made of local gravel materials with a central loamy core connected to a concrete grouting gallery. The height of the dam above the base of its foundation is 47 m, and the length of the dam crest is 391.5 m. The slope of the upstream face is 1:3.3 and of the downstream face 1:2.2–2.4. The intake part of the bottom outlet and the intake structures for water supply are placed in an intake tower. Water intake takes place from three levels. An emergency spillway is installed in front of the water tower; it is designed as a shaft spillway with no gates, with its spillway crest having a diameter of 10.50 m. A reinforced-concrete two-level construction 231 m long runs from the intake tower; in the lower part, it is used as an outlet conduit to transfer flood discharges from the emergency spillway to the area downstream below the dam, and the pipelines of the bottom outlets run in the upper part. It is also used as a communication corridor between the valve chamber of the bottom outlets at the downstream base of the dam and the intake tower.

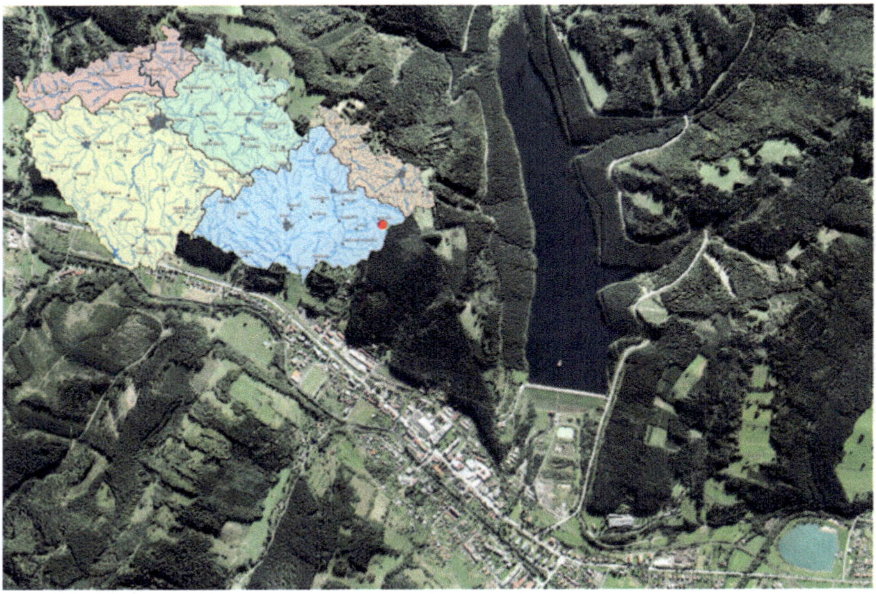

Fig. 13.5 Situation of the Karolinka dam and reservoir

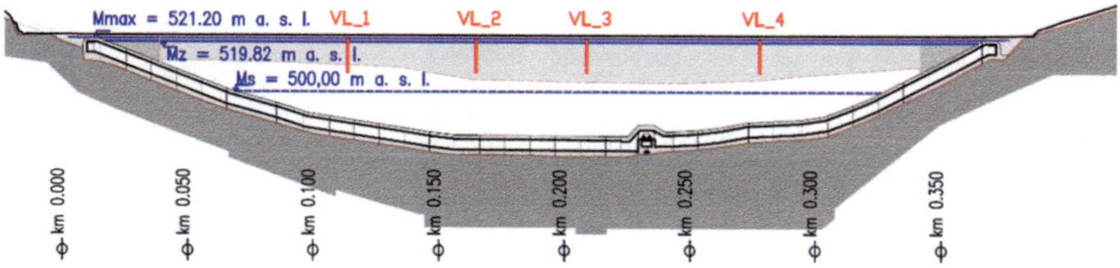

Fig. 13.6 Longitudinal section through the grouting gallery—the situation of the dam sealing wall and EIS monitoring probes

The purpose of this hydro-engineering structure is to deliver raw water for supplying the Vsetin area and the Vlara area, to protect the downstream area from floods, to maintain a minimum discharge in the river below the dam and to use it for power generation.

For the reason of seepage detected on the downstream face of the dam, the water reservoir was operated in a limited regime practically from its beginning. Based on investigations performed, it was stated that seepage is caused by the insufficient permeability of gravel sand used and technological indiscipline when the individual zones of the dam body were filled. The stabilisation zone of the downstream face was filled with very heterogeneous layers that differ particularly by the content of silt particles (the designed abundance less than 4%, but the real one reaching even 11%), which essentially affected the permeability of the dam soil [45–47]. The findings pointed to a potential threat to the stability of the dam material during extreme water level stages in the reservoir. Based on the long-term measurement and observation of seepage conditions in the dam body, it was decided to build a vertical cut-off diaphragm wall in the earth-fill core of the dam reaching a depth of about 16 m on average (Fig. 13.6). The decision came out carried out within dam safety supervision [20] and the basic results of the geotechnical survey of the dam. Its construction was carried out in 2013.

In March 2011, a monitoring system was supplemented by four probes to the downstream part of the earth-fill sealing core of the dam, which monitors changes in the electrical conductivity of soil using the EIS method, and thereby indirectly also changes in the degree of saturation of pores by water. The total length of probes is 13 m and the length of measuring electrodes 0.05 m; always 20 sensors are on one probe. The position of the sensors resulted from the position of observed inrushes [48], the knowledge of the characteristics of materials used [49] and the technical documents of dam construction [45], it means that it was predefined vertically and horizontally. The measuring rods were mounted into the dam core from the dam crest using a static penetration set PAGANI of the company GEO*test*, a.s (Fig. 13.7). The system enables the monitoring of changes in the electrical conductance, $G = 1/R$ of soils of the dam core and hence indirectly the changes in water content.

The correlation of the results obtained from the whole system is not easy. It seeks to systematically link the measured data with changes in the level of water in the reservoir, weather conditions, seepage and other standardised variables [50, 51]. All

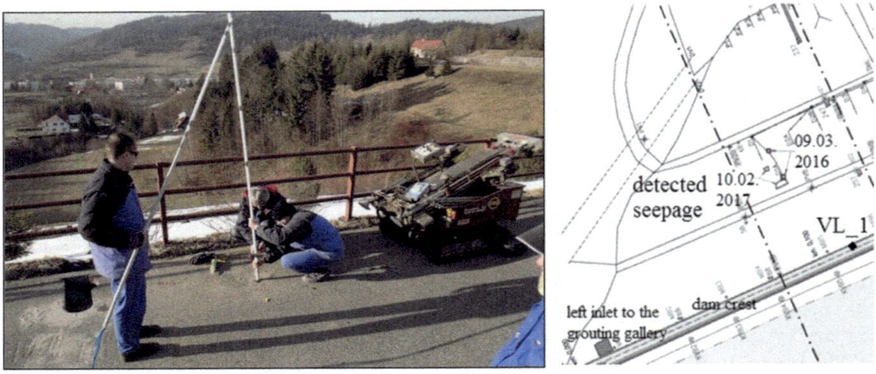

Fig. 13.7 Installation of probe VL_4 of the EIS method into the dam's sealing core (left), detected seepage sites after 2013—an outline of the situation (right)

discussions about the measured values of the electrical conductance G are made under the assumption that the other parameters, which could influence the electrical conductance of the soils (concentration of ions, temperature, water chemistry and the like), remain unchanged.

Primary information can be used for comparisons of measurement and visual observation. One of the possibilities of the evaluation of monitoring is the dependence of the conductance G of the soils on the position h of the sensor (Fig. 13.8 up). The trend of G in comparison with the water level in the reservoir with time for measured levels 10 and 13 m is shown in Fig. 13.8. The first EIS electrode is at the height $h = -2$ m (2 m below the dam crest), and the last electrode is mounted at the height $= -13$ m. The results of monitoring by the EIS method correspond with visual observation when, in February 2015 and 2017 and in January 2016, inrushes of water reappeared on the downstream face in the left part of the dam between the upper and lower berms (Fig. 13.7 right).

The complexity of the construction of the earth-fill dam of the Karolinka water reservoir and of the ongoing processes shows that the assessment only by monitoring by the EIS method is not clear, but it gives a certain idea about the changes in the water saturation of soil in the monitored area of the dam. It will be necessary to determine in the overall concept of the efficiency of the implemented measures how the given fact correlates with seepage measured in the base of the dam and other monitored parameters such as the water level in the reservoir, the temperature of water and soil and the pattern of pore pressures. The technology of construction and the process of ageing of the clay–cement mix of the cut-off wall could obviously have an important effect on the measured values of soil conductance.

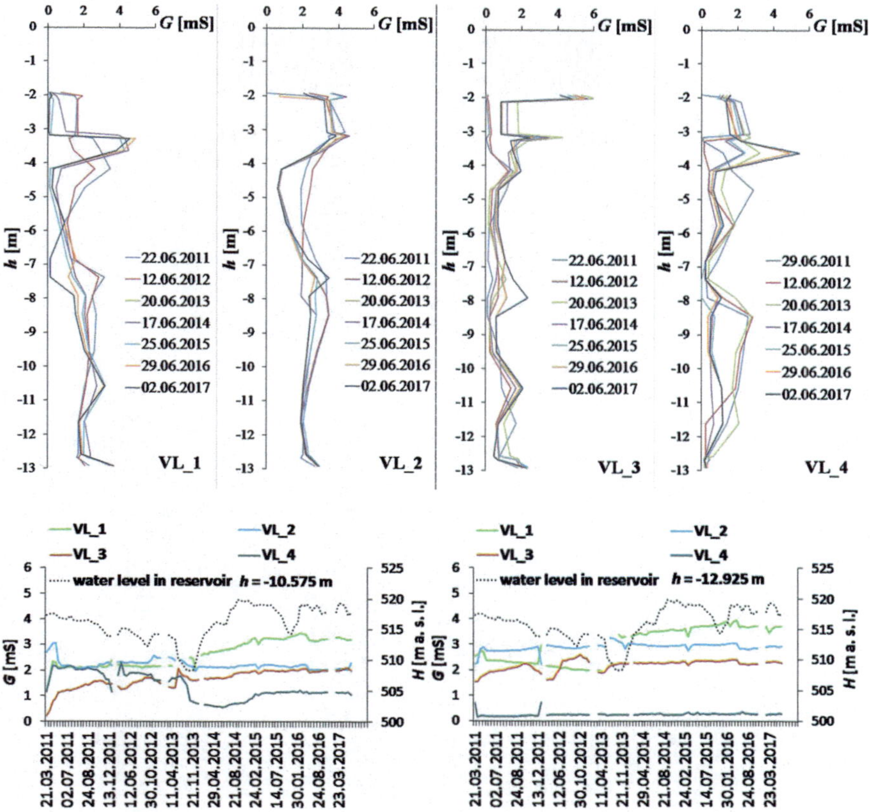

Fig. 13.8 Electrical conductance *G* measured in June 2011–2017 (top), the trend of *G* in comparison with the water level in the reservoir (bottom)

13.5.2 Hornice Earth-Fill Dam

The Hornice earth-fill dam and small water reservoir lie in the CR, south-west of the City of Brno, north of the municipality of Hornice (Fig. 13.9). The dam and water reservoir was approved on 31 May 1985 and originally served as an irrigation reservoir. Water from the reservoir was collected through an accumulation tank to an irrigation station located below the dam, and from there it was taken for irrigating adjacent plots of land. However, it does not serve this purpose anymore. At present, it is used only for the retention and accumulation of surface water and for fish farming and duck breeding by local unions and associations. Since 2012, when property settlement was agreed on, the owner of the dam and water reservoir is Zemědělské družstvo Dešov (Farming Cooperative Dešov).

The height of the dam above the original ground is 8.22 m, and the height of the dam above the base of the dam is 11.20 m. A gravel road runs on the dam crest; its

Fig. 13.9 Situation of the location

average width is 4.25 m. According to the project [52], the dam at its base is 46.80 m wide. The upstream face has a slope of 1:2.5 and is fortified with stone packing. The downstream face of the dam has a designed slope of 1:2.5 as well, but when installing an EIS monitoring system, a change in the slope was detected, which is probably a result of deformation of the downstream face due to the formation of an inrush area. The downstream face is provided with maintained grass cover. For the remediation of seepage of the earth-fill dam, the BioSealing method was used, based on the stimulation of the biological activity of bacteria, which occur naturally in soils [53–55]. The number and type of these bacteria are governed by the properties of soils, such as pH value, the temperature of water in soil or salt content. Their activity is stimulated using the injection of nutrients close to the place of seepage. Water flowing through soil transports nutrients towards the place of the highest seepage

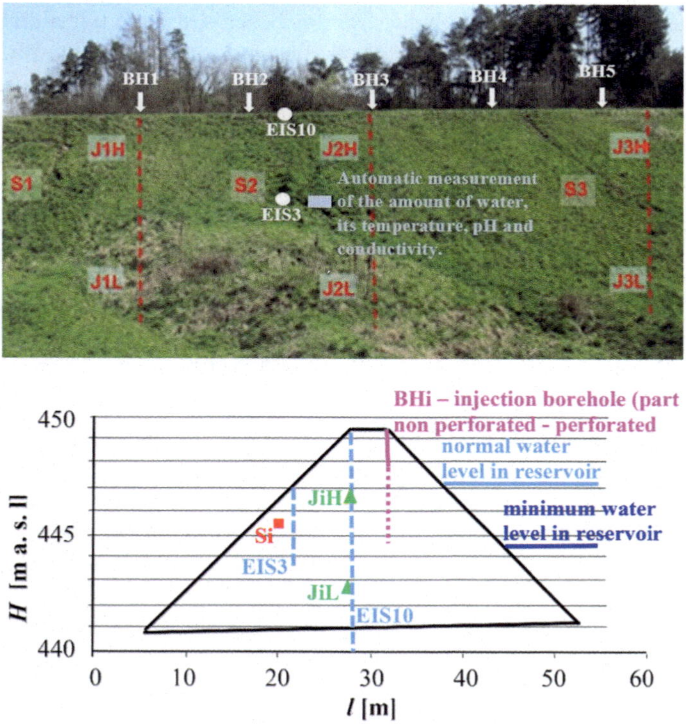

Fig. 13.10 Situation of the monitoring system and the position of injection boreholes

where the number of bacteria present and their products will rise due to nutrition. During the process, the chemistry of the soil can change, as well as the transport of bacteria and the capture of fine particles transported by the flowing water. Bacteria produce a slime that surrounds particles, and the soil is thus sealed. Approximately eight weeks after the injection of nutrients, the bacterial activity begins to decrease and returns to the original state. The fine particles captured in the bacterial slime, however, form a solid seal which remains even after the death of the bacteria.

A few types of sensors (Fig. 13.10) were used for documentation of the BioSealing process. There are pore pressure sensors in boreholes JiH and JiL, and Si in a trench on the downstream side, a flowmeter for seepage, a pH meter, thermometers and conductometers. As well as two active EIS probes with fully automatic data transmission were used. Probe EIS10 was installed from the dam crest, being 10 m long and containing 22 sensors, with the first sensor being placed at a level of 1 m beneath the dam crest. Probe EIS3 was 3 m long, containing 16 sensors, with the first sensor being placed beneath the dam crest on the downstream face in the area of the formation of the inrush area at a depth of 3.03 m beneath the dam crest, with the first electrode being 0.03 m beneath the surface.

An aqueous nutrient solution Nutrolase was dosed daily into injection boreholes BHi with a perforated part to the level of normal water level in the reservoir for four weeks. The solution Nutrolase in an amount of 0.040 m^3 per borehole per day was poured into the borehole. The concentration of the Nutrolase solution was 0.005 m^3 of Nutrolase + 0.200 m^3 of water for a daily dose to five boreholes. The amount of rinsing water was 0.100 m^3 per borehole, and the rinsing phase was split into two subphases 0.050 m^3 each with a certain time delay between them (about 1 h) avoiding overflow from boreholes during rinsing (in order not to affect/distort the measured amount of seepage).

A few results of the BioSealing pilot test carried out at the Hornice earth-fill dam in a period of 9 May–3 June 2016 are as follows (Figs. 13.11, 13.12 and 13.13). The water level in the reservoir was measured using special EIS probes in correlation with the values of a standard hydrostatic water level sensor placed in the water reservoir and in borehole BH3. During the experiment, the water level in the reservoir (gradually) rose by 0.17 m. After the termination of the experiment, the water level in the dam rose by another 0.09 m (due to intensive precipitation—on 6 June). The total rise of the water level in the reservoir since the beginning of the experiment was 0.26 m (Fig. 13.11). This rise of the water level was accompanied by an increase in daily seepage from about 0.050 m^3 per day to (0.200–0.250) m^3 per day. A similar result was also measured in borehole BH3 (Fig. 13.12). The line discontinuity of the water level was caused by the temporal rise of the water level during the rinsing; during the application of the solution and rinsing (two phases), the automatic measurement of the water level was interrupted (for about 2–3 h), and the measurement was performed manually. After the application of BioSealing, an almost immediate sudden drop of seepage was measured—from 3.200 m^3 per day to (0–100) m^3 per day, regardless of the fact that the water level in BH3 rose by 0.40 m during this period.

A very good correlation was also obtained with other measured parameters [56]. It was primarily about the daily pattern of the development of pore pressures in selected piezometers (placed in trenches, at upper levels—those most affected by the experiment), in which during the first week of injections, an immediate intensive response to pore pressures was observed. Another main outcome was addiction the daily pattern of the development of seepage and conductivity (of leachate).

The pilot application of the BioSealing technology in the Czech Republic, used on the earth-fill dam of a small water reservoir, has proved the capability of efficiently sealing seepage in the earth-fill dam. In comparison with the use of conventional remedial methods for dams by means of jet grouting or the construction of a sealing blanket, BioSealing has significantly lower costs and requires only smaller interventions into the dam body. But as it is documented in Fig. 13.13, the effectiveness of the measures at the site was short-lived [57–59]. After a higher rainfall episode on 6 June, the process of seepage reoccurred. In the daily values of electrical resistance on both curves EIS10 and EIS3, it is possible to see the start of the BioSealing process (a decrease in the values of electrical resistance means an increase of water content in pores of the soil) on 9 May 2016.

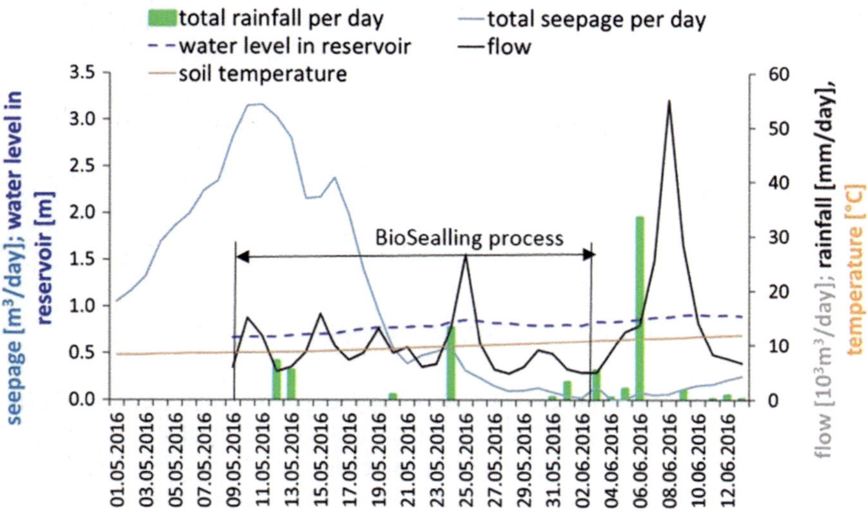

Fig. 13.11 Monitoring of the BioSealing process

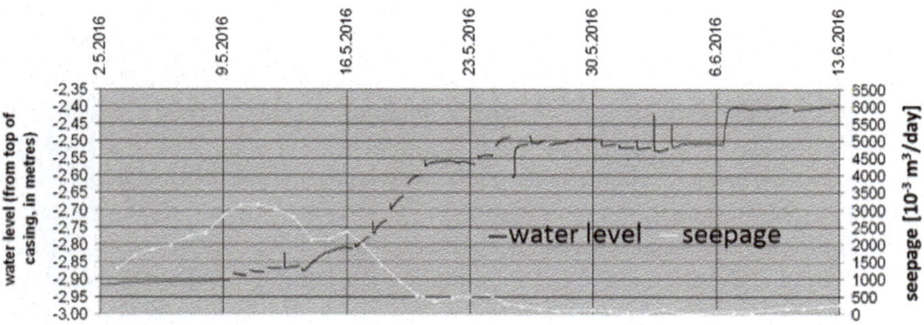

Fig. 13.12 Water level in BH3 and daily seepage—during and after BioSealing [56]

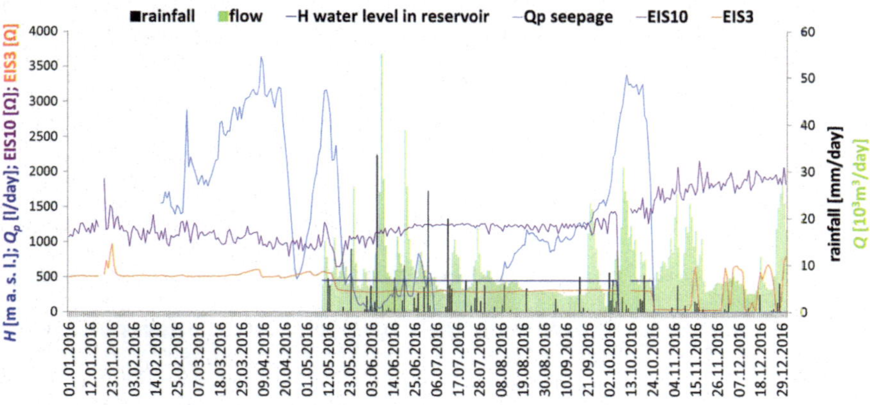

Fig. 13.13 Average daily values of monitored parameters

The EIS3 probe, which is placed directly in the inrush area, shows in the following period that a rise of the electrical resistance values *R* was not observed, and the values *R* remained low, which indicates a soil with higher water content.

The EIS10 probe, which is installed in the core of the dam, shows in the following period that water from the profile flowed away and the values *R* gradually reached the original level, which, however, they did not exceed. A further increase in the values *R* since the end of October is related to the discharge of the reservoir.

13.6 Conclusions and Recommendations

Remediation of seepage of earth-fill dams is currently most often carried out using construction procedures of special foundation engineering. Sealing is carried out from the dam crest. Used are technologies of construction of a grout curtain, as in the Karolinka dam, or innovative BioSealing as in the Hornice dam.

There are a whole number of methods of monitoring the regime of seepage water in earth-fill dams, one of which being the EIS method ever more often used at the present time. The above-given results of monitoring the process of seepage using the EIS method show the suitability of its application in documenting the processes which take place in the soil before, during and subsequently after the remediation. A significant advantage is the fact that the measuring probes remain in the same position over the entire time. Thus, it is possible to monitor changes still at the same measured levels and to evaluate them relatively. Because it is an indirect geophysical method of measurement, it is necessary for measurement to eliminate the effect of parasitic parameters (temperature, chemistry, etc.) on the monitored variables. The measurement was carried out using a unique measuring apparatus constructed within the solution of international project E!7614 in the EUREKA applied research programme.

Acknowledgements This paper has been created within the sustainability of international project E!7614 supported by the Ministry of Education, Youth and Sports within the special-purpose support of the EUREKA programmes and FAST-S-16-3757 "Increasing the Safety and Reliability of Selected Hydrotechnical Structures".

References

1. DharmaGaia translated from Braj Bhasha by Miltner V (1994) Krishna and eight seals. Praha, p 224. ISBN: 80-85905-03-5 (in Czech)
2. Cílek V (2009) People have forgotten why a house has a pedestal. Hospodářské noviny HN.IHNED.CZ 3.7.2009 (in Czech)
3. Matějíček J (1996) Water management in a river basin. Povodí Moravy, a.s., Brno (in Czech)
4. Water Management Section of the Ministry of Agriculture of the Czech Republic (2006) Water in the CR into a pocket, p 99. Available at http://www.vitejtenazemi.cz/archiv/voda_cs/voda_cr.pdf, ISBN: 80-7084-498-1 (in Czech)

5. Ministry of Agriculture of the Czech Republic (2013) Facts on water in the Czech Republic, p 36. Available at http://eagri.cz/public/web/file/211903/Fakta_o_vode_final.pdf, ISBN: 978-80-7434-048-2 (in Czech)
6. Arshad MA, Coen GM (1992) Characterization of soil quality: physical and chemical criteria. Am J Altern Agric 7:5–12
7. Smith N (1972) A history of dams, 1st edn. Citadel Press, New York, USA. ISBN 978-080-6502-915
8. Lukáč M, Bednárová E (2006) Design and operation of hydraulic structures. In: Embankment dams and dikes. Publishing House Jaga group, s. r. o., Bratislava, p 183. ISBN: 80-8076-015-2 (in Slovak)
9. Horský O, Bláha P (2011). The application of engineering geology to dam construction, or, what experience has taught us, 1st edn, rev and extended ed of Czech version. Repronis, Ostrava, p 296. ISBN: 978–80–7329–278–2
10. USACE (2006) ETL 1110-2-561. Reliability analysis and risk assessment for seepage and slope stability failure modes for embankment dams
11. Kearey P, Brooks M, Hill J (2002) An introduction to geophysical exploration, 3rd edn. Blackwell Publishing, Oxford. Available at http://fac.ksu.edu.sa/sites/default/files/AN_INTRODUCTION_TO_GEOPHYSICAL_EXPLORATION_brooks_0_0.pdf, ISBN: 978-0-632-04929-5
12. Sheriff RE (1989) Geophysical methods. Prentice Hall, Englewood Cliffs, NJ. ISBN 01-335-2568-6
13. Pařílková J, Zachoval Z, Műnsterová Z, Van De Walle B, Gomboš M (2015) Detection of the size of the effective grain using the EIS method. In: EUREKA 2015—3rd conference and working session, EUREKA. Brno University of Technology, VUTIUM, Brno. ISBN: 978-80-214-5149-0, ISSN: 2464-4695
14. Műnsterová Z, Pařílková J, Gomboš M, Pavelková D (2017) An estimation of the size of the effective soil grain using the EIS method in the field conditions. Acta Hydrologica Slovaca 18(2):244–252 (in Czech)
15. Pařílková J (2010). Monitoring the morphology of the bottom of streams and reservoirs using electrical impedance spectrometry method. J Civ Eng Archit 4(6):60–70, (serial number 31), USA. ISSN: 1934-7359
16. Gomboš M, Pařílková J, Tall A, Pařílek L, Kandra B (2011) Assessment of electrical impedance measurements at different soil moisture values in silty-loamy-clayey soils. In: EUREKA 2011. Štramberk, pp 65–77. ISBN: 978-80-214-4325-9
17. Johanson S (1997) Seepage monitoring in embankment dams. Royal Institute of Technology, Stockholm, p 50. ISBN: 91-7170-792-1
18. Kamshilin AN (2005) The results of physical modeling experiments with the close dry suffosion. Geophys Res Abstr 7, Sref-ID:1607-7962/gra/EGU05-A-06746
19. Jandora J, Říha J (2002) Failure of embankment dams due to overtopping, 1st edn. ECON publishing, s.r.o., Brno, p 188. ISBN: 80-86433-15-5 (in Czech)
20. Hydraulic works—DSS. Dam safety supervision [online] [cit. 2015-01-26]. Available at http://www.vdtbd.cz/bezpecnost-vodnich-del
21. Žatecký S (2013) Failures of dams of small water reservoirs. In: A paper for the conference water reservoirs 2013, Brno, pp 26–30. Available at vodninadrze.pmo.cz/download/pmo-sbornik-2013-www.pdf (in Czech)
22. Justin JD (1932) Earth dam projects. Wiley, New York, p 345
23. Wahl TL (1998) Prediction of embankment dam breach parameters. A literature review and needs assessment. Dam Safety Research Report DOS-98-004, USBR, p 61
24. Laník J, Halada J (1960) A book on soil 1st part: soil and plants, 2st edn. Státní zemědělské nakladatelství, Praha, p 259 (in Czech)
25. Morrison-Saunders A (2004) Assessing impact: handbook of EIA and SEA Follow-up. Earthscann, London, p 338
26. Hillel D (1998) Environmental soil physics. Academic Press, San Diego, USA, p 771. ISBN: 0-12-348-525-8

27. Bell S (2004) Elements of design in the landscape, 2nd edn. E.&F.N. Spon, London
28. Horáková J (2000) Water analytics. University of Chemistry and Technology Prague. ISBN: 80-7080-391-6 (in Czech)
29. Pitter P (1999) Hydrochemistry. Publishing House of University of Chemistry and Technology, Prague (in Czech)
30. The Czech Hydrometeorological Institute—CHMI. Available at http://www.chmi.cz/portal/
31. The T.G. Masaryk Water Research Institute—WRI. Available at http://www.vuv.cz
32. Ministry of Agriculture (2010) Guideline No. 1/2010 to dam safety supervision of hydraulic works, Ref no. 37380/2010-15000. Available at http://www.dppcr.cz/prilohy/pravo/Souhrnny_MP_k_vykonu_TBD.pdf
33. http://www.eurekanetwork.org/content/eureka-network-projects
34. http://www.msmt.cz/vyzkum-a-vyvoj-2/program-eureka-cz-lf-1; http://www.msmt.cz/vyzkum-a-vyvoj-2/inter-eureka
35. Kšáda P (2008) The methods of streambed morphology monitoring on physical model. Bachelor Thesis, Brno University of Technology (in Czech)
36. Patočka M (2006) Physical modelling of water pollution and monitoring technique. Diploma Thesis, Brno University of Technology (in Czech)
37. The Editors of Encyclopaedia Britannica (2018) Electrical impedance. Available at https://www.britannica.com/science/electrical-impedance
38. Davis JL, Topp GC, Annan P (1977) Measuring soil water content in situ using time-domain reflectometry techniques. Geol Surv of Canada, Pap 77-1B, pp 33–36
39. Fenton GA, Griffiths DV, Cavers W (2005) Resistance factors for settlement design. Can Geotech J 42(5):1422–1436
40. Beneš V (2009) Geophysical methods suitable for a quick survey of the state of earth-fill dams. Company material available at www.gimpuls.cz (in Czech)
41. Pařílková J (2003) Non destructive methods of protective dikes monitoring. Final report of the grant project 103/01/0057, Grant Agency of Czech Republic (in Czech)
42. Pařílková J (2006) Optimisation of monitoring methods of free water level movement and its effects on earth dikes. Final report of grant project 103/04/0741, Grant Agency of the Czech Republic (in Czech)
43. Pařílková J, Radkovský K (2016) Z-meter IV—user's manual. Print Copy General, Ltd., 22p
44. http://www.pmo.cz/cz/o-podniku/vodni-dila/karolinka/
45. Jareš Z, Krejčí V (2015). Karolinka water reservoir—dam reconstruction. In: Meeting of water management departments, Karolinka, p 8 (in Czech)
46. Sílová T (2015) Karolinka dam and water reservoir—cutoff diaphragm wall. Final Seminar Paper, Olomouc, March, p 17 (in Czech)
47. Zapletalová M (2016) Monitoring of the seepage regime of the dam of the Karolinka water reservoir. Final Seminar Paper. Morava River Basin s.e., p 39 (in Czech)
48. Woznica L (1993) Karolinka, landslide, stage XVI. Final Report, Woznica, Brno, MS
49. Čištín J, Hálek V (1975) The dam and water reservoir on the River Stanovnice near Karolinka. Filtration stability of loose material. Final Report of the Scientific-Research Institute of Hydraulic Engineering and Water Management of BUT, Brno (in Czech)
50. Rupp D (2011) Monitoring of seepage conditions of the dam of the Karolinka reservoir. In: EUREKA 2011. Štramberk, pp 78–86. ISBN: 978-80-214-4325-9
51. Rupp D, Pazdírek O (2012) The Karolinka reservoir dam monitoring. In: EUREKA 2012, Brno, pp 97–103. ISBN: 978-80-214-4537-6
52. Hejtman Z (2012) DATA SHEET of the Hornice water reservoir on plot of land 124/33, Hornice cadastral territory. A.1. Summarised and technical report, p 8 (in Czech)
53. Rupp D (2013) Implementation of new technology preventing water seepage and leaks through the body of dams. The project carried out in the programme "Innovation" of the provider—the Ministry of Industry and Trade. Available at http://www.geotest.cz/sluzby/vyzkum-a-vyvoj/ (in Czech)
54. Lambert J, Veenbergen V, Van der Hoek E, Karstens S (2004) Environment BioSealing: how micro-organisms become our little allies in repairing leaks ingeokring newsletter 11(1):9–12

55. Liao H, Zhao K, Lambert JWM, Veenbergen V (2005) River management bureau Changjiang water resources commission wuhan 430010 China and geodelft 2628 CK delft the Netherlands experimental study on biosealing technology for seepage prevention
56. Rupp D, Pazdírek O (2016) BioSealing—additional injection works. In: EUREKA 2016, Lednice, pp 243–252. ISBN: 978-80-214-5338-8, ISSN: 2464-4595
57. Pařílková J, Pařílek L, Fránková H, Nehudek A, Pazdírek O (2016) Monitoring of the earth-fill dam of the Hornice reservoir using the EIS method. In: EUREKA 2016, Lednice, pp 253–268. ISBN: 978-80-214-5338-8, ISSN: 2464-4595
58. Pařílková J, Pařílek L, Nehudek A, Pazdírek O, Gjunsburgs B (2017) Experience using the Z-meter to measure earth-fill dam Hornice. In: EUREKA 2017, Brno, pp 56–59. ISBN: 978-80-214-5573-3, ISSN: 2464-4595
59. Pařílková J, Gjunsburgs B, Zachoval Z, Veselý J, Műnsterová Z (2017) An earth-fill dam monitored by the EIS method during the application of nutrient aqueous solution used in the biosealing method. In: IOP conference series: materials science and engineering, vol 251. Available at http://iopscience.iop.org/article/10.1088/1757-899X/251/1/012130/meta

Part VI
Water Structures

Chapter 14
Flood Protection in the Czech Republic

D. Duchan, A. Dráb and J. Říha

Abstract Floods as extreme hydrological phenomena can influence many spheres of human life due to their destructive effects and significant expenses on required measures taken. Variable parameters of floods and their irregular occurrence unfavourably complicate risk management. This fact leads to complications in the systematic accomplishment of flood control measures. Flood control measures never provide absolute protection, and certain risk has to be always considered in flood-prone areas. However, negative flood effects may be reduced by implementing relevant flood control measures in the threatened area. Since the year 2007 in Europe, the flood protection concept is harmonized with the Directive 2007/60/EC of the European Parliament and the Council on the assessment and management of flood risks (https://www.eea.europa.eu/policy-documents/directive-2007-60-ec-of). During the past decade, the Czech Republic has prepared flood management tools which have been verified on hundreds of case studies. Currently, the Directive has been completely implemented, and flood management program is part of river basin plans. In this chapter, the main flood protection and flood risk-related issues are discussed. Firstly, the historical background is mentioned and discussed. Special attention is paid to the process of flood risk methodology and its applications at practical flood protection solutions. The present state of flood protection in the Czech Republic is briefly mentioned too.

Keywords Flood · Flood protection · Flood risk analysis · European flood directive

D. Duchan (✉) · A. Dráb · J. Říha
Institute of Water Structures, Faculty of Civil Engineering,
Brno University of Technology, Veveří 95, 602 00 Brno, Czech Republic
e-mail: duchan.d@fce.vutbr.cz

M. Zelenakova et al. (eds.), *Management of Water Quality and Quantity*,
Springer Water, https://doi.org/10.1007/978-3-030-18359-2_14

14.1 Introduction

14.1.1 General Background

Flood is defined as a significant temporary increase in the level of water in watercourses or other surface waters. During the flood, water is overbanking and flooding the areas along the streams like rivers and brooks and may cause material damage and injury to humans. The flood management is of fundamental importance also for dam safety. Overviews mentioned in Bulletins [1, 2] indicate that flood events have been the main reasons for dam failures.

According to the concept of Ministry of Agriculture in Czech Republic [3], the floods are classified as:

- *Natural (hydrologic) flood* is caused by natural conditions at the river catchment at a given locality. The main causes of natural floods are considered steady heavy rainfalls at large regions (regional floods) or intense local rainfall (flash floods), sudden snowmelt, ice jamming or the formation of jams at bridges, weirs, etc.
- *Man-made floods* are caused by the failure or malfunction of a hydraulic structure or by the management of critical state at a hydraulic structure. Man-made floods are divided into three types:
  - Type 1—flood caused by the collapse of the hydraulic structure (e.g. dam break),
  - Type 2—flood caused by the collapse of the gate or valve of bottom outlets, water intake or spillway,
  - Type 3—flood caused by the emergency release of water during the management of a critical state at the hydraulic structure endangering the safety of the installation.

Floods may have negative effects on human health, property, on the environment and cultural heritage. The mitigation of these harmful effects can usually be achieved by the realization of appropriate flood protection measures (FPM). However, it must be taken into account that such measures cannot guarantee absolute protection against a flood. The experience obtained from extreme flood events in the Czech Republic (CR) in recent years leads to the confirmation of the need for a systematic approach addressing issues related to flood protection. Methods based on the theory of flood risk management provide the most effective solutions. The basic principles of such methods are anchored in the Directive of the European Parliament and of the Council 2007/60/EC of 23 October 2007 on the assessment and management of flood risks [4]. The adoption of the Directive has been an impulse for the application of methods for managing flood risks in the countries of the European Union, and in the CR. The procedures above have been developed in the Czech Republic and applied in accordance with general European and international trends since the catastrophic flood in 1997. From this perspective, the Czech Republic was very well prepared to meet the requirements of the Directive [4], both in terms of available tools for implementing flood risk management methods and in terms of their practical verification on a number of pilot sites followed by the design of suitable flood protection measures.

The concept of FPM must be differentiated according to the land cover of flood-prone areas and should be supported by cost-benefit analysis and the assessment of other aspects such as cultural, historical, environmental, social and other.

The special approach to flood protection is required for cultural heritage, both in the process of flood risk management and the implementation of FPM. Cultural heritage localities are defined as immovable and movable property or complexes being important evidence about historical development, lifestyle and environment of the society from its earliest times to the present, as manifestations of creative skills and labour of man in various fields of human activity, for their revolutionary, historical, artistic, scientific and technical values. Significant effort to improve flood protection of cultural heritage at the international, regional and local level dates since 1997 of the twentieth century.

14.1.2 Floods in the Czech Republic

During the last two decades, the Czech Republic experienced numerous floods. The floods were documented in the reports of the Ministry of Environment [5–11] and the Water Research Institute [12, 13]. The summary of the floods since 1997 is shown in Table 14.1.

1997—Regional flood

In July 1997, severe floods occurred in the Moravian and Silesian part of the Czech Republic with further flood wave propagation to Poland, Germany, Slovakia, Austria and Hungary [12]. 59 losses of human life and about 2 billion USD of material losses were caused by the extreme flood wave in 1997, which at some river reaches exceeded 500-year flood discharge.

Table 14.1 Floods in the Czech Republic since 1997

Flood event Date	References	Type of flood	Number of fatalities	Material loss [USD]
1997—July	[12]	Regional	59	1.91E+09
1998—July	[10]	Flash	10	6.18E+07
2000—March	[11]	Small region	2	1.03E+08
2002—July		Flash	2	2.00E+07
2002—August	[13]	Regional	17	2.32E+09
2006—March, April	[5]	Regional—snow melt	11	2.74E+08
2009—June	[6]	Flash	18	3.21E+08
2010—May, June	[7]	Regional	3	2.45E+08
2010—August	[8]	Small region	5	5.23E+08
2013—May, June	[9]	Regional	16	8.00E+08

Southern Poland and the east-northern Czech Republic experienced two periods of extensive rainfall, the first occurring from 3 to 10 July and the second 17–22 July. The precipitation was caused by a low-pressure system moving from northern Italy to Moravia and Poland. The unusual development occurred when the zone of higher air pressure between the Azores Islands and Scandinavia was obstructed. The centre of the low-pressure zone stagnated over the east of the Czech Republic and southern Poland for a long period.

The precipitation was extremely high, measuring 300–600 mm, and corresponded to several months' average rainfall over a few days. The waters rose about 3 m above the previously recorded averages and were so high that they flooded over the gauging stations and staffs. It was one of the heaviest rainfalls in the recorded world's history [14]. The flood was referred as the Millennium Flood or Great Flood of 1997 [15].

Flooding started on July 5 in the CR and spread to Poland on July 6. Those early floods were rapid flash floods. Water levels rose by up to 4 m in half a day. During the flood, several levees were damaged with some total collapsed. The most significant losses were indicated in case of Kvasice levee collapse (Fig. 14.1) when the city of Otrokovice was completely flooded with the water depth exceeding 4 m (Fig. 14.2). Another serious collapse was the right-bank levee of the Odra River, which caused ecological catastrophe caused by extensive spills of oil stored at the tanks at the floodplain behind the levee.

Fig. 14.1 Collapsed levee close to Kvasice on the Morava River

Fig. 14.2 Flooded statue of Tomas Bata in Otrokovice

2002—Regional flood

In August 2002, a flood developed by continuous heavy rains ravaging Europe and caused damage of billions of USD in the Czech Republic, Austria, Germany, Slovakia, Poland, Hungary and other countries [13]. The flood was of a magnitude expected to occur roughly once a century (100-years flood), locally even once in 500 years.

Flooding resulted from the passage of two low-pressure systems bringing warm moist air from the Mediterranean northwards. The floods started with heavy rainfall in the Eastern Alps, gradually moved to the east along the Danube. The flood damage on the shores in the large cities was not severe and was greater in the areas affected by the floods later [16]. The rainfall moved northeast to the Bohemian Forest and the source areas of the Elbe and Vltava rivers resulting in catastrophic water levels in the Czech Republic. Several villages in the Northern Bohemia were more or less destroyed by rivers changing their courses or massively overflowing their banks.

Prague, the capital of the Czech Republic, was significantly damaged, it is considered to be the worst flood ever to hit the capital. The discharge in the Vltava culminated at 5300 m^3/s, 20% more than during the flood of 1896. Most of Prague's artwork was saved due to the warning of high water levels. However, there was significant damage to the Prague subway which was completely flooded at some sections. The successful evacuation was the reason for relatively little loss of life in the capital. About 40,000 people were evacuated from the Prague centre and dwelling along the

Vltava river. In total, 16 people lost their lives in the Czech Republic due to the flood, and damage from the flood was estimated at 2.3 billion USD.

2006—Flood from rainfall and snow melting

From February to April 2006, many rivers across Europe, especially in the catchments of the Elbe and Danube rivers, swelled due to heavy rain and melting snow [5]. The Elbe River also rose 13 cm higher than in 2002 downstream of the junction with the Vltava river, corresponding to 150-years event. The Czech Republic was not so much affected because during four years after the flood in 2002, efficient levee system along the Elbe was constructed.

Many dikes and levees breached because of unusually long and hard winter in Central Europe resulting in extremely high snow cover namely in mountainous regions and due to poor conditions of some levees (typically old farmer levees) which did not sustain the long-lasting load. An example of levee failure is the Horka dike which collapsed due to the internal erosion along the buried, not-properly embedded conduit (Fig. 14.3).

2009—Flash floods

The 2009 floods [6] were a series of natural disasters that took place in June 2009 in various places in Central Europe. In the Czech Republic, the heavy local rains

Fig. 14.3 Collapse of the Horka levee at the Morava river

caused overflowing of the streams and propagated to the large rivers like Oder, Elbe and Danube.

Starting on June 25 and 26, the precipitation concentrated in the area around Belgrade (Serbia) and Southern Hungary. In Austria and the Czech Republic, the situation eased. On June 27 and 28, a front moved towards Southern Poland and Baltic states. Further, heavy air masses struck the Czech Republic. After June 29, heavy thunderstorms repeatedly caused further local floods during the following days. Local areas of heavy rain of up to 50 mm in a few hours were recorded across the Czech Republic.

In the CR, persistent heavy rainfall beginning on June 22 led to the rise of smaller Vltava tributaries like the rivers Malše, Blatnice and Černá. At the same evening, the Rožnovská Bečva in Moravian part rose about 1.2 m in Valašské Meziříčí and its water level at the estuary into the Bečva significantly rose ten times above the normal depth.

The floods in the North Moravia and Silesia proceeded differently. In two hours on June 24, strong rainfall brought flash floods with up to 80 l/m^2 of precipitation at the streams Jičínka and Zrzávka. The depth of the Jičínka brook swelled up to 5.5 m at about 30 min and exceeded the 1997 flood by 2 m. In Jeseník nad Odrou, the stream Luha rose up to 2 m in the span of a half-hour; four people died in that community, three by drowning.

2010—Regional flood

During May and June 2010, a devastating series of weather events occurred across Central Europe, namely Poland which was the worst affected country. The Moravian and Silesian regions were the most affected parts in the CR. Heavy rainfall during 17–22 May 2010 reasoned in the increase in water levels of the Bečva river and its tributary the Rožnovská Bečva, where return period $N = 50$ years was reached during the days 17–19 May. At the same time, the flood hit the rivers like Morava and Dyje and also smaller streams Litava and Kyjovka where lower degrees of flood activity was declared.

Saturated soils at the catchment and another heavy rainfall on 2–5 July caused an increase in discharges and water stages at the entire Morava river basin, where the most critical situation occurred at the lower Morava in the cities of Kroměříž and Lanžhot. The flood danger situation also arose the Dřevnice, Olšava and other smaller streams.

2013—Regional flood

Extreme flooding affected namely western regions of the Czech Republic, namely its Bohemian part [9, 10]. Heavy rainfall event in the region 9 years after the 2002 event was reported, a state of emergency was declared in a total of 302 municipalities across the region. The spring weather preceding the floods had been wet in the Bohemian region, May 2013 had been one of the three wettest months during the last century. Low-pressure zone formed over the northern Adriatic and tracked north towards central Europe. On May 30–June 1, upper areas of the Elbe and Vltava river basins experienced rainfall total varying from 150 to 200 mm, locally reaching even 250 mm, which in just a few days was the equivalent normally occurring over two and half months on average.

In the Czech capital Prague, floodwater covered the esplanades along the Vltava with the discharge 3200 m^3/s on June 3, compared to the almost 5300 m^3/s during the floods of 2002. Parts of all subway lines were closed. Heavy machinery was brought in to protect the historic Charles Bridge in the city to remove floating debris accumulating at the upstream side of the bridge. One thousand troops from the Czech Army were called up to assist at erecting mobile flood defences.

14.2 Historical Background

Individual and local flood protection was built by property owners since long ago. The remainders of farmer dikes protecting agricultural land may be found at the flat floodplains along the Morava or Elbe river. Dwelling at the riverside was locally protected by quay walls or by movable locks installed into the grooves at the door-frames.

The first attempts of systematic flood protection date back to the era of Austro-Hungarian Empire of which the Czech Republic was the northern part. All activities of the great regulation of Danube and its principal tributaries were focused on the flood protection and navigation. Historically, the most extensive engineering measures were undertaken in order to secure the capital of the Empire—Vienna. Since 1565, the attempts to force the main arm of the Danube into the old river bed were undertaken. The flood protection activities were always linked to the improvement of the navigability of the river. The first systematic flood protection measures were established between 1775 and 1792 which enabled urban development in the area of the former floodplain and wetlands. In 1831, plans had been drafted at the initiative of the Hungarian politicians István Széchenyi and Gábor Baross, who succeeded in financing the large-scale navigation and flood protection project. In Hungary, the regulation of the Tisza river started in August 1846 and ended in 1880. The resultant length of the flood-protected rivers in Hungary comprises 2940 km.

The construction of flood protection measures of modern conception began in the Czech Republic at the end of the nineteenth century and the construction is still going on now. In the legislation, the development was primarily focused to increase the safety and reliability of flood protection systems. In the seventies of the twentieth century, the legal basis for the monitoring and supervision was established where some of the internationally accepted recommendations were adopted [17].

The most intensive development of flood levees in the Czech Republic dates to the period between two world wars (1918–1938) during the 'First Czechoslovak Republic'. The extensive regulation of the Morava and Elbe Rivers was realized including the system of flood levees following the streams and protecting inhabited and agricultural areas in the wide floodplains along the streams.

One of the important tributaries of the Danube is the Morava River which is draining the largest catchment in the area of the Czech Republic (Fig. 14.7). Here, the most dramatic changes started in the 1930s [18]. The most significant regulation realized in 1934–1938 included the Baťa navigation canal, connecting Otrokovice and Rohatec and following the Morava River almost parallel to its channel. For the flood protection of the adjacent area, flood levees were built along the channel, cutting former meanders of the Morava River in many places. The abandoned blind arms were obstructed by the fills on which flood defence levees were constructed. At the same time, the Morava river channel was replaced and re-routed together with corresponding tributaries. The shortened river course resulting in increasing slope together with the navigation canal called for numerous hydraulic structures—steps, weirs and navigation locks located along the stream.

The flood protection management had been accompanied by more systematic hydrological observations which had gradually developed during the nineteenth century. In the first half of the nineteenth century, started the first permanent observations and measurements of hydrological variables and the network of rainfall observation stations was established as well. In 1875, the Hydrological Committee of the Czech Kingdom was founded containing the departments of rainfall and river hydrology. Later on, during the First Republic, the State Hydrological Institute was founded in 1920. Since 1930, the hydrology was considered to be an autonomous discipline. During the Second World War, the rainfall observation was commended to the meteorological services. In 1954, the meteorological and hydrological services were linked together and the Hydrometeorological Institute was established and transformed into the Czech Hydrometeorological Institute (CHMI) after splitting Czechoslovakia into the Czech and Slovak Republic. Since 1950, the mathematical methods and models have been in use and help with the weather forecasts and prognosis of extreme hydrological events.

An example may be mentioned in the case of the water-gauge station at Rohatec, which was established in 1886 and started systematic hydrological observations on the River Morava in the Strážnické Pomoraví region [18].

Additional information about recent practices is mentioned in Sect. 14.3.3.

14.3 State Regulation, Decrees and Guidelines

14.3.1 State Administration

In the Czech Republic, the central water authority is the Ministry of Agriculture (MA). The responsibility of the MA extends to water management activities and care for water management infrastructure. According to the Water Act, the Ministry of Environment has a wide range of competence, which covers flood protection or hydrometeorological forecasts.

The principal owners of flood protection facilities are usually the River Board state agencies and only minority of those facilities belongs to municipalities. The Forests of the Czech Republic manage the lower important watercourses and small dams. Only exceptionally local municipalities or other owners (water supply companies, etc.) can manage watercourses and small dams [17].

At the regional and local scales, the regional and local environmental offices are part of the local administration.

14.3.2 General Legislative Framework for Flood Management

The general framework for flood protection management, design, construction, maintenance and safety assessment of related structures is defined by the obligatory statements of corresponding laws and decrees. More detailed requirements are anchored in the statements of technical standards and guidelines. Unfortunately for the user, it is quite difficult to implement the complex system of standards and follow up its development and all cross-references between individual standards and legislative documents. Moreover, due to the time factor over the course of the elaboration and publication of laws, decrees and standards some inconsistency occurs in terminology and sometimes also in individual statements.

Two principal legal documents concerning construction, administration and control in water management and also in flood protection in the Czech Republic are the Civil Engineering Act [19] and the Water Act [20].

The general rules for landscape and urban planning, development, construction and maintenance of civil structures, including water structures and flood protection measures can be found in the Civil Engineering Act (Building Code). Since 2007, the Civil Engineering Act defines restrictions, responsibilities and ownership demands and regulates preparation procedures and building permits [17].

The administration, ownership, control, and use of water are defined in the Water Act. The important part of the Water Act is dealing with watercourses and water structures, the responsibility of their owners and the responsibilities to perform the supervision and inspection of hydraulic structures. The improved extensive part including numerous paragraphs deals with various aspects of the flood protection. The last amendment of the new Water Act (since 2001) was issued in late summer in 2010 into which the implementation of European directives was done. Since this year, only a few minor changes were implemented into the Water Act [17].

Decrees have lower force in the Czech legislation hierarchy than laws; however, their statements are also obligatory. The following are closely related to the flood protection:

- Decree 471/2001, Coll. concerning technical-safety supervision of water structures.
- Decree 367/2005, Coll. concerning technical requirements for water structures.

- Decree 216/2011, Coll. concerning particulars of the handling rules and operational rules for water structures.
- Decree 79/2018, Coll. concerning the method and extent of the proposal for delimitation of flood zones and their documentation.
- The set of laws and decrees concerning emergency activities during floods.

Technical standardization in Czechoslovakia was organized since the beginning of the twentieth century. In 1951, the responsibility for technical standardization was taken over by the state administration through the Institute for Standardization. The previous voluntary standards became obligatory under the law. The new legislation defined technical standards as non-obligatory from 1995 again. The exceptions are the statements of laws and decrees referring to technical standards or the decrees of bodies entrusted by law to order standards to be obligatory.

Technical standards in flood protection structures (namely levees) have been elaborated since the 1960s. The set of standards dealing with flood protection included topics like river engineering, weirs, calculation of wave effect, construction of embankments, the design of outlets and intakes, stability computations for embankments, measurements and observations for water structures, the set of standards for dam engineering, etc. Most of the standards were improved and accepted as the Czech National Standards. Approximately, since the year 1987, those Czech National standards were extensively improved and updated. Some of those improved standards have been accepted as 'Technical standards for water management'. By requirements for new types of structures (e.g. dry reservoirs, etc.), some new standards have been issued as well [17].

Dam standardization in the Czech Republic has a long tradition, which began from the end of the nineteenth century. The knowledge and experience accumulated over time are dispersed across a great number of documents [17].

14.3.3 Recent Practices in Flood Protection

Recent practices in flood protection in the Czech Republic are anchored in Water Act [20]. Here, the protection against floods is understood as measures aimed at avoiding and preventing losses of human lives and damages to material property of the population, as well as to the environment, performed through systematic preventive measures increasing the flood attenuation capacity of the river basins and affecting the progression of floods. Protection against flood is assured by flood protection plans and, in the case of announcing a critical situation, by emergency plans. The Act defines behaviour and duties of population, enterprises, authorities and rescue services during the flood.

For the indication of flood arrival and danger and for the initiation of rescue activities, three degrees of flooding activity are defined. Flooding is considered at the second or third degree. Its end is at the revocation of the second degree of the activity.

According to the Water Act [20], *flood protection measures* are divided into

- Preventive measures are containing a determination of floodplain areas, the specification of flood activity degrees, flood protection plans and inspections, flood forecasting and reporting, organizational and technical activities and development of flood reserve stock. An important issue is the training of personnel participating in flood protection activities, flood reporting and warning.
- Measures were taken during flood like flood attenuation measures via hydraulic structures, flood protection and rescue activities, activities ensuring functions and services in affected territories.
- Flood documentation and assessment, identification and assessment of damage caused by flood, evaluation of factors affecting the flood and adopted measures.
- Construction, maintenance and repairs of flood protection structures.

The recent practice consists in the delimitation of *flood* zones according to the Decree 79/2018, Coll. [21] concerning the proposal for delimitation of flood zones. Existing documentation concerns flood extent maps at which flood scenarios corresponding to return periods of 5, 20, 100 and 500 years, extreme historical flood extents and the extents of dam break floods are displayed.

In some cases, so-called *active zones* are demarcated in the existing floodplain documentation. These sub-areas in the floodplain represent zones with high hazard and extensive flood damage potential. This is demarcated at all relevant flood-prone areas where flow velocity exceeds 1.5 m/s or water depth exceeds 1.5 m or the product of velocity, and the depth exceeds 0.75 m^2/s. An example of flood zones with marked active zones is shown in Fig. 14.4.

At the determined and declared flood plain areas, namely at the active zones, following restrictions are prescribed. It is forbidden to locate, permit and build structures inside the active zone of the flood plain area except hydraulic structures aimed for water management and flood protection purposes. In the active zones, it is prohibited to exploit raw materials and soils, to store materials and substances and to localize fences, hedges, campgrounds, etc. negatively influencing water flow. It is also prohibited to store materials, substances and other objects that could be washed away by water flow.

An important part of the flood protection system is *flood protection plans*. These documents summarize information necessary for flood protection of a facility, municipality, river basin or another territorial unit. Flood protection plan contains part of with names, addresses and communication links to participants involved in the flood protection, the warning and watching services. In the graphical part, maps and plans show flood plain areas, evacuation routes and meeting sites. These plans are compiled at the municipal, regional, republic and catchment level. In some cases, flood protection plans are developed for individual facilities like industrial plants.

The flood forecasting service shall provide reliable information to the public, flood protection authorities and other participants involved in the flood protection about the potential occurrence of flood and hydrometeorological conditions indicating the evolution of floods. In the Czech Republic, this service is provided by the Czech

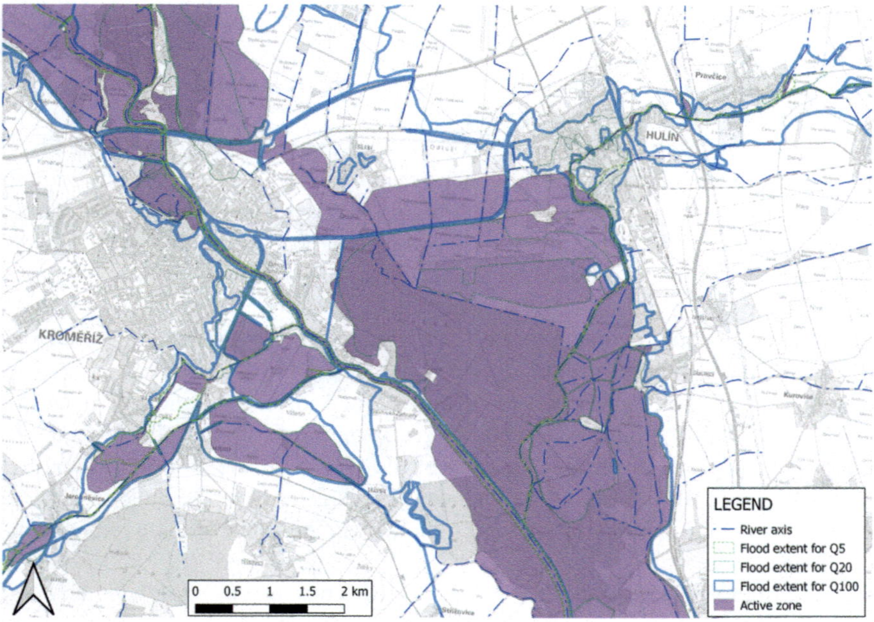

Fig. 14.4 Example of the flood zone with the marked active zone in the city of Kroměříž, Czech Republic

Hydrometeorological Institute (CHMI) in cooperation with the river administration agencies [22].

The Flood Risk Directive [4] (FRD) signifies that flood risk analysis methods are gaining ground in EC Member States and thus also in the Czech Republic. Procedures of flood risk analysis have been developed in the Czech Republic since the catastrophic floods of 1997 in line with worldwide trends and have been tested and applied in hundreds of case studies to date. Currently, the FRD Guideline based on experience with flood risk analysis applications is being processed (see the following sections).

14.4 Flood Risk Management

The FRD 2007/60/EC [4] is calling for the development of effective tools for the prioritization as a basis of technical, financial and political decisions in flood risk management. The FRD requires three stages of flood risk management:

- Preliminary flood risk assessment.
- Creation of flood hazard and risk maps for various scenarios.
- Development of flood risk management plans.

At the implementation of the individual stages defined by the FRD [4] for the conditions in the Czech Republic, the proposed and applied methods have been based primarily on existing available data such as flood extent maps, cadastral maps and data provided by the Czech Statistical Institute.

In this section, the philosophy of FRD implementation in the Czech Republic is introduced, and the way in which the general requirements have been met is presented. Brief information is given about the preliminary flood risk assessment method, flood hazard and risk map compilation for the given set of flood return periods, and also about the development of flood risk management plans using economic efficiency criteria and multi-criteria assessment techniques.

The FRD [4] only defines general requirements for the three stages mentioned above. Member states themselves decide about the appropriate methods needed for its implementation as geographical, hydrological and social differences demand specific approaches. Therefore, the working group on floods was established in 2007 primarily to support the implementation of the FRD [4]. The working group provides a platform for information exchange through a series of thematic workshops addressing particular issues of the FRD to help member states with the implementation process.

The terminology of flood risk analysis is still not uniform worldwide and also not within Europe. In the Czech Republic, the concepts of hazard, exposure and vulnerability as components of flood risk have been accepted.

It is essential that the flood risk assessment methods in the Czech Republic have been founded on existing and well-established data and procedures. This prerequisite principle has resulted in the necessity of modifying and adapting existing techniques [43].

14.4.1 Preliminary Flood Risk Assessment

The aim of preliminary flood risk assessment is the identification and definition of areas where more detailed risk-based methods should be applied. The initial basis for preliminary flood risk assessment has been existing documentation, namely flood extent maps. These documents provide a starting point for preliminary risk assessment and the identification of watercourses in the Czech Republic with potential significant flood risk. The following aspects have been considered:

- Number of residents in the floodplain area.
- Property in the urbanized areas.
- Pollution sources.
- Transport infrastructure.

Necessary data were taken from the register of census districts [23], structures and ZABAGED [24]. By combining the aspects above using GIS techniques, it was possible to preliminarily estimate the endangered property and the number of affected inhabitants in existing areas with the potential for flooding [25]. The procedure

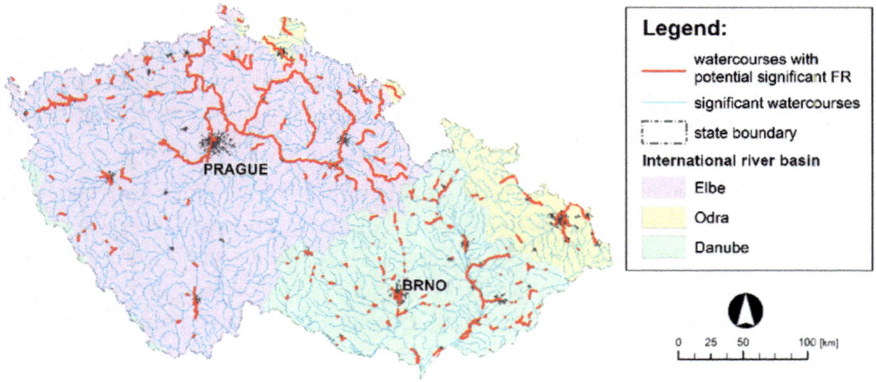

Fig. 14.5 Watercourses in the CR with high-flood risk potential [25]

resulted in the identification of 2965 km of watercourses in the Czech Republic which have high flood risk potential and for which more detailed evaluation is proposed (Fig. 14.5).

14.4.2 Flood Hazard and Risk Mapping

The second stage in the FRD implementation is focused on flood hazard, flood danger and flood risk maps. For the creation of the maps, a semi-quantitative method has been proposed based on a matrix for the determination of the danger level [26]. The procedure originates from Recommendations from Federal Office of Water Management in Germany [27], which was adapted for the conditions in the Czech Republic. In the context of the matrix approach, the concept of 'flood danger'—exposure to hazard—has been adopted using the following steps:

- The quantification of flood hazard via hydraulic calculations and the evaluation of flood intensity *IP* [27], Eq. (14.1).
- Determination of exposure to flood hazard using the 'danger matrix' [26].
- The construction of flood risk maps by combining flood danger with the vulnerability of the area.

At the first step, the flood hazard is expressed in terms of flood intensity IP which is a measure of the destructive ability of a flood and is defined as a function of water depth h and velocity v as follows [27]:

$$IP(h, v) = \begin{bmatrix} 0, & h = 0 \\ h, & h > 0\,\mathrm{m},\ v \leq 1\,\mathrm{m/s} \\ h \cdot v, & v \geq 1\,\mathrm{m/s} \end{bmatrix} \quad (14.1)$$

Flood hazard maps are developed using existing flood extent maps based on previous hydraulic calculations. The majority of older studies used one-dimensional (1D) hydraulic model, and only a few were elaborated using two-dimensional (2D) simulations. The crucial problems with the use of existing flood extent maps are as follows:

- Existing flood extent maps have been prepared for flood scenarios corresponding with return periods of 5, 20 and 100 years as required by previous Czech legislation and did not include the scenario defined by [4, 21] for floods with a low probability or extreme event scenarios. In the Czech Republic, this scenario is defined as the event with return period $N = 500$ years, for which further hydraulic calculations are being required.
- Flood extent maps do not generally contain information on water depth, which has to be subsequently evaluated using GIS tools. Usually, the information about local flow velocity is not adequate or is completely missing.

Solving the problems mentioned above is easier in cases where an updated hydrodynamic model is available. This allows the additional calculation of the Q_{500} scenario and the completion of missing data on water depths and velocities. However, in the year 2013, it was decided to upgrade all existing models and to apply 2D shallow flow models [28] for all significant streams with wider floodplains and inundation areas. This enables to evaluate local water depth and velocity for further processing.

Based on the calculated flood intensity IP from Eq. (14.1), the flood danger Ri is evaluated using the so-called danger matrix of from Eq. (14.2). In the Czech Republic, the method is applied for the relevant flood scenarios related to flood discharges Q_5, Q_{20}, Q_{100} and Q_{500}. Resulting danger level R is considered as the maximum danger level obtained in individual flood scenarios according to Eq. (14.4). The maps express the flood danger across the whole floodplain regardless of the land use.

Recommendations according to Table 14.2 are taken into account during the determination of such derived hazard values.

Table 14.2 Danger classification and verbal description consistent with [26, 27, 29]

Danger	Danger level	Recommendation
$R \geq 0.1$ or IP ≥ 3	(4) High (red)	**Do not permit** new built up areas. It is necessary to put forward designs for flood protection measures for existing buildings
$0.01 \leq R < 0.1$	(3) Medium (blue)	New construction is possible **with restrictions**. The location of sensitive structures such as hospitals, fire departments, etc. is unsuitable
$R < 0.01$	(2) Low (orange)	Construction **is possible**, but land parcel owners must be warned of the potential flood hazard. It is necessary to employ special flood measures for sensitive buildings
$p < 0.0033$ ($N > 300$)	(1) Residual (yellow)	Questions associated with flooding should be solved using urban planning, taking into account sensitive structures

Flood danger R_i for a flood scenario i with the exceedance probability p_i and a return period of N_i years is obtained from the following formula [25–27, 29]:

$$R_i = (0.3 + 1.35 \cdot \mathrm{IP}_i) \tag{14.2}$$

where the exceedance probability p of flood scenario i is expressed as follows:

$$p_i = 1 - e^{-\frac{1}{N}} \approx \frac{1}{N_i} \tag{14.3}$$

Resulting danger R is expressed as the maximum values of the danger R_i related to flood scenarios represented by the return period $N \in [5, 20, 100, 500]$ years:

$$R = \max_{0<i\leq n} R_i \tag{14.4}$$

where $n = 4$ is a number of assessed flood scenarios. The obtained flood danger values R are classified according to Table 14.2. The specific problem was the definition and delimitation of sub-regions corresponding to 'residual danger'. The extent of real past extreme floods with high return periods ($N = 500$) was the basis for the estimation of the residual danger area. Furthermore, the extent of alluvial loams in the area and the surface inundated by a potential dam break flood were also taken into account. The outer envelope of the flood extents mentioned was defined as the 'residual danger' zone in danger maps.

The results of the described analysis in the area of interest are maps of flood danger (Fig. 14.6). The categorization of danger enables the assessment of the suitability of existing or planned land use and provides the recommendation of restrictions on activities or the development of corresponding areas with higher danger rates (Table 14.3). The method described can be used in the process of urban planning, during the preliminary proposals for flood protection measures, etc. Risk maps combine data about danger and vulnerability in the exposed area.

Table 14.3 An example of selected land use zones

Land use	Acceptable risk
Residential	(2) Low
Public services	
Transportation and utility	
Industrial and manufacturing	
Agriculture	
Sport and recreation	(3) Medium
Water area	(4) High
Parks and open spaces, gardens, woods	
Arable land, meadows, pasture land	

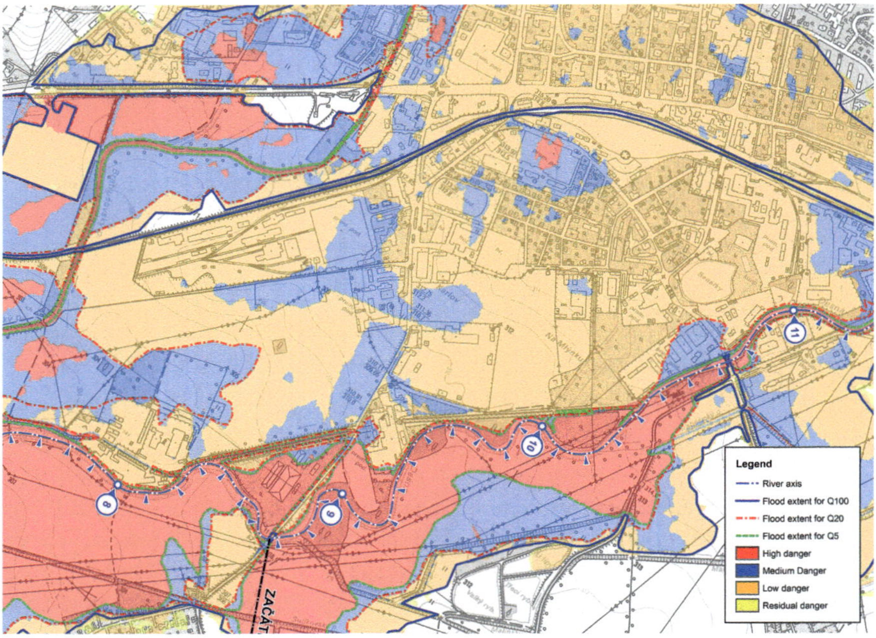

Fig. 14.6 An example of a flood danger map at locality Šumperk, CR

The vulnerability data can be derived from urban plans and maps and should be verified by site investigation. Based on the available urban plans, it is possible to define classes of land use (Table 14.3—column 'Land use'). Vulnerability maps contain locations with sensitive facilities (social care, hospitals, rescue services, police, etc.), historical monuments and potential pollution sources due to flooding.

A value of the maximum acceptable risk is assigned to each class according to Table 14.3—column 'Acceptable risk'. The maps of the areas thus classified according to land use (vulnerability maps) are 'overlaid' by danger maps and are processed by GIS analytical tools into risk maps in which existing or anticipated areas with exceeded acceptable risk are highlighted using a spectrum corresponding to Table 14.2. The following logical step is a more detailed analysis of 'risky areas' from risk management and risk attenuation views to an acceptable level. The described method using tailor-made software tools for GIS applications has been elaborated, tested and applied in the CR since the year 2001 at hundreds of localities which contain more than 1500 km of watercourses. The results of the described analysis in the area of interest are maps of flood risk (see Fig. 14.7).

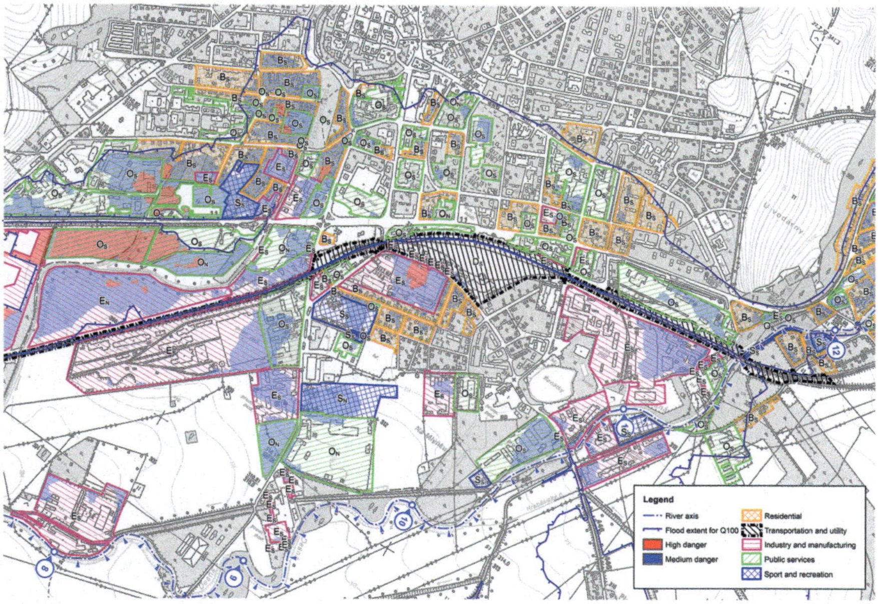

Fig. 14.7 An example of a flood risk map at locality Šumperk, CR

14.4.3 Flood Risk Management

The FRD [4] requires that flood risk management plans be established by hazard and risk maps. Flood risk management plans are primarily focused on prevention, protection and preparedness. During the development of flood risk management plans, existing hydraulic and feasibility studies are used as basic documents. Following risk-based methods have been implemented in the CR in the studies mentioned:

- The matrix for danger level determination [26, 27] is being used for the assessment of floodplains from the viewpoint of flood hazard, vulnerability and risk. The method enables the general identification of areas and structures where the acceptable risk has been exceeded.
- The method based on flood loss estimates described below serves for an assessment of the economic efficiency of structural FPMs. For the estimation of flood losses, the available geographical data and data from the census register [23] are employed.

According to the requirements of the FRD, the processing of flood risk management plans stems from the results of spatial analysis using the matrix method (Sect. 14.4.2) in which recommendations for further analysis and flood protection in areas with exceeded acceptable risk are introduced. For the complex assessment of FPM, variants according to [4] economic, environmental and social viewpoints are taken into account. For this purpose, the multi-criteria analysis is frequently used. At present, the evaluation of the economic effectiveness of FPMs is elaborated in detail while the risk-based analysis related to other loss categories is still subject to research activities. The evaluation of the economic effectiveness of FPMs consists of following steps:

- an estimation of the extent of endangered property in the floodplain,
- an estimation of material flood losses and quantification of economic risk,
- a determination of quantitative economic criteria using cost-benefit analysis.

The endangered property in the floodplain is estimated for the existing state and the state after the implementation of flood protection measures. Firstly, the processing of flood hazards is carried out for flood scenarios corresponding with at least the Q_5, Q_{20}, Q_{100} and Q_{500} discharges.

The estimate of flood damage to property in the floodplain is performed using so-called damage functions which express the relation between water depth and the percentage of damage. The damage functions were developed and tested for the floods in the years 1997, 1998, 2002 and 2006 for residential buildings, service buildings, industrial buildings, roads, railways, bridges, paved areas, infrastructure, sport and recreation areas, farmland and forest land. The functions were verified for the consecutive floods in 2010 and 2013 (Sect. 14.1.2). For the elements mentioned, the economic value of assets was estimated based on official census data. Material flood losses in monetary units are obtained by multiplying the economic value of assets and the corresponding percentage of damage. The total potential economic flood loss D for a given flood scenario with exceedance probability p is calculated by summarizing the losses across all elements in the analysed flooded area. Based on the potential flood losses D, the economic flood risk is expressed as follows (Fig. 14.8) [30]:

$$RI = \int_{0}^{p_{\mathrm{H}}} D(p)\mathrm{d}p \tag{14.5}$$

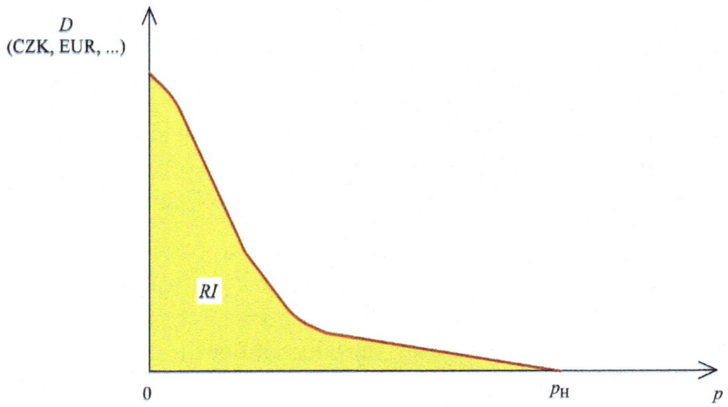

Fig. 14.8 Schematic interpretation of a flood loss exceedance curve *D*(*p*)

where $RI = E(D)$ is average annual flood risk in monetary units, p is the exceedance probability of the corresponding peak flood discharge determined by Eq. (14.3), $D(p)$ is the dependence of potential losses on flood exceedance probability p and p_H is the exceedance probability of harmless discharge corresponding to the flood protection level in the area.

Equation (14.5) may be rearranged by transforming probabilities into discharges:

$$RI = E(D) = \int_{Q_a}^{Q_b} D(Q) \cdot f(Q)\mathrm{d}Q \tag{14.6}$$

where $D(Q)$ is flood loss at the discharge Q, $f(Q)$ is probability density function of peak discharges, Q_a is harmless flood discharge and Q_b is the discharge with very low probability approaching zero.

A practical procedure for the average annual economic risk estimate is as follows:

- The exceedance probability for individual flood scenarios represented by selected N-year flood discharges is estimated using Eq. (14.3).
- The hydraulic characteristics of the flood, namely water depth, are determined by hydraulic modelling (1D, 2D).
- The direct economic losses in the area of interest are derived using the damage curves and water depth for selected flood scenarios. A graphical interpretation of the obtained function $D(p)$ can be seen in Fig. 14.8.
- For practical applications, Eq. (14.5) is rearranged into Eq. (14.6) which is figured out by the numerical integration. It gives an estimate of the annual average risk. In case of the proposal of flood protection arrangement, Eq. (14.6) is applied for the present state $Q_{a,ORIG}$ and new increased harmless discharge $Q_{a,NEW}$ after the improved flood protection level. Annual economic risks RI_{ORIG} and RI_{NEW} before and after the application of flood protection measures are obtained. In Fig. 14.9, the shaded area shows the difference between present and future risks.

The method described has been implemented into the tailor-made GIS application.

The final value of RI serves for the evaluation of the economic efficiency of proposed variants of structural flood protection measures. Primarily, the relative efficiency RE was used for the cost-benefit analysis [30]:

$$RE = \frac{RI_{ORIG} - RI_{NEW}}{I \cdot DR} \tag{14.7}$$

where I is the investment cost and DR is the discount rate (in CR estimated by the value $DR = 0.03$). The flood protection arrangements are profitable if $RE > 1$.

The method described has been widely applied in the assessment of hundreds of anticipated flood protection measures in the CR financed from the funds of the European Investment Bank. During this assessment, the economic risk was the principal quantitative criterion. At the assessment, more indicators have been used like the absolute efficiency, repayment period, or semi-quantitative indicators for technical viability, safety and reliability, environmental indicators, etc.

According to [4], the flood protection measures and their subsections related to individual areas should be prioritized using multi-criteria analysis during which other important partial risks are taken into account. For the flood risk management plans in the Czech Republic, the partial risks RI_i are related to the following loss categories:

- RI_1—annual average loss—economic risk according to Eqs. (14.5) and (14.6) (CZK/yr.);
- RI_2—annual average affected population (inhabitants/yr.);
- RI_3—annual average number of affected sensitive buildings (buildings/yr.);
- RI_4—annual average affected area with historical monuments (m^2/yr.);
- RI_5—annual average number of affected potential pollution sources (sources/yr.).

Quantification of partial risk RI_j can be performed using Eq. (14.5) modified for the aforementioned partial risks:

$$RI_j = \int_0^{p_H} D_j(p)\mathrm{d}p \tag{14.8}$$

where RI_j is the risk for consequence (loss) category j and $D_j(p)$ represents the functional dependence of corresponding potential loss on flood exceedance probability p.

The final assessment is usually done using multi-criteria analysis. A more detailed description of these methods is outside the scope of this paper [31–33], and some recommendations are noticed in Sect. 14.7.

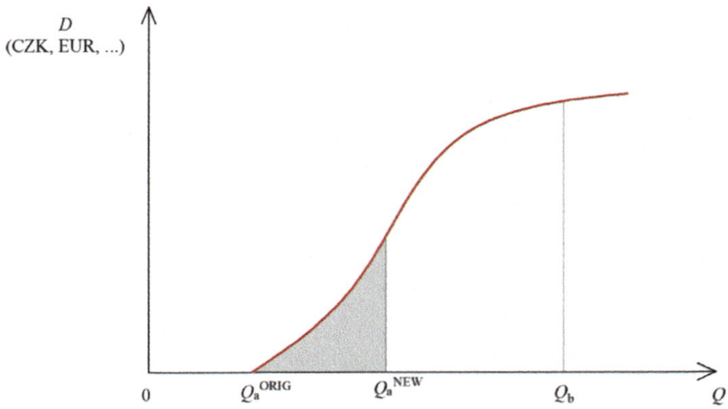

Fig. 14.9 Flood loss depending on flood discharge *D*(*Q*)

Table 14.4 Characteristic error in provided hydrologic data [34]

Hydrological data	Accuracy class			
	I	II	III	IV
	Standard error [%]			
Average annual discharge (Q_a)	8	12	20	30
M—day's discharges (Q_{30d}–Q_{300d})	10	15	25	40
M—day's discharges (Q_{330d}–Q_{364d})	20	30	45	60
N—year's discharges (Q_1–Q_{10})	10	20	30	40
N—year's discharges (Q_{20}–Q_{100})	15	30	40	60

14.4.4 Uncertainties

At the flood risk analysis, uncertainties should be taken into account. The most significant uncertainties are:

- hydrologic uncertainties,
- uncertainties in the Digital Terrain Model (DMT) and geodetical survey,
- uncertainties in the hydraulic model applied.

Hydrologic uncertainties concern accuracy of the flood discharges and design flood hydrographs related to given probability expressed via return period *N*. In the Czech Republic, these are quantified according to the 'accuracy class' at the standard ČSN 75 1400 [34] via standard error (Table 14.4).

The uncertainties in *land topography and geometry* are given by the accuracy of the geodetical measurements (about 10 mm), and in more extensive floodplains, the Digital Terrain Models (DTM) are used. Recently, the DMR 5G [35] is used for the topography modelling. Experience shows that the accuracy of the relief model [35] is about 0.1 m; however, in the case of vegetated surface, the error may locally exceed 0.2 m. The levels in the river channel, at hydraulic structures and important singularities (water diversions, junctions, etc.), should be determined using standard geodetic land surveying.

The use of the appropriate *hydraulic model* is crucial for obtaining relevant results. Concerning dimensions, 1D and 2D models are worldwide used [36–39]. Sometimes branching 1D models are applied, and here, we denote them as 1.5D models. For the flow in an open channel, 1D and 1.5D models are the best options. When the flow spreads over the wide flat floodplain, it is necessary to use 2D models. However, it is known that 2D models do not give good results in narrow channels where the size of the computation grid does not fit the dimensions of the river bed, slopes, curvature, etc. This shortcoming is presently overcome by hybrid 1D and 2D models where the channels are modelled by the 1D algorithm and shallow inundation flow is considered to be 2D. The interaction between two models is governed by so-called transition structure which assures continuity of the flow and water level at the transition between 1D and 2D models. The mesh size and selected time step are crucial characteristics influencing numerical stability and accuracy. The uncertainties in hydraulic models are also due to applied model parameters like surface roughness, contraction/expansion coefficients, method of embodying singularities like single buildings, water sources/sinks, obstructions like fences, etc. All these uncertainties may be reduced by careful model calibration for the past floods (Sect. 14.1.2). At the model calibration, the inaccuracy of calibration data (estimated discharge, measured water depth) must also be taken into account.

All uncertainties mentioned and the importance of the flood protection measures are governing the height of the *levee freeboard*. It may vary from 0.3 to 1.0 m, and in the Czech Republic, the common freeboard height accounts 0.5–0.6 m. As mentioned above, consequences in case of levee breaching should be taken into account together with other factors mentioned. When designing freeboard height, another important factor is the expected settlement of the embankment which from experience could reach 0.4% to extremely 1% of levee height. At the undermined areas, the levee crest has to be designed including overelevation related to the expected land subsidence or considering the possibility of a rapid increase of its crest level. It is recommended to propose levees with a wider crest with the corresponding stock of appropriate earth. The subsidence forecast should be verified periodically by visual inspections and geodetical surveying. In the case of international or boundary streams, the freeboard height has to be coordinated based on trans-boundary negotiations and agreement.

14.5 Flood Protection in the Czech Republic

14.5.1 Requirements on the Flood Protection Level

There are no strict requirements on the design flood (return period) for individual land use categories. However, historically, some guidance is anchored in older and also in presently valid (however not obligatory) national standards. In Table 14.5, recommended design flood return periods are linked to particular land uses according to the older Czech technical standard.

Table 14.5 Design flood discharges for river treatment purposes [42]

Land use	Design discharge
Historical town centres, historical monuments	$\geq Q_{100}$
Continuous urbanization, industrial area, important roads, infrastructure	$\geq Q_{50}$
Dispersed built-up area (residential, industrial), continuous recreational areas (cottages, etc.)	$\geq Q_{20}$
High value land like orchards, hop gardens, etc.	$\geq Q_{20}$
Arable land (according to its basification)	Q_5–Q_{20}
Pastures and forests	Q_2–Q_5

Table 14.6 Design flood discharges according to [40]

Land use	Design discharge
Historical monuments	$\geq Q_{100}$
Public roads	According importance Q_{20}–Q_{100}
Continuous urbanization, important industrial area	Q_{100}
Larger built-up areas or manufacturing, services	Q_{50}–Q_{100}
Smaller urban areas	Q_{20}–Q_{50}
Local roads	Q_{10}–Q_{50}
Orchards, gardens, hop gardens	Q_{10}
Arable land	Q_{5}
Meadows, forests, pastures	Q_{30days}–Q_{1}

In the year 2000, the Strategy for the Flood Protection [40] with the documentation 'Support of the prevention against floods' was issued by the MA of the Czech Republic. According to [17, 40], the following factors should be taken into account when deciding about the design flood return period *N*:

- the necessity of flood protection with respect to the lowering of damage and loss on human lives and material property,
- flood protection level should be determined using the risk-based methods with respect to the following criteria:
  - population at risk,
  - the value of property and potential material losses,
  - location of important structures and strategic facilities (principal highway, railway, gas duct, power plants, etc.),
  - available flood arrival time and warning time,
  - information about sediment transport,
  - information about water sources and water supply infrastructures,
  - information about water pollution sources (wastewater treatment plants, industry handling with dangerous toxic substances),
  - general feasibility of FPM,
  - impact on nature, landscape and cultural heritage,
  - economic effectiveness of FPM.

Recommendations according to the Strategy for the Flood Protection for the territory of the Czech Republic are shown in Table 14.6.

14.5.2 Technical Arrangements

Basic flood protection structures applied in the Czech Republic are as follows:

- Structures increasing channel capacity:
  - dikes, levees,
  - floodwalls,
  - levee spillways (usually side spillways),
  - mobile structures (log gates, inflatable bags, etc.).
- Structures for the transformation of flood hydrograph:
  - large dams with significant flood attenuation volume,
  - dry reservoirs, polders,
  - managed floodplains.
- Additional arrangements for the management and disposal of inner waters:
  - drainage system (dikes, drain pipes),
  - pumping stations.

All mentioned arrangements are standard structural tools applied for flood control and flood protection worldwide. In the CR, until now, there is no central database of existing flood protection measures. This is probably because of various bodies which have constructed and presently own structural flood protection measures. The technical level and maintenance of flood protection structures correspond to the owner. In general, the owners of flood protection measures can be:

- River Agencies and state enterprises are responsible for overall river basin management, conceptual planning of flood protection measures, etc. There are five River Agencies in CR corresponding to five major river basins—The Elbe, Vltava (Moldau), Oder, Morava and Ohře rivers. These enterprises provide enough professional staff to manage flood protection issues. They are the major owner of flood levees and similar structures in the CR.
- Municipalities may be owners of local flood protection arrangements like levees, floodwalls, flood attenuation reservoirs including small dams, etc. An example may be flood protection in Prague consisting of more than 20 km of levees, floodwalls, mobile walls including several pumping stations and appurtenant structures (gates, sewer closures, etc.). Smaller municipalities usually do not possess enough technical staff and operation, and maintenance of flood protection measures may be a problem.
- Forests of the Czech Republic, state enterprise are the administrators of smaller streams and relevant flood protection arrangements related to these streams. They have their departments for water management, even if these are only minor and "less important" part of the enterprise and thus underfinanced.
- Private bodies are owners of arrangements protecting private property against floods. These may be private companies, factories, individual inhabitants, etc.

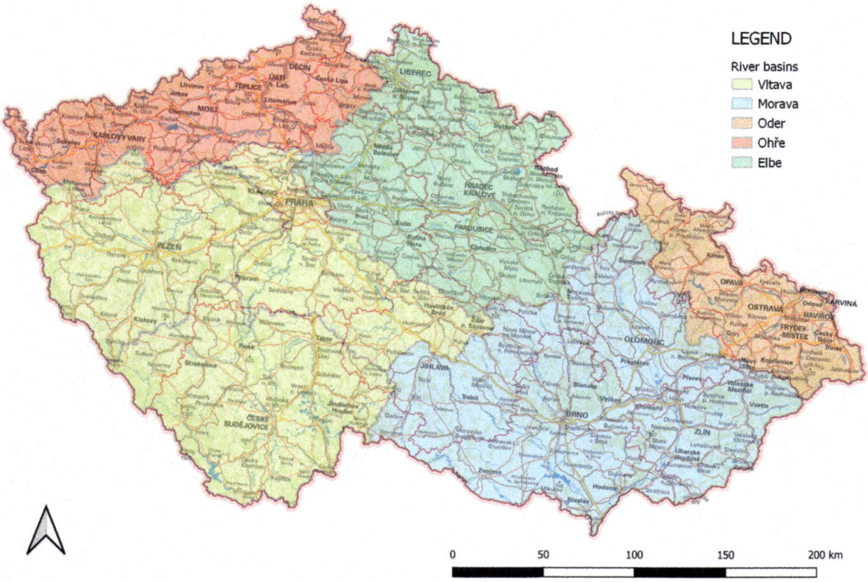

Fig. 14.10 Scheme of river basins in the Czech Republic. Blue—Morava, brown—Oder, yellow—Vltava, green—Elbe and red—Ohře [41]

Table 14.7 Summary of levee's length including floodwalls at individual catchments [km]

Levees including floodwalls	Before 1997	After 1997
The Morava river basin	1014	39
The Oder river basin	80	103
The Vltava (Moldau) river basin	65	26
The Elbe river basin	107	72
The Ohře river basin	29	4
Total	1295	244
Total	1539	

Based on the questionnaires answered by the River Board Agencies (Fig. 14.10) in the year 2017, the summary was compiled on the total length (in km) of *levees* including floodwalls owned and managed by individual agencies (Table 14.7).

In the Czech Republic, vital discussion about the application, function, risks and profits of *levee spillways* is in progress. Its design and parameters issue from the fact that levees are hydraulic structures damming water and should be protected against breaching due to overtopping. The problem consists in the fact that the area behind levees would be 'artificially' flooded before reaching the levee crest and so channel capacity would not be fully utilized. This shortcoming may be solved using movable gates on the side spillways which call for more demanding performance activities of the levee owner's staff, gate testing and operation during the flood.

Fig. 14.11 Levee spillway at the Bulhary weir

Another option is the location of levee spillway upstream of the movable weir which enables manipulation of water level at the backwater. An example of such levee spillway is at the weir Bulhary at the Dyje River (Fig. 14.11).

Large dams and reservoirs play a crucial role in flood routing and are one of the most important structural measures for the reduction of the flood damage within the integrated flood management system. The flood protection purpose is anchored in most of the operation orders and manipulation rules of large dams with multi-purpose function. Their role was confirmed at the extreme flood in 2002 and 2013 (Sect. 14.1.2).

During the last two decades, more than one hundred *dry reservoirs* for trapping local, mostly flash floods were constructed in the CR. These are mostly so-called small reservoirs, i.e. dams up to 9 m in height, volume up to 2 million m^3. These installations serve mostly for the attenuation of flash floods occurring at the upper parts of river basins.

There are several *polders* located at the floodplains adjacent to the streams with significant flood attenuation volume. In case of polders, the care must be taken to protect the surrounding dikes which are mostly built from locally available materials with its risk factors like variable geological conditions at and sub-base, earth material of poor quality, limited maintenance, supervision, etc. Polders are commonly located in river deltas and junctions (e.g. Polder Soutok, Lednice) or former fenlands.

14.6 Discussions and Conclusions

In this section, the flood risk techniques practised in the Czech Republic are briefly mentioned. The recent procedures used for the development of flood hazard and flood risk maps are described together with the development of danger and vulnerability maps. The assessment of economic efficiency should be a necessary part of flood risk management plans. The methods described are by the FRD [4]. From this viewpoint, the Czech Republic is well-fitting requirements on flood protection harmonization prescribed by the European Committee.

Parallel with the permanent flood protection planning, the measures against the floods are being designed and constructed. Foremost, these arrangements are levees and dry reservoirs. Lessons learned from previous floods show the importance of large dams with significant flood attenuation volume in flood routing. The most efficient is the management and real-time operation of the system of dams based on real-time rainfall measurement and data processing using adaptive models.

The important issue is the technical surveillance of flood protection structures. The particularity of levees, floodwalls and dry reservoirs is that they cannot normally be impounded and that their trial operation is usually firstly performed during the flood event.

14.7 Recommendations

Even if the flood risk procedures have been well established in CR, further research into flood risk techniques should continue to improve and refine results and outputs obtained. Special attention should be focused on:

- the more detailed multi-criteria flood risk assessment,
- non-structural flood protection measures and the evaluation of their effectiveness,
- inclusion of risks from the exposure of inhabitants to flood hazards,
- assessment of environmental risks,
- assessment of risks due to the flooding of sensitive facilities and historical monuments,
- estimation of indirect losses,
- more comprehensive uncertainty analysis in risk management.

The persisting issues are environmental aspects of all flood protection measures located at the floodplains, which are often declared to be environmentally protected areas. Further attention should be focused on incorporation and accommodation of flood protection structures into the natural environment of the riverine zone.

It is also recommended to perform continuous informing, education and training of the population at flood-prone areas together with the flood rescue services (fire brigade in CR). Floods in 1997 proved that 'flood memory' may be a critical phenomenon influencing the behaviour of the population at risk before and during the flood.

References

1. ICOLD (1995) Dam failures statistical analysis. ICOLD Bull 99:76
2. ICOLD (2003) Dams and flood. Guidelines and cases histories. ICOLD Bull 125:229
3. Ministry of Agriculture (2010) The concept of flood protection in the Czech Republic using technical and natural measures. Ministry of the Environment of the Czech Republic, Prague, p 29
4. FRD (2007) Directive 2007/60/EC of the European parliament and of the council of 23 October 2007 on the assessment and management of flood risks
5. CHMI (2006) The spring flood of 2006 in the Czech Republic. Final Report. Ministry of the Environment of the Czech Republic, p 158 (in Czech)
6. CHMI (2009) Assessment of the floods of June and July 2009 on the territory of the Czech Republic. Final Report, Ministry of the Environment of the Czech Republic, Prague, p 131 (in Czech)
7. CHMI (2010) Assessment of the floods of May and June 2010. Report, Ministry of the Environment of the Czech Republic, Prague, p 167 (in Czech)
8. CHMI (2010) Assessment of the August 2010 flood. Report, Ministry of the Environment of the Czech Republic, Prague, p 131 (in Czech)
9. CHMI (2013) Assessment of the floods in June 2013. Final Report, Ministry of the Environment of the Czech Republic, Prague, p 257 (in Czech)
10. ERA (1998) The catastrophic 1998 flood in the Orlicke hory region, 10 years after the flood. Elbe River Agency, Hradec Králové (in Czech)
11. ERA (2000) Final report on the March 2000 flood in the Elbe River basin. Elbe River Agency, Hradec Králové, p 25 (in Czech)
12. WRI (1997) Evaluation of the flood in July 1997. CD ROM 1–8 (in Czech)
13. WRI (2002) Assessment of the catastrophic August 2002 flood in the Czech Republic. Final Report, Ministry of the Environment of the Czech Republic, p 160 (in Czech)
14. Szamalek K (2000) The great flood of 1997 in Poland: the truth and myth. In: Petts G, Amoros C (ed) Fluvial hydrosystems, pp 67–74
15. Kundzewicz ZW (2007) Summer 1997 flood in Poland in perspective. In: Vasilev OF (ed) Extreme hydrological events: new concepts for security, pp 97–110
16. European Floods (2002) everipedia.org/wiki/2002_European_floods. Accessed 27 Aug 2018
17. Central European Flood Risk Assessment and Management in CENTROPE (2003) CEframe 363. Current standards for flood protection. www.central2013.eu/fileadmin/user_upload/Downloads/outputlib/CEframe_363_Current_standards_for_flood_protection.pdf. Accessed 27 Aug 2018
18. Brázdil R, Máčka Z, Řezníčková L, Soukalová E, Dobrovolný P, Grygar TM (2011) Floods and floodplain changes of the River Morava, the Strážnické Pomoraví region (Czech Republic) over the past 130 years. Hydrol Sci J 56(7):1166–1185. https://doi.org/10.1080/02626667.2011.608359
19. The Act No. 183/2006 Coll. (2006) Act on town and country planning and building code (The Building Code). Czech Republic
20. The Act No. 254/2001 Coll. (2011) Act on water (The Water Act). Czech Republic
21. Decree 79/2018, Coll. (2018) Concerning the method and extent of the proposal for delimitation of flood zones and their documentation. Czech Republic
22. Flood forecast service. hydro.chmi.cz/hpps. Accessed June 2018
23. Census District and Building Register (2009) Czech statistical office, Prague. www.czso.cz/csu/rso/registr_scitacich_obvodu. Accessed 16 Sept 2010 (in Czech)
24. ZABAGED (2009) Fundamental basis of geographical data. Czech office for surveying, mapping and cadaster, Prague
25. Drbal K, Štěpánková P (2008) Preliminary flood risk assessment in the Czech Republic. In: Das Magdeburger Gewässerschutzseminar 2008, Magdeburg, pp 97–99
26. Zimmerman M, Pozzi A, Stoessel F (2005) Vademecum—hazard maps and related instruments, The Swiss system and its application abroad. In: PLANAT, Bern, Switzerland, p 19

27. FOWM (1997) Empfehlungen: Berücksichtigung der Hochwassergefahren bei raumwirksamen Tätigkeiten. EDMZ, CH-3000. Federal Office for Water Management, Bern (in Dutch)
28. Vreugdenhil CB (1988) Numerical methods for shallow water flow. Kluwer Academic Publishers, p 261
29. Říha J et al (2005) Flood risk analysis. In: Monographs of the water structures institute, FCE Brno University of Technology, CERM, 7, p 286
30. Čihák F, Satrapa L, Fošumpaur P (2006) Guidelines for the assessment of flood protection proposed to the IInd period of the programme "prevention against floods". Czech Technical University, p 12
31. Brans JP, Mareschal B (2005) PROMETHEE methods. In: Figueira J, Greco S, Ehrgott M (eds) Multiple criteria decision analysis: state of the art surveys. Springer, New York, pp 163–195
32. Meyer V et al (2007) GIS-based multicriteria analysis as decision support in flood risk management. Report T10-07-06, FLOODsite Consortium
33. Zeleňáková M, Zvijáková L (2017) Using risk analysis for flood protection assessment. Springer International Publishing, Berlin, p 128
34. ČSN 75 1400 (1990) Hydrologic data of surface waters. Czech national standard, Czech Office for Standards, Metrology and Testing, Prague (in Czech)
35. Digital model of relief of the fifth generation, DMR 5G. geoportal.cuzk.cz. Accessed July 2018
36. Fan Y, Ao T, Yu H, Huang G, Li X (2017) A coupled 1D-2D hydrodynamic model for urban flood inundation. Adv Meteorol 2017(2819308):12
37. Patel DP, Ramirez JA, Srivastava PK, Bray M, Han DW (2017) Assessment of flood inundation mapping of Surat city by coupled 1D/2D hydrodynamic modeling: a case application of the new HEC-RAS 5. Nat Hazards 89(1):93–130
38. Davidsen S, Lowe R, Thrysoe C, Arnbjerg-Nielsen K (2017) Simplification of one-dimensional hydraulic networks by automated processes evaluated on 1D/2D deterministic flood models. J Hydroinformatics 19(5):686–700
39. Al Amin MB, Sarino, Haki, H (2017) Floodplain simulation for Musi river using integrated 1D/2D hydrodynamic model. In: Sriwijaya international conference on engineering, science and technology, Sicest 2016, p 6
40. Strategy for the Flood Protection for the Territory of the Czech Republic (2000) Ministry of Agriculture of the Czech Republic
41. Water portal of Czech republic. voda.gov.cz/portal/cz/aplikace/sap_toky.htm. Accessed June 2018 (in Czech)
42. ODN 75 2103 (1993) River management. Czech technical standard (in Czech)
43. Dráb A, Říha J (2010) An approach to the implementation of European directive 2007/60/EC on flood risk management in the Czech Republic. Nat Hazards Earth Syst Sci 10(9):1977–1987

Chapter 15
Small Hydropower Plants in the Czech Republic

D. Duchan, A. Dráb and O. Neumayer

Abstract This chapter aims to present selected small hydropower plants realized in the Czech Republic for about 16 years, which are interesting both from a technical point of view and from the point of view of the architectural design. To improve the economic efficiency of the investment, most of the newly built hydroelectric power plants were implemented using existing weirs or dams. Besides the new buildings, a number of general reconstructions of the technological and construction part of the small hydropower plants were also carried out, aiming to increase the efficiency of these sources. In the end, the foreseen future development in the field of power plants in the Czech Republic is indicated.

Keywords Small hydropower plant · Design · Weir · Dam · Reconstruction · Kaplan turbine

15.1 Introduction

In the Czech Republic, all hydroelectric works with an installed power of up to 10 MW belong to the category of small hydroelectric power plants (SHPP). In 2018, a total number of 1600 SHPPs were in operation, with a total installed power of 351 MW [1]. Over the last 16 years, the number of SHPPs operated has increased by about 162% and installed power by 93% (see Fig. 15.1). Due to the geographical location of the Czech Republic, where most of the large rivers in our country originate, the SHPPs are an important part of hydro-energetics.

Most of the power plants built during this period use existing weirs or dams to provide the head. The reason for this approach is substantial savings in investment costs. Another significant group of power plants realized during this period is the gen-

D. Duchan (✉) · A. Dráb · O. Neumayer
Faculty of Civil Engineering, Institute of Water Structure, Brno University of Technology, Veveří 95, 602 00 Brno, Czech Republic
e-mail: duchan.d@vutbr.cz

O. Neumayer
Aquatis, a.s., Botanická 834/56, 602 00 Brno, Czech Republic

M. Zelenakova et al. (eds.), *Management of Water Quality and Quantity*, Springer Water, https://doi.org/10.1007/978-3-030-18359-2_15

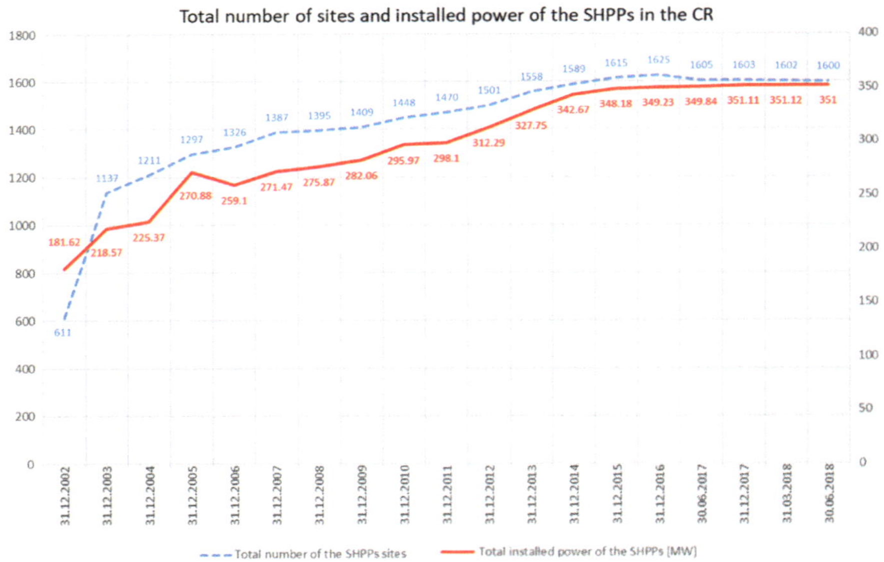

Fig. 15.1 Time progression of the total number of sites and installed power of the SHPPs in the Czech Republic [1]

eral reconstructions of the structures and technological parts of the historical power plants, mostly from the first half of the twentieth century. From construction layout point of view, the SHPPs realized in this period can be divided into three main groups:

- weir SHPP,
- diversion SHPP,
- dam SHPP.

SHPPs in the Czech Republic are usually designed for continuous operation (run off the river power stations). Exceptions are the dam SHPP that uses the influenced discharges from the reservoirs. SHPPs are commonly designed for unattended automatic operation with occasional monitoring and the possibility of automatic flow control by turbines using water-level operation control. The design of the SHPP is based on the results of hydro-energetic calculations, which make it possible to determine the installed capacity of the SHPP, the average annual electricity generation, and serve as a basis for determining the type and number of sets. When designing a powerhouse of the power station (i.e. substructure and superstructure), the emphasis is placed on optimizing dimensions in terms of space options, investment costs, but ensuring good operating conditions. The intake section of the SHPP is designed with the emphasis on minimizing hydraulic losses and in the case of weir power plants, to ensuring a favourable distribution of flow velocities in front of turbine intakes. Other principles in the design of the intake structure are related to the limitation of clogging by sediments and floating objects. The elevation arrangement of the intake and outlet sections is made taking into account the anticipated bed level of the upstream and downstream apron. An important criterion is also the position of the minimum operating water levels concerning the limitation of the intake swirls and the required immersion of the upper edge of the outlet from the draft tube. The height position of the powerhouse is chosen to minimizing the costs of foundation (substructure) and the necessary protection against flooding during design flood discharges (superstructure). From the operating point of view, the design of the SHPP mainly concerns the assembly, dismantling and maintenance of the technological equipment, installing intake and draft tube stop logs, connecting of individual parts of the MVE (staircases, ladders, etc.), trash rack cleaning, service roads, etc. At the design stage, the process of execution is also considered and, last but not least, the power output from the power plant and the connection to the existing power grid. The solution has to take into account the possibilities of placement of fish pass and include the necessary discharge into the hydro-energetic calculations. The main principles of the engineering design of power plants are included in technical standards ČSN 75 2601 (Small hydropower plants—Basic requirements) [2] and ČSN EN 61116 (Electromechanical equipment guide for small hydroelectric installations) [3] in the Czech Republic.

15.2 Main Principles of the Construction and Technical Solution of the SHPP

15.2.1 Input Data for the Design of the SHPP

The range of potentially useful data is broad and includes, in particular:

- Exploration of interest site, which serves mainly for the acquisition of photo documentation, video recording, purpose measurements, completion of drawing documentation and other documents.
- Hydrological data according to the technical standard ČSN 75 1400 (Hydrological data on surface water) [4] in the form of m-day and N-year discharges. In special cases, this may be, for example, a chronological time series of discharges to capture more complex flow rates variations (e.g. influencing the operation of a peak-load water power station, etc.).
- Geodetic data that includes geodetic measurement of interest site, maps, etc.
- Drawing documentation of existing building objects. For this type of documents, it is necessary to compare the content of the documentation with the actual status. In the case of paper documentation, it is necessary to consider its subsequent digitization.
- Handling regulations for hydraulic structures that provide the necessary data on the operation, hydrological data and basic drawing documentation.
- The documentation of the manufacturers of the technological equipment necessary for the design of the type and number of the sets, the determination of the installed power, the efficiency of the equipment, etc.

- Geological data used to design the building part of the power plant and the realization of structures, which can significantly affect the total investment costs for construction (especially in the case of cofferdams, complicated foundation conditions, etc.).
- Data for designing power output from a power plant that includes information on power lines, options for connection to a high- or low-voltage grid, the need to implement a transformer station, etc. Connecting a power plant in more remote areas can be a significant part of total investment costs.

The scope of the necessary documents is, of course, proportional to the level of documentation, the time and financial possibilities of the investor.

15.2.2 Suitable Types of Turbines

In general, it is possible to declare that in today's times, the use of horizontal axial Kaplan turbines is proposed for the SHPP. The main reason is the overall profitability of the realization of the power plants and the possibility of lowering the investment costs of works with low values of heads and relatively large discharges. Their hydraulic profile is generally axially symmetrical with the axial distributor. The most common types of construction of axial Kaplan turbines are as follows:

- PIT turbine,
- Belt-driven or bevel gear bulb turbine,
- S-type turbine (this type of Kaplan turbine often occurs in a modification with a fixed vanes distributor labelled as semi-Kaplan).

In the case of dam and diversion power plants, we can mainly meet axial Kaplan and Francis turbines, depending on design discharges and heads.

15.2.3 Powerhouse of the SHPP

In the case of new structures, the substructure of the SHPP's powerhouse is usually box-shaped, made of reinforced concrete and placed below the level of the surrounding terrain. The dimensions of the substructure must allow the proposed number of sets with accessories to be accommodated. The internal configuration of the SHPP's powerhouse is maximally adapted to the used technological equipment. In the substructure, there are usually other technological equipment, including oil separators, leakage water pumping set, lubricating oil pumping set, control system, etc. The location of turbine generators, switchboards of the power plant and other electrical equipment is possible, according to the chosen concept, both in the substructure and the superstructure of the powerhouse. The superstructure of the powerhouse is merged directly to the substructure. The design is dependent on how the powerhouse is protected against flooding during design floods. The typical design is a monolithic reinforced concrete or masonry structure. The superstructure serves primarily as a supporting structure for a crane track of a bridge crane or other types of lifting equipment. It also protects the SHPP's technological equipment from weather conditions. In justified cases, the powerhouse can be designed without the superstructure, with the assumption that it will be overflowed during the flood. Disassembly of the technological equipment from the powerhouse is done either by a mounting hole in the roof of the structure using a crane or using a door. All entries into the powerhouse and all openings (ventilation, mounting hatches, etc.) must be designed to provide protection against flooding.

15.2.4 Intake and Outlet Structures of the SHPP

The main task of the intake structure at the weir and diversion SHPPs is to let water from the upstream apron to the intakes on the turbines. For the weir type of power plants, these are intake structures in which water flows predominantly with the free surface. From the operational point of view, these objects belong to the most critical parts of the hydraulic circuit of the SHPP. The intake structures should meet a number of requirements that are relevant to design. This is mainly about:

- Ensure that the required amount of water is taken under all operating conditions and throughout the lifetime of the work, with minimal operator, energy and complex automation features.
- Ensure that sediments and mechanical impurities do not enter the intake structure where they could endanger the operation of the power plant.
- Reduce the occurrence of intake swirls.
- Ability to prevent clogging and destruction of trash rack, during floods, ice run and in normal operation.
- Reduce the freezing of all exposed ones, i.e. trash racks incl. cleaning machine, stop log grooves, etc.
- Minimize hydraulic losses and ensure a favourable distribution of flow velocities in front of turbine intakes.

Complying with all the above recommendations can be very problematic. For this reason, the resulting design of the intake structure is usually an appropriate compromise.

Correct hydrotechnical design of intake structure, respecting these conditions, is therefore very difficult and, for each particular case, more or less unique. The success of the design of the inflow objects depends, in addition to the amount of theoretical knowledge required, on the experience and the hydraulic and design sensitivity of the designers. An important tool in this context is numerical or physical modelling [5, 6] (see Fig. 15.2). This fact is evident especially in the case of power plant construction at existing barrages where it is necessary to respect existing building structures and not to limit their functionality.

Fig. 15.2 Physical model of the weir SHPP

The basic dimensions of the intake structure of the weir and diversion power plant influence the recommended maximum flow velocities at the intake sill (approx. 0.5–0.6 m/s) and on machine-cleaned trash racks (approx. 0.8–1.0 m/s). The above speeds must be assessed for the worst operating condition. The ground plan of the intake structure must be coordinated with the design of the elevation arrangement. Here is the determining elevation of the intake sill at least 0.5 m above the upstream apron bed and the bottom of the intake to the turbines. This intake bed transition is usually designed in the shape of a warped ruled surface. The height design of the inflow object is further related to the assessment of the intake swirls. On the rivers with significant sedimentation movements, a second intake sill is usually proposed before turbine intakes, about 0.3–0.5 m high. In these cases, it is necessary to design an effective rinsing of settled material.

15.3 Examples of Realized SHPP in CR

This section presents selected SHPPs realized in the Czech Republic for about 16 years, which are interesting both from a technical point of view and from the architectural design point of view. These are the following SHPPs:

- SHPP Litoměřice,
- SHPP Mířejovice,
- SHPP Rozkoš,
- SHPP Troja,
- SHPP Železný Brod.

The location of the power plants is shown on the map of the Czech Republic in Fig. 15.3. The following text provides a more detailed construction and technical description of each site.

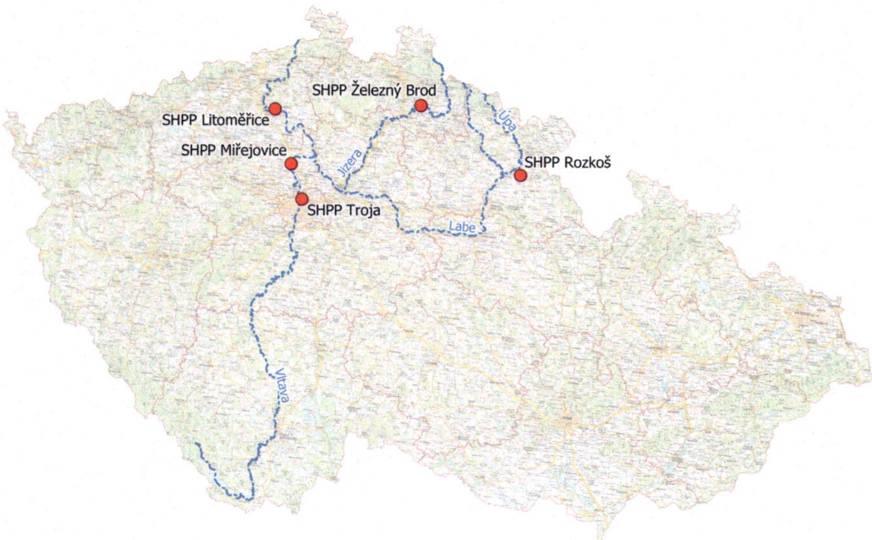

Fig. 15.3 Location of the selected SHPPs on the map of the Czech Republic

15.3.1 SHPP Litoměřice

The realization of the SHPP Litoměřice is the result of many years of efforts for the energetic exploitation of the České Kopisty barrage, which is located at the Labe River near the village of České Kopisty and the town of Litoměřce (Table 15.1). The SHPP is designed beside weir and is situated on the right bank of the Labe River next to the České Kopisty weir at river 795.688 km (see Figs. 15.4, 15.5).

The original České Kopisty needle weir was built in 1914 in the framework of river training on the Labe and Vltava rivers. Between 1971 and 1976, it was replaced by a new drum gate weir with three sluices, located at the foot of an original weir. Part of the barrage is also a small and large navigation lock.

SHPP Litoměřice was put into trial operation at the end of 2012 and finished in June 2013. The completed work was soon verified by the unplanned and unexpected load test that was the flood event in June 2013 (see Fig. 15.9). The construction in this test has suffered without much damage.

Table 15.1 Basic technical parameters of the SHPP Litoměřice

Location	At river 795.688 km of the Labe river
Type of turbines	2 × axial Kaplan PIT turbines
Head	1–2.95 m
Discharge	2 × 150 m^3/s
Power output	6.46 MW

An important factor influencing the process of construction works on the SHPP Litoměřice was the discovery of the artesian water during the execution of the cut-off walls. The digging of the foundation pit had to be interrupted, and the pit stability had to be strengthened through the grouting of rock mass and relief wells. As a part of the measures against the Artesian water, 35 relief wells plus 442 injection wells were carried out, and the total length of the anchoring piles reinforcing the subsoil reached almost 33.4 km.

The main parts of SHPP are:

- intake structure,
- powerhouse,
- outlet structure,
- fish pass.

The construction of the SHPP has a total length of more than 300 m and a width of about 70 m. The powerhouse itself has 55 × 30.2 m ground plan; the footing bottom is positioned vertically in the deepest parts almost 18 m below the surrounding terrain (Figs. 15.4, 15.5 and 15.6).

The intake structure was built on the right bank of the Labe River over the existing České Kopisty weir and serves to supply water from the upstream apron to the turbine intakes. The structure consists of intake sill, quay wall, bottom, weir-separating pier, intake guide piers (see Figs. 15.7 and 15.8).

The intake sill has a length of about 88 m and is height elevated 0.5 m above the upstream apron bed. The bottom of the intake structure consists of a reinforced concrete slab in the shape of a warped ruled surface and ensures a smooth transition from the level of the intake sill to the bottom level in front of the turbine intakes. The quay wall has a suitable hydraulic shape in the plan view, composed of adjacent circular arcs. The weir-separating pier between the weir and the intake structure has a width of about 21 m and is designed with a streamline shape. Two reinforced

Fig. 15.4 Powerhouse of the SHPP Litoměřice from the outlet side

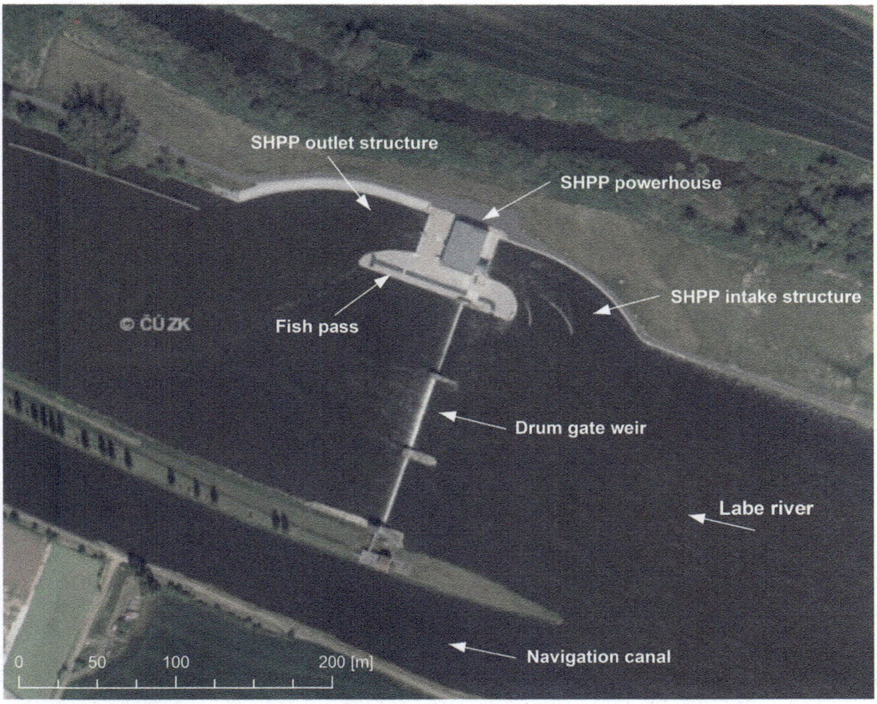

Fig. 15.5 Situation of the SHPP Litoměřice (based on data [10])

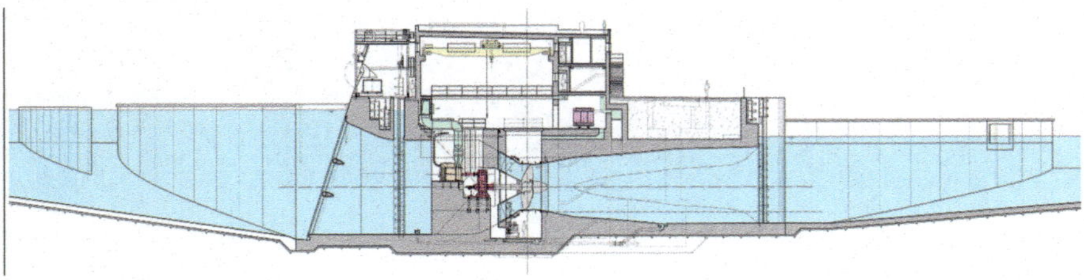

Fig. 15.6 Cross section through the SHPP Litoměřice (*source* Aquatis, a.s.)

concrete intake guide piers of hydraulically appropriate shape were realized to direct the flow in the area of the inflow to the intake structure (Fig. 15.8).

The substructure of the SHPP's powerhouse has a ground plan dimension of 30.2 × 55 m. In the substructure of the SHPP, there are two axial PIT Kaplan turbines with runner diameter 5.10 m. Turbines actuate horizontal synchronous generators through the gearbox. Turbines are designed with basic parameters listed in Table 15.2.

The synchronous generators have horizontal construction and are located in the PITs. Technical parameters of generator are listed in Table 15.3.

The superstructure of the powerhouse is an industrial building with a flat roof (see Fig. 15.4). The machine room is a 14.8 × 29 m high hall, equipped with a bridge

Fig. 15.7 Intake part of the SHPP Litoměřice under construction

Fig. 15.8 Intake part of the SHPP Litoměřice under construction (left) and after completion (right)

Table 15.2 Basic technical parameters of turbines (parameters are given for one turbine)

Number of turbines	2
Runner diameter	5.10 m
Number of runner blades	3
Range of heads	1.00–2.95 m
Range of discharges	45–150 m^3/s
Rated speed	54–60 rpm
Maximum shaft power	3500 kW

Fig. 15.9 SHPP Litoměřice during the flood in 2013 (*source* Povodí Labe, s.p.)

Table 15.3 Basic technical parameters of generators (parameters are given for one generator)

Number of generators	2
Installed power	3230 kW
Voltage	6.3 kV
Rated speed	750 rpm

crane track and a removable 8 × 4 m roof hatch. The superstructure of powerhouse also includes waterworks, dispatching, archive, repository and sanitary facilities. The entrance to the powerhouse is located above the design flood water level and accessible via an external staircase. The second entrance is possible with watertight doors from the terrain level. The architectural design is adapted to the location of the power plant in the flood plain. The outlet structure is hydraulically and structurally similar to the intake structure (see Fig. 15.4). A separate structure, which was a prerequisite for the realization of the SHPP, is a fish pass (see Fig. 15.5). This structure located in the weir pier allows migration of salmonid and carp fish, incl. eel. The fish pass width is 3.0 m, and length is approx. 90 m.

15.3.2 SHPP Mířejovice

The Mířejovice hydraulic structures were built in the early twentieth century. The original needle and sluice gate weir was approved in 1905. In 1921, a partial reconstruction of the weir was permitted to replace the needle weir with rolling gate weir. In 1922, the construction of the SHPP was started and into operation was put in 1928. Based on the 1927 permit, the sluice gate part of the weir was then replaced by "Stoney" gates (Table 15.4).

The hydraulic structures include a gated weir, SHPP (Fig. 15.10), two navigation locks (small and large) and a slalom canal that is used for sports purposes. The structures of SHPP are located on the left bank of the Vltava river, next to the existing large navigation lock (Fig. 15.11).

The current owner and operator of the SHPP Miřejovice have decided to undertake a complex reconstruction of the existing technological equipment. It was implemented in the years 2009–2012 due to the state of the technological equipment that was more than 70 years old and morally obsolete. SHPP Miřejovice consists of the following structures (see Fig. 15.11):

- supply canal including intake structure,
- powerhouse including intake and outlet structure.

Table 15.4 Basic technical parameters of the SHPP Miřejovice

Location	At river 18.0 km of the Vltava river
Type of turbines	4 × vertical Kaplan + 1 × vertical Francis turbines
Head	1.60–4.07 m
Discharge	4 × 36 m^3/s + 1 × 30 m^3/s
Power output	4 × 1.3 MW + 1 × 0.7 MW

Fig. 15.10 Powerhouse of the SHPP Miřejovice from the intake side

Fig. 15.11 Situation of SHPP Miřejovice (based on data [10])

Supply canal of the SHPP is constructed as an open canal with a total length of 750 m, a bottom width of 23 m and slopes 1:1.5. The intake part of the supply canal has a funnel shape and is protected by the coarse screen. Supply canal is connected to the intake structure of the SHPP (see Figs. 15.12 and 15.14), which includes:

- trash racks with automatic cleaning machine,
- stop logs,
- sluice gates.

The water intake to the turbines is provided by the concrete spirals which are located in the substructure of the powerhouse (see Figs. 15.12 and 15.13). Five vertical Francis turbines were originally installed in the SHPP. The SHPP includes five vertical sets—four newly installed Kaplan turbines (labelled as TG1-TG4) and the original Francis turbine (labelled as TG5) after general repair (Fig. 15.13).

The reconstruction of the SHPP's technological equipment, implemented in 2009–2012, involved in the first stage the reconstruction of the high-voltage switch room, the reconstruction of the steel platform above the intake part of the SHPP,

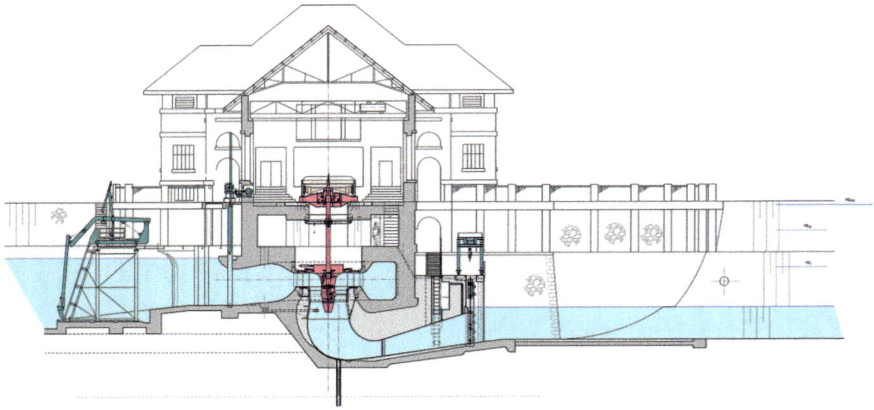

Fig. 15.12 Cross section through the SHPP Miřejovice (*source* Aquatis, a.s.)

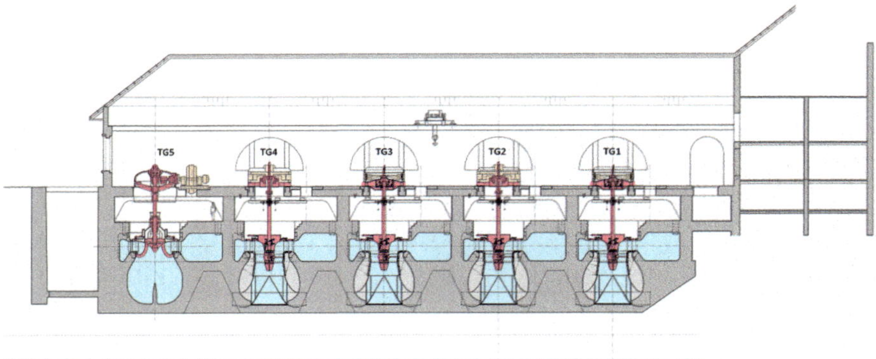

Fig. 15.13 Section through the SHPP Miřejovice (*source* Aquatis, a.s.)

including the replacement of the trash rack and the installation of a new automatic cleaning machine (see Fig. 15.14).

The next stage of the reconstruction was the overhaul of one of the five original sets (TG5). Upon the agreement of the power plant operator with the National Monument Institute, it was decided that one of five sets (TG5) with Francis turbine will be maintained during the reconstruction and will continue to operate in the original arrangement (see Figs. 15.15 and 15.16; Tables 15.7 and 15.8).

The reconstruction of the remaining four sets was realized between 2010 and 2012 and included the replacement of turbines TG1 to TG4 including generators and related building modifications.

Five vertical sets with Francis turbines with a 3.0 m runner diameter were originally installed in the SHPP (see Fig. 15.16), with a maximum discharge of 30 m^3/s (i.e. the total discharge of five sets was 150 m^3/s). The total installed power of the SHPP was 5×0.7 MW = 3.5 MW. The torque transmission from the vertical turbine

Fig. 15.14 Steel platform above the intake part of the SHPP with a new automatic cleaning machine

Fig. 15.15 Historical set with vertical Francis turbine in the generator hall of SHPP Miřejovice (bevel gear, generator and regulator)

Fig. 15.16 Historical photograph of a Francis turbine runner for a SHPP Mířejovice

Fig. 15.17 New sets with vertical Kaplan turbine in the generator hall

shaft to the horizontal shaft of the generator was secured to the turbines by bevel gear fitted with generators and regulators (see Fig. 15.15) on the generator floor in the generator hall.

When selecting new power plant equipment, the effort was made to minimize the building modifications of the substructure of the powerhouse. This corresponded to the selection of new sets TG1-TG4 consisting of a vertical Kaplan turbine with a runner diameter of 2.85 m and a maximum discharge of 36 m^3/s. The total discharge of the sets TG1-TG5 is 174 m^3/s after the reconstruction. The installed power of one new set has increased to 1.3 MW, i.e. the total installed power of the SHPP reaches around 5.9 MW after the reconstruction. Kaplan turbines are directly connected to newly installed vertical synchronous generators located on the generator hall (see Fig. 15.17).

The original concrete spirals of turbines were not changed in any way in the fitting of new turbines with Kaplan turbines (see Figs. 15.12 and 15.13). On the floor of the generators were made mainly the construction adjustments necessary for the installation of new vertical generators. The largest range of construction work from the entire reconstruction of the SHPP was the modification of the draft tubes of turbines and installing new draft tube stop logs (see Fig. 15.18). The superstructure of the powerhouse has internal dimensions of approximately 11.8 $\times$ 60 m and is covered by a saddle roof. A bridge crane is installed in the top superstructure above the floor of the generators. Electrical equipment of SHPP is located next to the generator room on the left side. Technical parameters of TG1-TG5 sets are listed in Tables 15.5, 15.6, 15.7 and 15.8.

Fig. 15.18 Construction works on the outlet part of the SHPP including modification of the draft tubes of turbines and installing new draft tube stop logs

Table 15.5 Basic technical parameters of turbines TG1-TG4 (parameters are given for one turbine)

Number of turbines	4
Type of turbine	Vertical Kaplan, directly connected to the generator
Runner diameter	2.85 m
Number of runner blades	4
Range of heads	1.60–4.07 m
Maximal discharge	36 m^3/s
Rated speed	107.14 rpm
Maximum shaft power	1320 kW

Table 15.6 Basic technical parameters of generators TG1-TG4 (parameters are given for one generator)

Number of generators	4
Nominal output	1333 kVA
Rated speed	107.14 rpm

Table 15.7 Basic technical parameters of historical turbine TG5

Year of manufacture	1928
Type of turbine	Vertical Francis
Runner diameter	3.0 m
Range of heads	1.1–4.1 m
Maximal discharge	30 m^3/s
Rated speed	49.98 rpm
Maximum shaft power	700 kW

Table 15.8 Basic technical parameters of historical generator TG5

Year of manufacture	1928
Nominal output	850 kVA
Rated speed	300 rpm

15.3.3 SHPP Rozkoš

Rozkoš reservoir was built in the 80s of the twentieth century as a side reservoir with a water supply from the river Úpa and two smaller streams (Šonov and Rozkoš stream). The purpose of the Rozkoš reservoir is to influence the discharge in the Labe River, to provide water supply for irrigation, to transform the flood discharges in the Úpa River, to provide recreation and fish breeding. The dam is constructed with earth core. The length of the dam in the crest is 412 m, and the maximum height above the base of excavation is 26.4 m. The dam structure includes bottom outlets which consist of a tower structure at the foot of the upstream slope, two pipelines with the diameter 1.4 m underneath the dam and the outlet structure at the foot of the downstream slope (see Figs. 15.19 and 15.20) (Table 15.9).

Table 15.9 Basic technical parameters of SHPP Rozkoš

Location	Rozkoš reservoir
Type of turbines	1× axial Kaplan S-turbine
Range of heads	8.0–14.0 m
Discharge	5.0 m^3/s
Power output	0.6 MW

SHPP Rozkoš is situated next to the left stilling basin wall of the outlet structure. The SHPP uses the power potential of the given site, i.e. it exploits the influenced discharges from the bottom outlet valves and the head given by the difference in water levels in the reservoir and under the bottom outlets. The SHPP Rozkoš was built in the period from March 2007 to May 2008. The main construction parts of SHPP are:

- penstock,
- powerhouse,
- outlet structure.

Penstock with a diameter DN 1400/1600 serves to supply water to the turbine and is additionally connected to the original left bottom outlet pipe DN 1400. At the intake to the penstock, the horizontal trash rack is installed. The penstock has a total length of about 16 m. There is installed one axial Kaplan S-turbine with a direct connection to a synchronous generator in the powerhouse. The SHPP powerhouse (see Figs. 15.19, 15.21, 15.22) is most of the parts designed as an underground building made of reinforced concrete and adjacent to the left stilling basin wall of the outlet structure. The width of the structure is about 8.0 m, and the length is about 12.8 m. On the lower floor, there is a penstock with a butterfly valve and an

Fig. 15.19 Powerhouse of the SHPP Rozkoš from the outlet side (*source* Hydrohrom, s.r.o.)

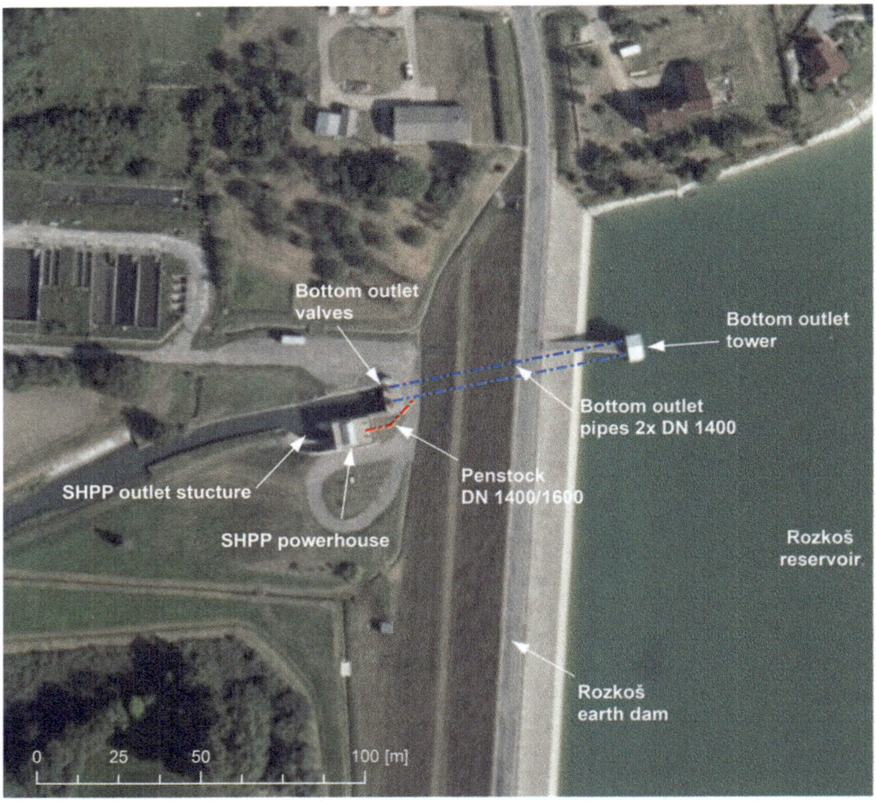

Fig. 15.20 Situation of SHPP Rozkoš (based on data [10])

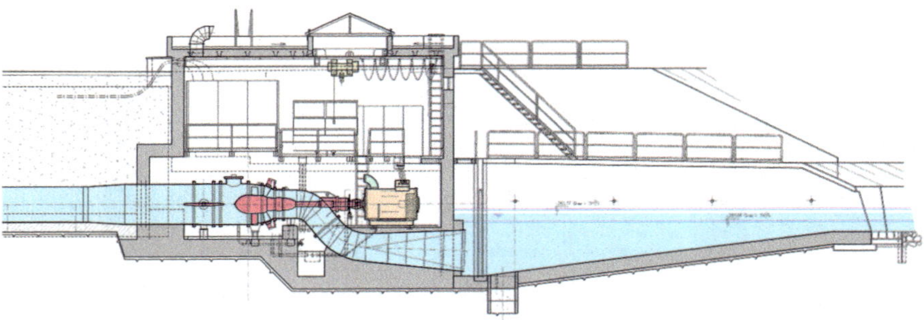

Fig. 15.21 Cross section of SHPP Rozkoš (*source* Aquatis, a.s.)

axial Kaplan S-turbine with a direct connection to the generator (see Fig. 15.23 and Table 15.10). On the upper floor, there is an entrance and all electrical equipment components. The engine room with minimized mounting space is equipped with a single girder crane.

The outlet is a reinforced concrete structure with a bottom in the shape of a warped ruled surface. Behind the outlet from the draft tube are located draft tube stop logs.

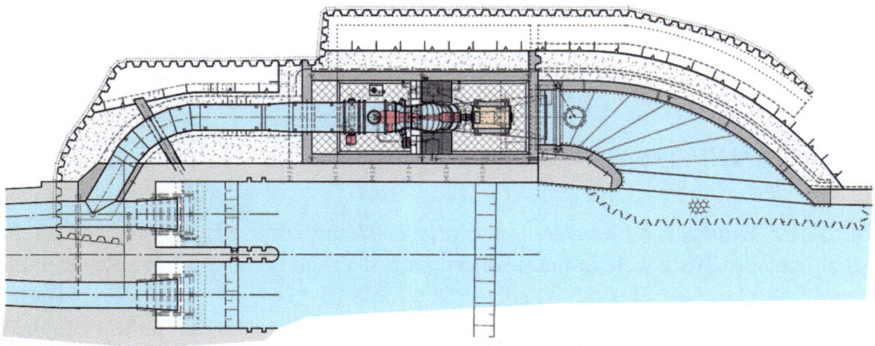

Fig. 15.22 Horizontal section through the SHPP Rozkoš (*source* Aquatis, a.s.)

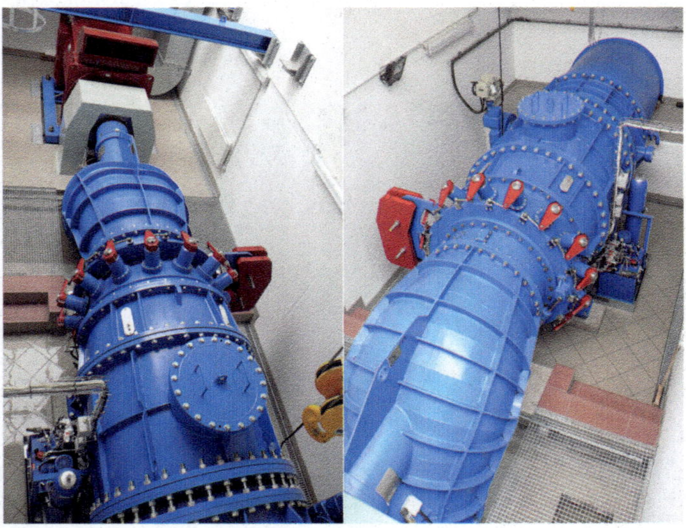

Fig. 15.23 Horizontal axial S-type Kaplan turbine in the SHPP Rozkoš (*source* Hydrohrom, s.r.o.)

Table 15.10 Basic technical parameters of turbine and generator of SHPP Rozkoš

Number of turbines	1
Type of turbine	Axial S-type Kaplan turbine, directly connected to generator
Runner diameter	1.00 m
Range of head	8.0–14.0 m
Maximal discharge	5 m^3/s
Rated speed	500 rpm
Voltage	420 V
Installed power	568 kW

15.3.4 SHPP Troja

The SHPP Troja was built between the year 2007 and 2009. SHPP is located on the left side of the Troja weir in the northern part of Prague on the Vltava River in the river 45.580 km (see Figs. 15.24 and 15.25; Table 15.11).

The purpose of the weir, which is part of the Troja–Podbaba hydraulic structures, is mainly the maintenance of water level for the navigation locks in Prague–Podbaba and the operation of SHPP Troja. The construction of the Troja–Podbaba hydraulic structures (i.e. weir, navigation canal and navigation lock) dates back to the years 1899–1902.

Fig. 15.24 View of SHPP Troja (*source* Povodí Vltavy, s.p.)

Table 15.11 Basic technical parameters of SHPP Troja

Location	At river 45.580 km of the Vltava River
Type of turbines	2 × axial bevel gear Kaplan turbines
Range of head	1.50–3.60 m
Discharge	2 × 40 m^3/s
Power output	2 × 1030 kW

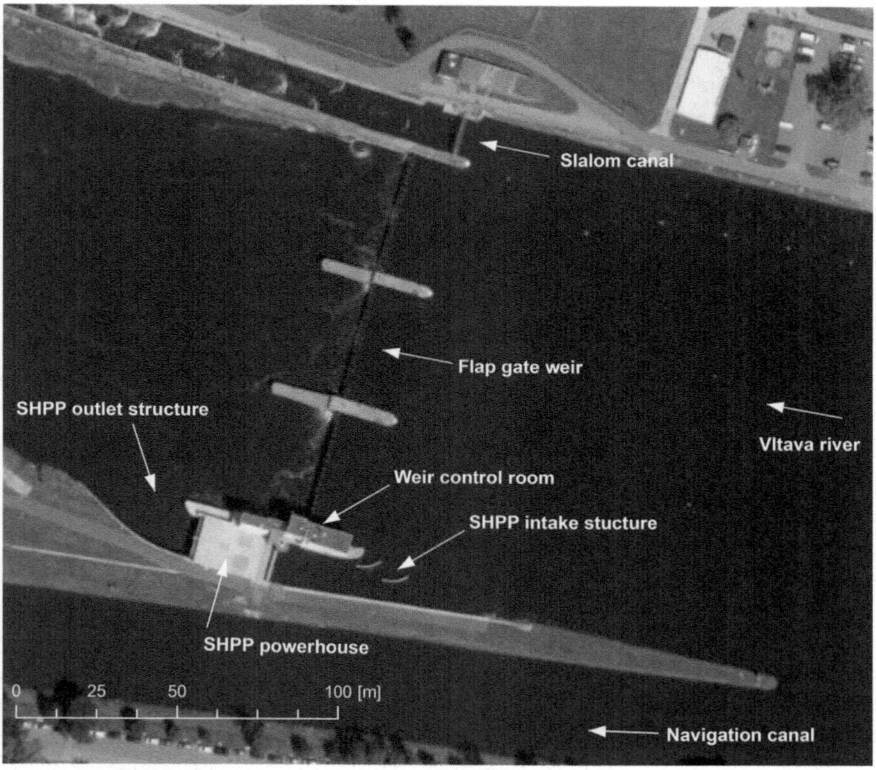

Fig. 15.25 Situation of SHPP Troja (based on data [10])

The Troja weir (after the reconstruction completed in 1978) has three gates of the same width. The bottom structure is in a shape of "Jambora" weir. There are flap gates with a height of 3.30 m mounted on weir bottom structure in each gate. To the left of SHPP Troja, about 150 m above the weir, there is an entrance to a 3 km long Troja–Podbaba upstream navigation canal. SHPP is located in a very cramped area on the left bank of the Vltava River between the Troja weir-separating pier and the Troja–Podbaba navigation canal.

The construction of the SHPP has a length of more than 115 m and a width of about 18 m (see Figs. 15.25, 15.26 and 15.27). The intake structure serves to supply water from the upstream apron to the turbine intakes. The structure consists of intake sill, quay wall, bottom, weir pier, intake guide piers.

The intake sill is height elevated 1.0 m above the upstream apron bed. The bottom of the intake structure consists of a reinforced concrete slab in the shape of a warped ruled surface and ensures a smooth transition from the level of the intake sill to the bottom level in front of the turbine intakes. The quay wall has a shape adapted to the local confined conditions in the ground plan and respects the direction of the navigation canal. The weir-separating pier between the weir and the intake structure is designed with a streamline shape. Two reinforced concrete intake guide piers of hydraulically appropriate shape were realized to direct the flow in the area of the inflow to the intake structure. The substructure of SHPP powerhouse is a waterproof reinforced concrete structure in shape of the box. There are installed two axial bevel gear Kaplan turbines. The basic technical parameters of the set are as shown in Tables 15.12 and 15.13.

The substructure of the SHPP powerhouse also includes all the necessary equipment (bridge crane, high-voltage and low-voltage switch rooms, transformer room, hydraulic circuit for pumping drainage and leakage water and access staircase, etc.). An interesting feature of the design of the SHPP powerhouse is that it is designed to

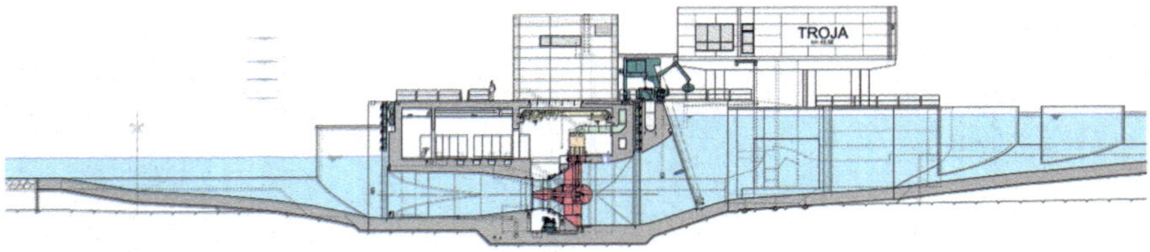

Fig. 15.26 Cross section through the SHPP Troja (*source* Aquatis, a.s.)

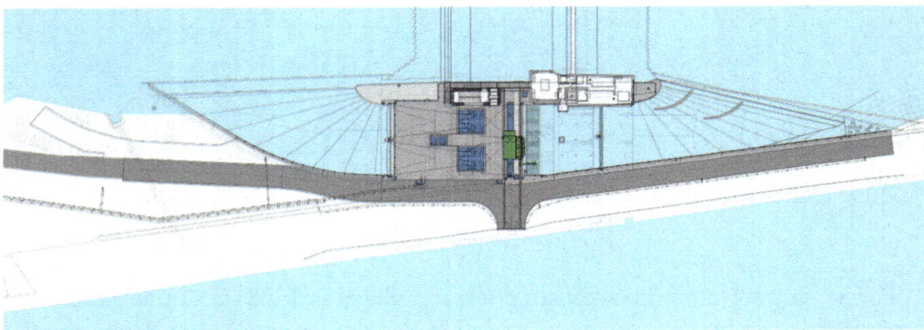

Fig. 15.27 Horizontal section through the SHPP Troja (*source* Aquatis, a.s.)

Table 15.12 Basic technical parameters of turbines (parameters are given for one turbine)

Number of turbines	2
Type of turbine	Axial bevel gear Kaplan turbine
Runner diameter	2.6 m
Range of head	1.50–3.60 m
Maximal discharge	40 m^3/s
Maximum shaft power	1030 kW

Table 15.13 Basic technical parameters of generators (parameters are given for one generator)

Number of generators	2
Generator type	Synchronous, vertical
Installed power	1030 kW
Voltage	6.3 kV
Rated speed	600 rpm

be overflowed during flood events. The reinforced concrete roof of the substructure is designed to withstand a water level of 4 m above its upper edge. This represents the discharge of over 100-years flood. The roof includes four steel watertight mounting hatches.

The superstructure of the SHPP powerhouse is composed only of a reinforced concrete watertight hollow pier equipped with entrance staircase and the ventilation ducts. This structure is constructed to the same height as the adjacent existing control room of the weir. The entrance to the SHPP is located above the water level of 100 years of flood discharge. The outlet structure is hydraulically and structurally similar to the intake structure.

15.3.5 SHPP Železný Brod

SHPP Železný Brod is based on the open-flow diversion scheme and exploits the power potential of the river Jizera in the river 97.480 km. The aim of the construction of the new SHPP was to preserve the original layout and to restore to the maximum possible extent the structures of the historic SHPP built in this area in the second half of the nineteenth-century (see Fig. 15.28) and in the 60s of the last-century shutdown. At this time, the dismantling of all the plant's technological equipment and the gradual filling of the entire length of the diversion canal were also followed. The structures of the historic SHPP included a sluice gate weir with intake structure, diversion canal, powerhouse and outlet canal. Historic powerhouse contained four identical horizontal Francis twin turbines. The reconstruction of the historical SHPP was carried out between 2009 and 2010, approximately 45 years after the shutdown.

The goal of the reconstruction was the renovation of the structures of the historic hydroelectric power plant in the original layout, but with the new powerhouse in a different location. The original powerhouse of the historic SHPP was demolished. Changing the powerhouse location involved prolonging the diversion canal and shortening the outlet canal. The main structures of SHPP are (see Fig. 15.29):

- sluice gate weir with fish pass,
- intake structure,
- diversion canal,
- powerhouse,
- outlet canal.

Fig. 15.28 Historical view on SHPP Železný Brod (outlet side) from 1867 (*source* Vodní elektrárna Železný Brod, a.s.)

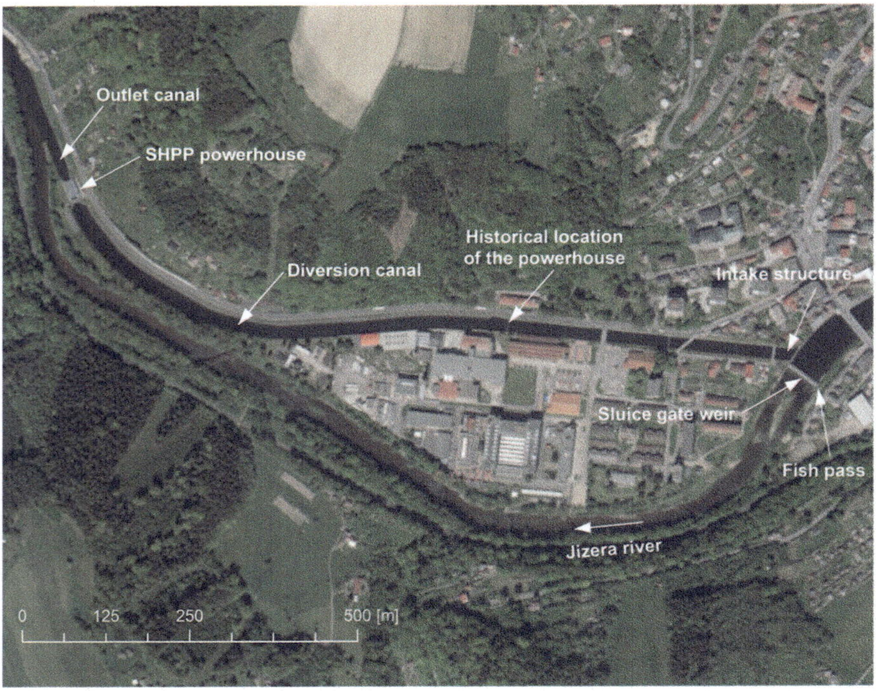

Fig. 15.29 Situation of SHPP Železný Brod (based on data [10])

The original sluice gate weir was completely renovated, and a new reinforced concrete construction of the fish pass was added to the left weir pier. Weir has three gates of width 14.7, 14.9 and 19.6 m. The reconstructed intake structure let the water into the diversion canal. The intake facility consists of ice breaker, reinforced concrete slab with a trash board supported on piers, an intake sill, a sand trap, a flushing canal with gate and intake gates. The original diversion canal of the historic power plant, extended from the original length of 480 m to 665 m, serves as the conduit for the reconstructed power plant. Reconstruction of the section of the original canal included mainly the extraction of the material through which the canal was practically covered and the subsequent rehabilitation of the original supporting walls and sides. The extended section of the canal partially utilizes the original supporting wall on the left bank, and in the section without supporting walls, the left bank has a slope of about 1:1.15. On the part of the canal are designed right-bank dikes.

The new SHPP powerhouse (see Figs. 15.30 and 15.31) is divided into three main operational parts:

- trash rack room (see Fig. 15.33),
- machine room (see Fig. 15.32) and service spaces for operation,
- switch and transformer rooms.

From a design point of view, we can divide the powerhouse into the substructure and superstructure. The substructure of the powerhouse is used to place a total of three axial bevel gear Kaplan turbines—one turbine with a diameter of 1.45 m and two turbines with a diameter of 1.1 m (see Fig. 15.32). The structure has a length of 30.5 m and a width of 12.9 m. The intake part includes an intake sill with a settler, inspection locks and trash racks with cleaning machines (see Fig. 15.33). On the outlets of the

Fig. 15.30 Powerhouse of the SHPP Železný Brod from the intake side

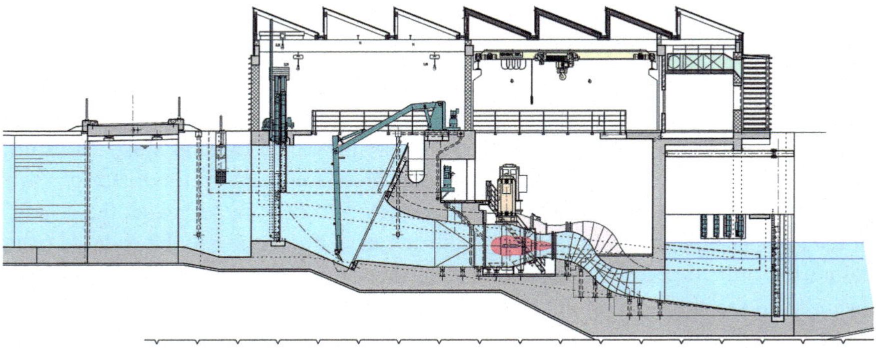

Fig. 15.31 Cross section through the SHPP Železný Brod (*source* Aquatis, a.s.)

Fig. 15.32 Machine room of the SHPP Železný Brod powerhouse (*source* Vodní elektrárna Železný Brod, a.s.)

draft tubes are proposed stop logs. The superstructure of the powerhouse is based on reinforced concrete substructure. The supporting system of the superstructure is made of reinforced concrete prefabricated columns in combination with supporting walls made of ceramic blocks. A bridge crane with a load capacity of 16 t is located in the superstructure. Great attention was paid to the architectural design of the powerhouse, which is situated in areas with a close connection to the complex of historical factory buildings and the natural environment of the Jizera watercourse. The outlet canal connects to the outlet part of the powerhouse and has a total length of 76 m. It let the water into the river Jizera. The basic technical parameters of the technological equipment are listed in Tables 15.14, 15.15, 15.16, 15.17 and 15.18.

Fig. 15.33 Trash rack room of the SHPP Železný Brod powerhouse (*source* Vodní elektrárna Železný Brod, a.s.)

Table 15.14 Basic technical parameters of SHPP Železný Brod

Location	Jizera, river 97.480 km
Type of turbines	3 × axial bevel gear Kaplan turbine with "S" draft tube
Head	5 m
Discharge	$2 \times 6.8\ m^3/s + 1 \times 10.8\ m^3/s$
Power output	2 × 245 kW + 1 × 496 kW

Table 15.15 Basic technical parameters of TG1 turbine

Number of turbines	1
Type of turbine	Axial bevel gear Kaplan turbine with "S" draft tube
Runner diameter	1.45 m
Head	5 m
Range of discharges	$1.9–10.8\ m^3/s$
Rated speed	247 rpm
Maximum shaft power	72–438 kW

Table 15.16 Basic technical parameters of TG1 generators

Number of generators	1
Generator type	Synchronous
Installed power	496 kW
Voltage	400 V
Rated speed	750 rpm

Table 15.17 Basic technical parameters of TG2 and TG3 turbines (parameters are given for one turbine)

Number of turbines	2
Type of turbine	Axial bevel gear Kaplan turbine with "S" draft tube
Runner diameter	1.15 m
Head	5 m
Range of discharges	1.2–6.8 m^3/s
Rated speed	326 rpm
Maximum shaft power	268 kW

Table 15.18 Basic technical parameters of TG2 and TG3 generators (parameters are given for one generator)

Number of generators	2
Generator type	Synchronous
Installed power	245 kW
Voltage	400 V
Rated speed	750 m

15.4 Conclusions

The paper provides an overview of the current use of hydropower potential in the Czech Republic in small hydropower plants. Between 2002 and 2016, we have seen a relatively significant increase in the number of small hydropower plants in the Czech Republic by 162% and installed power by 93%. The primary consideration in deciding on the realization of a SHPP is still the aspect of the economic efficiency of the investment invested. For this reason, most of the newly built hydroelectric power plants were implemented using existing weirs or dams. Besides the new buildings, a number of general reconstructions of the technological and construction part of the SHPPs were also carried out, aiming to increase the efficiency of these sources. The possibility of obtaining a financial subsidy was one of the factors that helped to improve the economic efficiency of the implementation of small hydroelectric power plants in that period. It is gratifying that a number of precisely technically and architectonically solved new constructions and reconstructions have been realized during the period, some of which are described in more detail in Sect. 15.3.

15.5 Recommendations

At present, most of the localities suitable for the economically efficient execution of the SHPPs are already exhausted in the Czech Republic. Most sites are now available offering under current conditions low economic efficiency for any investment in the SHPP. These are mainly the sections of streams with existing barrages with low heads of up to 2 m or locations requiring high investment in the construction of new barrages. Another group is the localities with complicated local conditions, for example, due to collision with the interests of nature conservation or other interested entities, complicated property rights relations, etc.

Further developments in the field of small hydropower plants can, therefore, be expected to be significantly slower than in the past 16 years. We expect the gradual redevelopment of new hydropower plants in previously unoccupied localities and the reconstruction of end-of-life historic hydroelectric power plants. Assessing the economic efficiency of a possible reconstruction of a power plant is based on a reliability assessment of its components [7–9].

Generally, in the Czech Republic, in the field of hydro-energy, the development of mainly pumping hydroelectric power plants in the future depends on the use of unstable electricity sources (wind and sun). The main obstacle to the implementation of pumped-water power plants is the fact that suitable locations are located in naturally valuable localities of the Czech Republic. It will be necessary to look for suitable technical solutions to minimize possible impacts on nature and landscape.

Acknowledgements This chapter was elaborated with the support of project FAST-J-18-5525 - Numerical modelling of overflowed bridge deck, which is a specific research project funded by Brno University of Technology. We would also like to thank those companies who provided the necessary data: Aquatis, a.s., Hydrohrom, s.r.o., Povodí Vltavy, s.p., Povodí Labe, s.p., Vodní elektrárna Železný Brod, a.s.).

References

1. Statistical data of the Czech Energy regulatory office. www.eru.cz/documents/10540/463106/VE_18_06.pdf. Accessed 27 Aug 2018
2. ČSN 75 2601 (2010) Small hydro-power plants—basic requirements. Czech national standard. Czech Office for Standards, Metrology and Testing, Prague
3. ČSN EN 61116 (1997) Electromechanical equipment guide for small hydroelectric installations. Czech national standard. Czech Office for Standards, Metrology and Testing, Prague
4. ČSN 75 1400 (2014) Hydrological data on surface water. Czech national standard. Czech Office for Standards, Metrology and Testing, Prague
5. Dráb A, Jandora J, Říha J, Neumayer O (2001) Numerical modelling of the intake parts of small hydropower plants. In: Hydropower in the new millennium, Proceedings of the 4th international conference on hydropower development. A.A. Balkema, Norway, pp 361–370
6. Sedláček J, Nowak P (2017) Kaplan turbines: the history and future prospects. In: Special Issue, 85th annual meeting of ICOLD, vol 24, no. 3/2017, pp 60–64
7. Špano M, Říha J, Dráb A, Bartoš J (2017) Reliability indicators for small hydro plants. Int J Hydropower Dams 24(3/2017):54–59
8. Špano M, Říha J, Dráb A (2016) Evaluation of the maintenance and operational dependability of small hydropower plant. Int J Sci Res Publ 6(12):484–489
9. Dráb A, Říha J, Špano M (2015) Failures and durability analysis of a small hydro power plant. In: ICOLD 25th congress/ICOLD 83rd annual meeting, symposium hydropower 15. ICOLD. Stavanger, NK: ICOLD, pp 138–144
10. Czech Office for Surveying, Mapping and Cadastre. ČÚZK Geoportal. geoportal.cuzk.cz. Accessed 27 Aug 2018

Part VII
Conclusions

Chapter 16
Updates, Conclusions and Recommendations for Management of Water Quality and Quantity

Martina Zelenakova, Petr Hlavínek and Abdelazim M. Negm

Abstract This chapter is devoted to the present and highlights the main conclusions and recommendations of the chapters presented in this book. This chapter contains information on the management of water quality and quantity mainly in the Czech Republic. It focuses on urban water management, pollution of water, modelling in water management. Also, it covers the subject of the water structures with a focus to flood protection, water supply and sewage systems and hydropower. In this chapter, a set of recommendations for future research is pointed out to direct future research towards the sustainability of water resources management which is one of the strategic topics of the Czech Republic.

Keywords Management · Quality · Quantity · Modelling · Water resources · Czech Republic · Pollution · Hydropower

16.1 Introduction

Management of water quality and quantity is an inevitable condition for sustainable development of water. Sustainable management of water resources is based on the principle that water as a natural resource should be utilized only to that extent which ensures future generations sufficient usable supplies. Fortunately, the Czech Republic

M. Zelenakova
Department of Environmental Engineering, Faculty of Civil Engineering, Technical University in Košice, Košice, Slovakia
e-mail: martina.zelenakova@tuke.sk

P. Hlavínek
Faculty of Civil Engineering, AdMaS Centre, Brno University of Technology, 612 00 Brno, Czech Republic
e-mail: hlavinek.p@fce.vutbr.cz

A. M. Negm (✉)
Water and Water Structures Engineering Department, Faculty of Engineering, Zagazig University, Zagazig 44519, Egypt
e-mail: Amnegm85@yahoo.com; Amnegm@zu.edu.eg

M. Zelenakova et al. (eds.), *Management of Water Quality and Quantity*, Springer Water, https://doi.org/10.1007/978-3-030-18359-2_16

is rich in water resources. Water is presented in the rivers, lakes, and reservoirs and that reserves contained in porous soil remain preserved in the same quantity and quality. It is evident that waters are vulnerable in terms of their hygienic quality and safety. For this reason, it is necessary to devote more attention to the protection and integrated management of water sources.

This book presents contributions on water issues from engineering point of view from experts in the field of water management from the Czech Republic. It is focused on stormwater management in urban areas, water quantity, hydraulics structures, hydrodynamic modelling and flood protection. The book brings state-of-the-art knowledge that can be effectively used for solving a variety of problems in integrated water resources management as well as the latest developments in the research area. The main goal of the book is put on water pollution, water management and water structures. The chapter presents a summary of the findings, updates, conclusions and recommendations of the studies on water quality and quantity.

16.2 Update

In the following, the national studies regarding the water resources in the Czech Republic concerned with climate change and extreme hydrological phenomena—-droughts and floods are presented. Also, some studies regarding the water management in buildings are mentioned. The brief results of the studies are introduced.

The assessment and management of surface water quality and quantity have a long tradition in the Czech Republic. Water quality was in the past it concentrates mainly on the evaluation of organic pollution using the saprobic system. Considering the modern trends of assessment of the ecological status of water bodies, a new assessment system named PERLA was developed by Kokeš et al. [1]. The system presents a complex of biological methods of ecological status assessment of rivers and connected activities in the Czech Republic. It involves 300 reference sites with respective biotic and abiotic data and a prediction model using a newly developed software HOBENT.

Although the management of surface water quality and quantity has a long tradition in the Czech Republic, the management of surface water bodies—fish ponds, water reservoirs and tanks—changed only little from the Middle Ages until the end of the nineteenth century. The intensification of fish production was initiated during the first half of the twentieth century [2]. This intensification includes liming and manuring of the ponds. A greater density of stocking led to the use of artificial feeds in the form of pellets and grain. These changes in pond management have led to an increase (up to ten times) in fish production in the last fifty years' decades. At the same time, the quality of the water and sediments has been deteriorating, and the functioning of the pond ecosystem has been disturbed. There are now massive blooms of phytoplankton, especially cyanobacteria, accompanied by great fluctuations in the oxygen concentration and pH. Fishponds exhibit high natural retention potential for phosphorus. Consequently, this phosphorus enters the ponds from non-point, diffuse and point sources, as well as from aquaculture management. Interesting results of phosphorus mass balance monitoring of nine large fishponds over 2010–2014 were presented in the study by Potužák et al. [3]. The results of long-term monitoring of sediments from more than 200 sites in the Czech Republic from 2011 to 2017 are introduced in the study by Baxa et al. [4]. Their study revealed that the total sediment volume of the Czech Republic, mainly from fishponds, is estimated to be 197 million m^3. Sediment may be used, e.g. for land application in case that contains pollutants are up to legislation limits given by Decree regulating the conditions for the application of sediments on agricultural land.

Anthropogenic activities are amongst the greatest factors producing global environmental pressures in terms of climatic and land use in the world and in the Czech Republic as well. Frélichová et al. [5] studied spatially explicit information at a national level on climate and land use change impacts in order to assess changes in the provision of ecosystem services (carbon sequestration, food production and soil erosion) in the agricultural sector of the country. This assessment shows that historical land use trends and land use under projected climate scenarios display some shared spatial patterns. Specifically, these factors both lead to a significant decrease of arable land in the border fringes of the Czech Republic, which is to some extent replaced by grasslands. An integrated approach to landscape functioning assessment which combines energy efficiency and hydrochemical balance studies are presented in the paper of Hesslerová et al. [6]. Six model sub-watersheds with different land use situated in southern Bohemia were chosen to show the influence of landscape management on landscape functioning. The river basins with higher human activity were shown higher surface temperature as well as higher electrical conductivity in the run-off surface water. Natural catchments with forests showed opposite trends. Their results proved that sustainable landscape management leads to lower temperature extremes and consequently to low matter and water losses. Presence of vegetation and water areas—water retention measures—green infrastructure in the countryside highly ensure its sustainability. Wetlands as one of these measures and exactly constructed wetlands with the horizontal sub-surface flow in the Czech Republic were studied by Vymazal [7]. These constructions are very effective in organics and suspended solids removal.

Water retention measures as part of rainwater management in urban areas are widely discussed in the edited book of Hlavinek and Zelenakova [8]. This book reviews the current state of the art in managing infrastructure, urban regions, industrial regions and inhabited areas with respect to flooding and water damage. It also gives a broad overview of the different manageable components that play a role in stormwater management. It includes chapters on planning infrastructure and regional development, modifications in urban regions. The authors review different hydrological conditions that make areas safer, or more prone to stormwater threats, and capture and store storm water. The sealed infrastructure and their role in stormwater hazards, including roads, public spaces, roofs and stormwater quality, are also discussed. An analysis of factors influencing the process of infiltration is carried out in [9], relevant uncertainties influencing the design of the storage volume of the infiltration facility are discussed in this paper.

The potential for wastewater reuse in the Czech Republic has discussed in Janosova et al. [10]. "On the basis of a country-wide analysis of the water management situation, different regions were identified with particular water stress problems" (https://www.sciencedirect.com/science/article/pii/S0011916405007241). Researchers from Institute of Municipal Water Management, Faculty of Civil Engineering, Brno University of Technology, AdMaS centre [11] describe the possibilities of drying sludge in order to reduce weight, improve the conditions for storage and management of sludge and, last but not least, pretreatment for the material transformation of waste in accordance with the principles of Circulation Economy in the EU called Circular Economy by means of: combustion, gasification or pyrolysis. The aim of their research in [12] is sludge management optimization by a mobile sludge dewatering press unit with optimal polymer combination for dewatering. The paper by Ševčík et al. [13] reviews microwave pyrolysis and presents a full-scale application of this technology to treat sewage sludge.

Experimental research concerning hydraulic conductivity for glass bead size and research of variable porosity was carried out in [14]. The validity of various published porosity functions and empirical formulae was verified with the use of the experimental data obtained from the glass beads. An interesting study is done in [15] where differences in terminology concerning soil deformation due to seepage are discussing.

Management of water quantity in the Czech Republic includes flood management. "Procedures of flood risk analysis have been developed in the Czech Republic since the catastrophic floods of 1997 in line with European and worldwide trends and have been tested and applied in hundreds of case studies to date" [16]. Dráb and Ríha [16] presented flood risk analysis procedures and techniques for the flood hazard and flood risk maps creation. Methods related to flood risk management plans are also mentioned in this paper. There are many local and regional studies in the Czech Republic related to flood protection and management, e.g. Blazkova and Beven [17] In this study, continuous simulation flood frequency predictions on the Skalka catchment in the Czech Republic are compared against summary information of rainfall characteristics, the flow duration curve and the frequency characteristics of flood discharges and snow water equivalent using the generalized likelihood uncertainty estimation limits of acceptability approach. In the paper [18], the approach to the dam safety assessment during floods is discussed together with problems faced especially in case of assessment of small dams.

16.3 Conclusions

Water plays a vital role in both the environment and the human life. Surface and groundwater resources of in the Czech Republic are fortunately rich enough to ensure current and prospective water needs although their effective management is inevitable. Assessment of the impact of climate variability on water resources is an essential activity because we consider water as a strategic raw material. The quantitative characteristics of renewable water resources of a region or river basin can be determined by two approaches: by using meteorological data or by using river run-off observations. In the next sections, some of the conclusions and recommendations of the chapters in this volume of the Earth and Environmental Sciences specified to water management are presented.

Chapter 2 "**Storm Water Management in Urban Areas**" presents designs corresponding to natural stormwater drainage. Stormwater management as a decentralized drainage system represents the point of natural processing of storm water and its return to in the natural water cycle. The acceptable solutions for stormwater management are evaporation, infiltration and regulated run-off to watercourses. These solutions can be implemented by the accumulation of storm water and its use for irrigation and flushing toilets. The reduced quality and yield of surface and groundwater due to droughts and changing climatic conditions lead to an increase of reusing storm water in the Czech Republic. Given the current continuing trend of reducing the cost of storm water collecting and treating technology, the stormwater management will be more often designed in practice.

Chapter 3 titled "**The Green Roofs and Facades as a Tools of Climate Cooling in the Urban Environment**" mentions many specific functions related to the improvement of the environment that green roof fulfil. It defines the basic function of the roof with vegetation cover, which is the protection of the construction and the indoor environment from meteoric water. It can be used for recreational purposes and at the same time fulfil the ecological function. The most important potential of green roofs though is their climate adaptation function. They help to moderate urban heat islands and measuring their retention capacity approved that even extensive green roof can significantly cut the peaks of run-off from extreme storm rains. By measuring on a particular green roof and calculations, it was verified that the vegetation reacts to solar radiation by evaporation of water, and thus by cooling of the surrounding environment, which is especially in the urban environment in mentioned summer months' great benefit.

Green roofs and infiltration facilities are tools for effective rainwater in urban areas. "**Infiltration of Rainwater in Urban Areas**" Chap. 4 presents the planning, design and performance of rainwater infiltration systems in urban areas. These facilities have to be part of integrated urban drainage-wastewater systems. As such the location of retention and infiltration facilities has to be planned and harmonized within the overall urban plan. Efficient arrangements may be installed at or close to individual estates such as private dwellings, blocks of flats, industrial, commercial and agricultural facilities. The effect on the technical drainage systems of rainwater such as sewerage and open channels may be considered if the retention and infiltration capacities are functionally planned, designed and performed. The main condition for the design of these facilities are professional hydrological, geotechnical and hydrogeological survey and analysis have to be performed before the design to determine soil profiles and infiltration coefficients. The modelling of groundwater flow is a very useful tool.

The modelling techniques of surface water are presented in Chap. 5 "**Stream Water Quality Modelling Techniques**". In this chapter, the principles of individual model types are mentioned. Their use is demonstrated in case studies carried out by the author during the last three decades in the territory of the Czech Republic. Particular problems results achieved, and questions arose at the modelling are briefly discussed together comments of interpretation of typical graphical and tabular outputs. The core issue for water quality modelling stream is comprehensive, complete, reliable and homogeneous data on catchment and channel characteristics, pollution sources and also water quality in streams obtained by monitoring. The reliability of the model may be significantly improved by its calibration and verification. When the stream water quality data are missing, sensitivity analysis should be provided. Generally, the last step of the modelling process should be feedback on the compliance between predicted values of water quality indicators by modelling and the real values obtained by monitoring after adopting proposed measures. For decision-makers, the results of the modelling should be properly summarized, depicted and interpreted.

Quality of water and specifically the topic of pharmaceuticals in the water cycle are discussed in Chap. 6 "**Pharmaceuticals in Urban Water Cycle**". The authors state the difficulty of the topic as the concentration of drugs in wastewaters is variable, depending on many factors such as seasonal period, urban sprawl, dilution of water spills due to heavy rain events, synergies with other pharmacological agents in wastewaters. Pharmaceuticals have diverse physicochemical nature, and a high degradation degree can be provided only on the basis of accurate analyses of the sewage. They solve the question of removal technology. Each removal technology has to be determined precisely to the specific medicinal substances and their concentrations. Another question is where to place a removal technology—on the sewage treatment plant or on the outflow from the hospital or decentralized for each larger territorial unit. It is necessary to have an idea of conjugation and metabolic processes taking place in the sewage to build efficient degradation technologies and to predict more accurately the concentration of drugs in the water cycle. Their effort is to preserve the environment purely for the future generations by investigating functional curative pharmaceuticals.

The water supply system from the catchment to the consumer should be studied and monitored as a complex with many interrelations. This principle is one of the key points of Water Safety Plans which are included in the new World Health Organisation (WHO) Drinking Water Quality Guidelines. To understand the sources and consequences of eutrophication and the pollution of both treated and untreated water could be done by microscopic analysis. Biological and hydrobiological audits are discussed step by step in the context of risk analyses—Hazard Analysis and Critical Control Points (HACCP), monitoring of water treatment technologies and the Water Safety Plans used in the Czech Republic in Chap. 7 "**Biological Audits in the System of Water Treatment Control**". Microscopic analysis can help to understand the sources and consequences of eutrophication and the pollution of both treated and untreated water, thus affecting the work of the designers, construction engineers and operatives, water managers, water suppliers, health authorities and decision-makers involved.

The reduced quality and yield of water mainly also due to changing climatic conditions lead to an increase in reusing wastewater in the Czech Republic. This topic is discussed in Chap. 8 "**Greywater Reuse in Urban Areas**". Reusing wastewater can be implemented at the wastewater treatment plant or in the building. The current water strategies in the Czech Republic do not strictly address the wastewater reuse because there are not needs to find new possibilities of water sources without conventional raw water sources. The problem is not only the technical and economic aspects but also the negative public opinion about the reuse of treated wastewater. Although in the Czech Republic, the system reuses greywater is presented only sporadically the occurring droughts and changing climatic conditions represent a challenge in the disposal of wastewater reuse management. In the Czech Republic, there is no direct regulation which would limit the wastewater reuse. It is expected that further increase needs of water and sewerage rates to allow a smooth recovery of predominantly obsolete utilities and the rehabilitation or construction of new water resources. This increase creates conditions for return on investment in the greywater treatment system in the selected buildings. The author points out that in Czech the system reuses greywater is used sporadically, but there is a challenge in the disposal of wastewater reuse management.

Chapter 9 "**The Necessary Documents for the Design Documentation for Water Supply and Sewerage Systems in the Czech Republic**" provides an overview of the preparation of design documentation for the construction of sewerage systems, sewer connections, water supply systems, and water-service pipes. It is based on the legislative and normative documents valid in the Czech Republic. This chapter explains the basic terminology used in fields of water supply and sewerage systems for easier orientation and understanding of the problematics. Attention is also focused on the responsibilities of the owners of sewerage systems, sewer connections, water supply systems and water-service pipes. It also describes the issue of unauthorized water consumption and the possible sanctions for administrative offences arising from the Water Act. It does not deal with the specific design of the above-mentioned constructions, but it introduces the minimum requirements resulting from the existing legislation in force in the Czech Republic. It also outlines the possibilities of various permission modes for these constructions and deals with the obligations of the owners of these constructions and the possible sanctions for administrative offences. Chapter 10 titled "**Evaluation of Technical Condition of Sewerage Systems Operated by Municipalities In The Czech Republic**" deals with the comparison of the technical condition of sewerage systems in the Czech Republic concerning water companies operating water management infrastructure and municipalities operating the sewerage systems by themselves. It presents comparison and classification of various defects in the sewerage systems. The assessed localities are municipalities up to 2000 inhabitants in the Czech Republic, which themselves operate sewerage systems. In these localities, the sewerage systems were monitored throughout 15 years, with the age of the system ranging from 40 to 50 years with regard to construction and development of the municipality in the relevant locality. This chapter offers a solution for assessing the structural and technical condition of sewer systems using a multi-criteria analysis based on the Failure Mode and

Effects Analysis (FMEA) method while keeping the relative assessment scale and comparing the results. The conducted analysis has determined the average defect rate of sewerage systems for selected municipalities operating their sewerage networks themselves.

Chapter 11 "**Numerical Modelling of the Fluid Flow at the Outlet from Narrowed Space for a Better Water Management**" presents the possibilities of numerical modelling of the fluid flow in changing flow space. In the water management, events of significant changes in the running fluid flow space occur very frequently. These might be effects on adjacent objects in close proximity such as walls of the designated space or effects on objects bypassed with the running fluid. The presented task is solved using Computational Fluid Dynamics based on a Finite Volume Method in ANSYS Fluent software. The authors explain the mathematical specification of fluid movements and principles of the correct choice of the numerical model. Total of four numerical models was selected, suitable for the solution of described problems. The mathematical approach to the solution and differences between individual models and results related to flow velocity and turbulence intensity are presented. Calculations and the experiment confirmed an assumption of accelerated flow field through a narrowed space.

In Chap. 12 "**Numerical Modelling of Fluid Domain Flow in Open Top Channel**" presents the influence of choosing mesh parameters on seismic response of fluid domains by numerical simulation of problems fluid-structure interaction during extremely loading. The type of mesh parameters does not have a significant influence on the response of the fluid domain, but it has an influence on a number of finite elements. The mesh density has a significant influence on the peak values of the wave. Resolution of numerical simulations of fluid domains has great importance for the reliability design of structures in interaction with fluid during extremely loading.

Chapter 13 "**Monitoring of Changes in Water Content in Soil Pores of Earth-Fill Dams**" provides a view of international cooperation and its achievements when dealing with a project E!7614 of applied research in the EUREKA programme. Within this project, implementation was constructed a unique measuring apparatus based on the method of electrical impedance spectrometry. The measurements were focused on seepage of earth-fill dams and were carried out at two dams in the Czech Republic: Karolinka and Hornice. There are a whole number of methods of monitoring the regime of seepage water in earth-fill dams, one of which being the electrical impedance spectrometry method ever more often used at present. The above-given results of monitoring the process of seepage using the electrical impedance spectrometry method show the suitability of its application in documenting the processes which take place in the soil before, during and subsequently after the remediation.

In Chap. 14 "**Flood Protection in the Czech Republic**" the main flood protection and flood risk-related issues are discussed. Firstly, the historical background is mentioned and discussed. Special attention is paid to the process of flood risk assessment and its applications at practical flood protection solutions. The present state of flood protection in the Czech Republic is briefly mentioned too. The recent procedures used for the development of flood hazard, and flood risk maps are described together with the development of hazard and vulnerability maps. The assessment of economic efficiency should be a necessary part of flood risk management plans. The chapter presents the implementation of the Flood Risk Directive in the Czech Republic. From this viewpoint, the Czech Republic is well fitting requirements on flood protection harmonization prescribed by the European Committee. Parallel with the permanent flood protection planning, the flood protection measures are being designed and constructed. Foremost, these arrangements are levees and dry reservoirs.

Chapter 15 "**Small Hydropower Plants in the Czech Republic**" presents selected small hydropower stations realized in the Czech Republic for about 16 years, which are interesting both from a technical point of view and from the point of view of the architectural design. To improve the economic efficiency of the investment, most of the newly built hydroelectric power plants were implemented using existing weirs or dams. Besides the new buildings, a number of general reconstructions of the technological and construction part of the small hydropower plants were also carried out, aiming to increase the efficiency of these sources. The possibility of obtaining a financial subsidy was one of the factors that helped to improve the economic efficiency of the implementation of small hydroelectric power plants in that period.

16.4 Recommendations

The current trends in population growth, industrialization and environmental pollution will lead to a significant negative impact on Planet Earth as a result of global warming and other (hardly predictable) negative consequences. It is still possible to change these trends and create conditions of ecological and economic stability which will be sustainable in the foreseeable future. The state of global balance can be established by satisfying the basic material needs of each individual on this planet, and each individual will have (at least approximately) the same opportunity to realize the individual human potential.

"What is ecological is also economical" should be the base principle in the management of water quantity and water quality. Flood mitigation measures, the measures to reduce energy consumption, construction of the elements of green infrastructure, revitalization of polluted watercourses, increasing municipal cleanliness, especially in the areas inhabited by marginalized groups, improving the technical and aesthetic level of surrounding forests, and implementing other measures should be in line with this principle. All of these contribute in the long run to harnessing the benefits, including the economic benefits through the creation of a positive synergy effect, for the inhabitants.

Comprehensive community care can be ensured through integrated management of water quality and quantity in line with the concept of sustainable development. A problem solved should not be understood as a one-off act, but as a step in a permanent process in line with the need to shift the management paradigm towards principled leadership, including the concept of mutual respect and understanding, limiting the deterioration of living conditions, reducing the burden on the countryside and promoting the rational use of natural resources. The benefits of such an approach will positively affect the quality of life in the rural community and will also affect those people who use nature's gifts and wealth in the use of their time.

The assessment of water resources, including studies on health, social, economic, technical and environmental state of water resources should be based on the preservation of the relevant scientific principles depending on the technology of their implementation. Research and development activities should, therefore, be based on a strategic analysis of the needs of the country. They should take into consideration and strengthen national expertise in applying sustainability principles to maintain the water resources for sustainable use.

References

1. Kokeš J, Zahrádková S, Němejcová D, Hodovský J, Jarkovský J, Soldán T (2006) The PERLA system in the Czech Republic: a multivariate approach for assessing the ecological status of running waters. In: Furse MT, Hering D, Brabec K, Buffagni A, Sandin L, Verdonschot PFM (eds) The ecological status of European rivers: evaluation and intercalibration of assessment methods. Developments in hydrobiology, vol 188. Springer, Dordrecht
2. Pechar L (2000) Impacts of long-term changes in fishery management on the trophic level water quality in Czech fish ponds. Fish Manage Technol 7(1–2):23–31
3. Potužák J, Duras J, Drozd B (2016) Mass balance of fishponds: are the sources or sinks of phosphorus? Aquacult Int 24(6):1725
4. Baxa M, Šulcová J, Kröpfelová L, Pokorný P, Potužák J (2019) The quality of sediment in shallow water bodies—long-term screening of sediment in Czech Republic. A new perspective of nutrients and organic matter recycling in agricultural landscapes. Ecol Eng 127:151–159
5. Frélichová J, Nelson E, Vačká D (2013) Past and future impacts of land use and climate change on agricultural ecosystem services in the Czech Republic. Land Use Policy 33:183–194
6. Hesslerová P, Chmelová I, Pokorný J, Šulcová J, Kröpfelová L, Pechar L (2012) Surface temperature and hydrochemistry as indicators of land cover functions. Ecol Eng 49:146
7. Vymazal J (2009) Long-term performance of constructed wetlands with horizontal sub-surface flow: ten case studies from the Czech Republic. Agric For Meteorol 149(6–7):913–919
8. Hlavinek P, Zelenakova M (2015) Storm water management: examples from Czech Republic, Slovakia and Poland. Springer Publishing House. https://doi.org/10.1007/978-3-319-25835-5
9. Duchan D, Říha J (2017) Stormwater infiltration facilities-philosophy, design. Int J Water Resour Arid Environ 6:33–40
10. Janosova B, Miklankova J, Hlavinek P, Winthens T (2006) Drivers for wastewater reuse: regional analysis in the Czech Republic. Presented at the International conference on integrated concepts on water recycling, Wollongong, NSW, Australia, Desalination, vol 187, pp 1–3, 103–114
11. Raček J, Ševčík J, Chorazy T, Hlavínek P, Kučerík J (2018) Zpracování čistírenských kalů prostřednictvím mikrovlnné pyrolýzy. Vodní hospodářství 68(1):16–20
12. Raček J, Anferova N, Hluštík P, Hlavínek P (2018b) Optimizing sludge management at the municipal solid waste incinerator—a case study. In: Proceedings, vol 2, p 651. https://doi.org/10.3390/proceedings2110651
13. Ševčík J, Raček J, Hluštík P, Hlavínek P, Dvořák K (2018) Microwave pyrolysis full-scale application on sewage sludge. Desalin Water Treat 112:161–170. https://doi.org/10.5004/dwt.2018.22108
14. Říha J, Petrula L, Hala M, Alhasan Z (2018) Assessment of empirical formulae for determining the hydraulic conductivity of glass beads. J Hydrol Hydromech 66(3):337–347. https://doi.org/10.2478/johh-2018-0021

15. Říha J, Petrula L, Alhasan Z, Popielski P, Dabska A, Fry J, Solski S, Perevoshchikova N, Landstorfer F (2018) Harmonisation of terminology and definitions on soil deformation due to seepage. In: Internal erosion in earthdams, dikes and levees: proceedings of EWG-IE 26th annual meeting 2018. Springer, Cham, Switzerland, s. 347–366
16. Dráb A, Říha J (2010) An approach to the implementation of European Directive 2007/60/EC on flood risk management in the Czech Republic. Nat Hazards Earth Syst Sci 10(9):1977–1987
17. Blazkova S, Beven K (2009) A limits of acceptability approach to model evaluation and uncertainty estimation in flood frequency estimation by continuous simulation: Skalka catchment, Czech Republic. Water Resour Res
18. Říha J, Špano M (2017) Evaluation of dam safety during floods. In: 85th ICOLD annual meeting international symposium "Knowledge-based Dam Engineering". ICOLD. Praha, CZ: Český přehradní výbor, z.o., p. 1–10

Printed by Printforce, the Netherlands